P9-AZX-463

World Records
in Chemistry

DISCARDED
FROM
UNH LIBRARY

World Records in Chemistry

Hans-Jürgen Quadbeck-Seeger (ed.)
Rüdiger Faust, Günter Knaus, Ulrich Siemeling

Translated by William E. Russey

Weinheim · New York · Chichester · Brisbane · Singapore · Toronto

Prof. Dr. H.-J. Quadbeck-Seeger
B1
BASF AG
D-67056 Ludwigshafen

Dr. R. Faust
Department of Chemistry
University College London
20 Gordon Street
UK-London WC1H 0AJ

Dr. G. Knaus
ZZS/OR – D 100
BASF AG
D-67056 Ludwigshafen

PD Dr. U. Siemeling
Fakultät für Chemie
Universitätsstr. 25
D-33615 Bielefeld

1st English edition, 1999

Library of Congress Card No.: applied for
A catalogue record for this book is available from the British Library.

Die Deutsche Bibliothek – CIP-Einheitsaufnahme

World records in chemistry / Rüdiger Faust ; Günter Knaus ; Ulrich Siemeling. Hans-Jürgen Quadbeck-Seeger (ed.). [Übers. William E. Russey]. – 1. Engl. ed. – Weinheim ; New York ; Chichester ; Brisbane ; Singapore ; Toronto : WILEY-VCH, 1999
Dt. Ausg. u.d.T.: Faust, Rüdiger: Chemie-Rekorde
ISBN 3-527-29574-7

Printed on acid-free and chlorine-free paper.

Cover Design: Gunther Schulz, D-67136 Fußgönheim
Composition: TypoDesign Hecker GmbH, D-69115 Heidelberg
Printing: Strauss Offsetdruck, D-69509 Mörlenbach
Bookbinding: Wilhelm Osswald & Co., D-67433 Neustadt

Printed in the Federal Republic of Germany.

Foreword

Chemistry faces a huge problem in public relations, with the general public, with the mass media, and with our political leaders. Chemistry is not only the Central Science, as the American Chemical Society describes it, but also the Useful Science and the Creative Science. Chemists invent new substances and new chemical transformations, and their contributions to human health and welfare and to the economic strength of advanced countries are enormous. In the U.S. more than one third of all manufacturing involves the chemical process industries broadly defined, those industries that perform chemical transformations to produce their products. No other segment of manufacturing comes close to this contribution, judged by the value added during production. The medicines that have so prolonged life expectancy and improved our general health, the materials that make modern life possible, all are the products of the inventiveness of chemists and their industry. Thus chemistry should be widely popular, hailed as the science that has done the most to permit the advance of human life from brief and brutish animal-like existence to our modern civilized life-styles.

Unfortunately, the reputation of chemistry is far from that which it deserves. Few of its contributions to human welfare are generally understood, and most of the news that the public receives about chemistry focusses on environmental problems or on toxic spills. The problem is so serious that some of the leading chemical companies have removed the word "chemistry" from their mottos. There are few books for the general public that describe exciting aspects of chemistry in non-technical terms, and until there are such books we can expect the negative image of chemistry to persist. This book by Hans-Jürgen Quadbeck-Seeger, a past president of the German Chemical Society, and his coauthors Rüdiger Faust, Günter Knaus and Ulrich Siemeling makes an important contribution toward solving the problem.

In an interesting and imaginative way, the book describes the chemical versions of world records. We are all fascinated with the analogs in other fields – the tallest building, the fastest runner, the facts listed in the Guiness Book of Records – and here we find brief and interesting descriptions of molecules with the longest and the shortest bonds, the greatest strain, the highest toxicity, the longest carbon chain, the worst smell, etc. We also learn about records in chemical industry, in Nobel prizes, and in scientific publications (what is the longest footnote?). The facts are fun and informative.

This isn't a textbook, and it isn't a novel. Instead it is a source book for all chemists, giving us ideas we can use in teaching and in presentations, and it is also a book that should fascinate young students who want a feeling for the chemical world. Finally, because it lays out the current records in many aspects of chemistry, it should serve as an inspiration to Beat That Record. I enjoyed it and recommend it to my fellow chemists and teachers, and to journalists who want to add spice to their writing about science.

October 1998

Roland C. D. Breslow, Columbia University New York

Table of Contents

Atoms and Molecules

Record Atoms

Bonding Records

Cluster Records

Hardness Records

Coordination Records

Solvent Records

Redox Hit List

$E = mc^2$

Acids and Bases

Superconductors

Highlights

Biotechnology

Catalysts

Chemical Industry

Appendices

The Largest Atoms

The **heaviest atom found in significant quantities in nature** is ^{238}U.[a] But this is by no means the largest of the natural atoms. Uranium, with an atomic radius of ca. 154 pm, is certainly no dwarf, but there are nevertheless sixteen other elements whose atoms are more voluminous. The size of an atom is determined not by the tiny nucleus, but by the surrounding shells of electrons: the diameter of a nucleus is roughly ten thousand times smaller than the atomic diameter. If the nucleus were the size of a cherry, then the entire atom would be so large that a church the size of Cologne Cathedral could be enclosed within it. The **largest atomic radius** is associated with cesium (272 pm), followed by rubidium (250), potassium (235), barium (224), and strontium (215). The **lightest atom**, the hydrogen atom, has the **smallest atomic radius**, a slender 37 pm. At the same time, a hydrogen atom is also the **largest atom ever observed experimentally** – although not in its normal state (the electronic ground state), but in a highly excited state. Hydrogen atoms have been detected in interstellar clouds with electrons of such high energy that their orbits – using Bohr terminology – are so far from the nucleus as to produce an atomic radius up to 0.339 mm. A hydrogen atom of that size would even be visible to the naked eye![1,b]

The Most Stable Atoms

Most atoms are stable under normal conditions, but the atoms of radioactive elements are unstable. The **radioactive nucleus with the longest lifetime** is ^{113}Cd; its half-life corresponds to an astronomical 9×10^{15} years (the age of the universe is estimated to be "only" 1.3×10^{10} years).[c] The **nuclei with the shortest half-lives**, including for example those of the most recently discovered elements 107–112, decay within a fraction of a second. The reasons for this instability have not yet been elucidated in every detail, but one significant factor is atomic mass. Using the "drop model" of the atomic nucleus, originally proposed by Bohr, it is easy to imagine that the large, heavy nuclei of the artificially produced elements would be wobbly, and that they would easily fall apart like big drops of water. However, this model is unable to explain, for example, why – among the ca. 2400 atomic nuclei now known – certain "magic" nuclei are especially stable. For this and many other reasons the two-centered shell model is the one that currently offers the most satisfactory explanations. Theoretical calculations predict that atomic nuclei of high atomic number could be relatively stable provided they contained 160 to 166 neutrons, and some experimental evidence supporting these claims has also been uncovered. Thus, the ^{266}106 nucleus containing 160 neutrons has a half-life of several seconds, roughly ten times that of the ^{265}106 nucleus with only 159 neutrons.

[1] D. B. Clark, *J. Chem. Educ.* **1991**, *68*, 454–455 and references cited.

Whether there will ever prove to be any truly stable heavy elements, so-called "superheavies," remains to be seen.

Analytical Highlights

Spectroscopic methods have become an essential part of chemical analysis. These methods are generally based on some interaction of the material to be analyzed with energy in the form of electromagnetic radiation – i.e., photons. Just as the strings of a musical instrument produce only very specific, characteristic tones when excited by energy in the form of bowing, striking, or plucking, so the excitation of an atom or molecule leads to characteristic "tones" in the electromagnetic spectrum, which are referred to as signals. Many spectroscopic methods are extremely sensitive: figuratively speaking, under some circumstances one can use them to "hear" extremely faint "tones," or to determine a "pitch" with great accuracy. For example, ESR spectroscopy "hears" the whispering of as few as 10^{11} spins, so this method can be used to detect less than 10^{-12} moles (1 picomole) of a substance.[1] The resolution available by this technique is quite low, however; a "pitch" cannot be determined with great accuracy. Possibly a record for **high resolution** (the equivalent of "perfect pitch" in analysis) belongs to Mössbauer spectroscopy, in which the relative accuracy $\Delta E/E$ for 14.4 keV photons (^{57}Fe resonance) is approximately 10^{-15}. Under especially carefully controlled experimental conditions it is possible to measure and quantify by this method even the influence of gravity on the photons employed (gravitational red shift).[2]

The non plus ultra in analytical chemistry is the **detection of individual molecules**: you can't get any better than that! Analysts are now in a position to "see" with a microscope the individual atoms and molecules on a surface, albeit not through a light microscope but with the help of a device called a scanning tunneling microscope, which is based on a completely different principle.[3] With such an instrument it is possible not only to detect individual atoms, but even to move them about.[4] Scientists at IBM created a stir some years ago when they used 35 xenon atoms to write the company logo on a nickel surface: in letters only 5 nm high![5] Techniques have in the meantime been developed for the

[1] See for example J. A. Weil, J. R. Bolton, J. E. Wertz, *Electron Paramagnetic Resonance*, Wiley, New York, **1994**, p. 501.
[2] See for example N. N. Greenwood, T. C. Gibb, *Mössbauer Spectroscopy*, Chapman and Hall, London, **1971**, pp. 80–81 and cited literature. P. Tourrenc, *Relativity and Gravitation*, Cambridge University Press, Cambridge, **1997**, p. 118. A "state of the art" description is provided by W. Potzel, C. Schäfer, M. Steiner, H. Karzel, W. Schliessl, M. Peter, G. M. Kalvius, T. Katila, E. Ikonene, P. Helistö, J. Hietaniemi, K. Riski, *Hyperfine Interact.* **1992**, *72*, 197.
[3] See for example J. Frommer, *Angew. Chem.* **1992**, *104*, 1325; *Angew. Chem. Int. Ed. Engl.* **1992**, *31*, 1298.
[4] See for example C. F. Quate, *Nature (London)* **1991**, *352*, 571 and cited literature.
[5] D. M. Eigler, E. K. Schweizer, *Nature (London)* **1990**, *344*, 524.

spectroscopic detection and study of individual molecules within a solid matrix or even in a stream of liquid (using laser-induced fluorescence).[6] These methods are currently beginning to find their way into the tool box of analytical chemistry.[7]

[6] See for example W. E. Moerner, T. Basché, *Angew. Chem.* **1993**, *105*, 537; *Angew. Chem. Int. Ed. Engl.* **1993**, *32*, 457. W. E. Moerner, *Science* **1994**, *265*, 46. With respect to biologically relevant systems see D. T. Chiu, R. N. Zare, *Chem. Eur. J.* **1997**, *3*, 335.

[7] N. J. Dovichi, D. D. Chen in *Single Molecule Optical Detection, Imaging and Spectroscopy* (Ed. T. Basché, W. E. Moerner, M. Orrit, U. P. Wild), VCH, Weinheim, **1996**, pp. 223–243.

[a] Plutonium has a higher atomic weight than uranium, but is present in nature only in the tiniest of traces.

[b] To be completely accurate (at the risk of splitting hairs) it should be recorded that quantum mechanics regards even the tiny hydrogen atom – in its ground state! – as infinitely large, because the probability of finding a 1s electron at any given distance from the nucleus never drops to zero no matter how great the distance.

[c] Estimating the age of the earth is extremely difficult. Earlier estimates were recently revised downward by about four billion years. See for example B. G. Börner, *Phys. Unserer Zeit* **1997**, *28*, 6.

The Longest Bonds

Molecules are held together by chemical bonds. The **longest bond** ever measured in a molecule has a mean length over time of ca. 6200 pm.[1] It was discovered while studying the van der Waals molecule ^4He–^4He, which is not stable under normal conditions (→ Atoms and Molecules, Bonding Records: The Strongest and the Weakest Bonds). Bond lengths in most stable molecules fall between ca. 100 and 300 pm, largely as a function of the radii of the bonded atoms. The distance between two specific atoms that are bonded together can vary considerably. As a rule, a longer distance corresponds to a weaker bond, a shorter distance to a stronger bond (one involving multiple-bond character, for example). This need not always be the case, however. For instance, theoretical considerations suggest that there is no bonding interaction between the two metallic centers in silver complex **1** (Fig. 1) despite their rather small separation [270.5(1) pm]; they are simply squeezed together as a consequence of the geometry of the bridging ligands.[2] It is impossible to identify a specific distance between two atoms within a molecule below which it is meaningful to speak of a "bond."[a] If the separation is greater than the sum of the van der Waals radii of the two atoms, then these are generally *not* bonded together. However, if the separation in question lies between the sum of the covalent radii and the sum of the van der Waals radii there may be a contact that under some circumstances could correspond to a weak bonding interaction. This caveat makes it difficult to proclaim a **longest known bond in a stable**[b] **molecule** (Fig. 1).[c] Especially long bonds are of course found between large atoms. It is therefore not surprising that the xenon–xenon bond in the cation of $Xe_2^+[Sb_4F_{14}]^-$ [308.7(1) pm] is one of the longest.[3] The rhenium–rhenium bond in $(OC)_5Re$–$Re(CO)_5$ is also very long, 304.1(1) pm,[4] but a significantly longer rhenium–rhenium bond, 309.1(2) pm, is present in the complex $[(OC)_5Re$–$Re(CO)_4\{C(OEt)SiPh_3\}]$.[5] Taking into account the reservations above, substantially longer contacts are found in large clusters.[d] Among transition-metal clusters the anion **2**, $[Os_{17}(CO)_{36}]^{2-}$, displays two exceptionally long metal–metal "bonds" according to the authors: The distance between Os(2) and Os(17) is 315.2(5) pm, and that between Os(10) and Os(17) is 314.0(4) pm.[6] A likely candidate for the record is the binuclear palladium complex **3** with a palladium–palladium distance of 433 pm,[7] which on the basis of a simple theoretical model is thought to represent a weak bond.[8]

---

[1] F. Luo, C. F. Giese, W. R. Gentry, *J. Chem. Phys.* **1996**, *104*, 1151.
[2] F. A. Cotton, X. Feng, M. Matusz, R. Poli, *J. Am. Chem. Soc.* **1988**, *110*, 7077. The molecular structures were determined by single-crystal X-ray analysis. Atomic distances cited in the further course of this section were also so determined unless otherwise noted.
[3] T. Drews, K. Seppelt, *Angew. Chem.* **1997**, *109*, 264; *Angew. Chem. Int. Ed. Engl.* **1997**, *36*, 273.
[4] M. R. Churchill, K. N. Amoh, H. J. Wasserman, *Inorg. Chem.* **1981**, *20*, 1609.
[5] U. Schubert, K. Ackermann, P. Rustemeyer, *J. Organomet. Chem.* **1982**, *231*, 323.
[6] L. H. Gade, B. F. G. Johnson, J. Lewis, M. McPartlin, H. R. Powell, P. R. Raithby, W.-T. Wong, *J. Chem. Soc., Dalton Trans.* **1994**, 521.
[7] P. Dapporto, L. Sacconi, P. Stoppioni, F. Zanobini, *Inorg. Chem.* **1981**, 20, 3834.
[8] M. Di Vaira, S. Midollini, L. Sacconi, *J. Am. Chem. Soc.* **1979**, *101*, 1757.

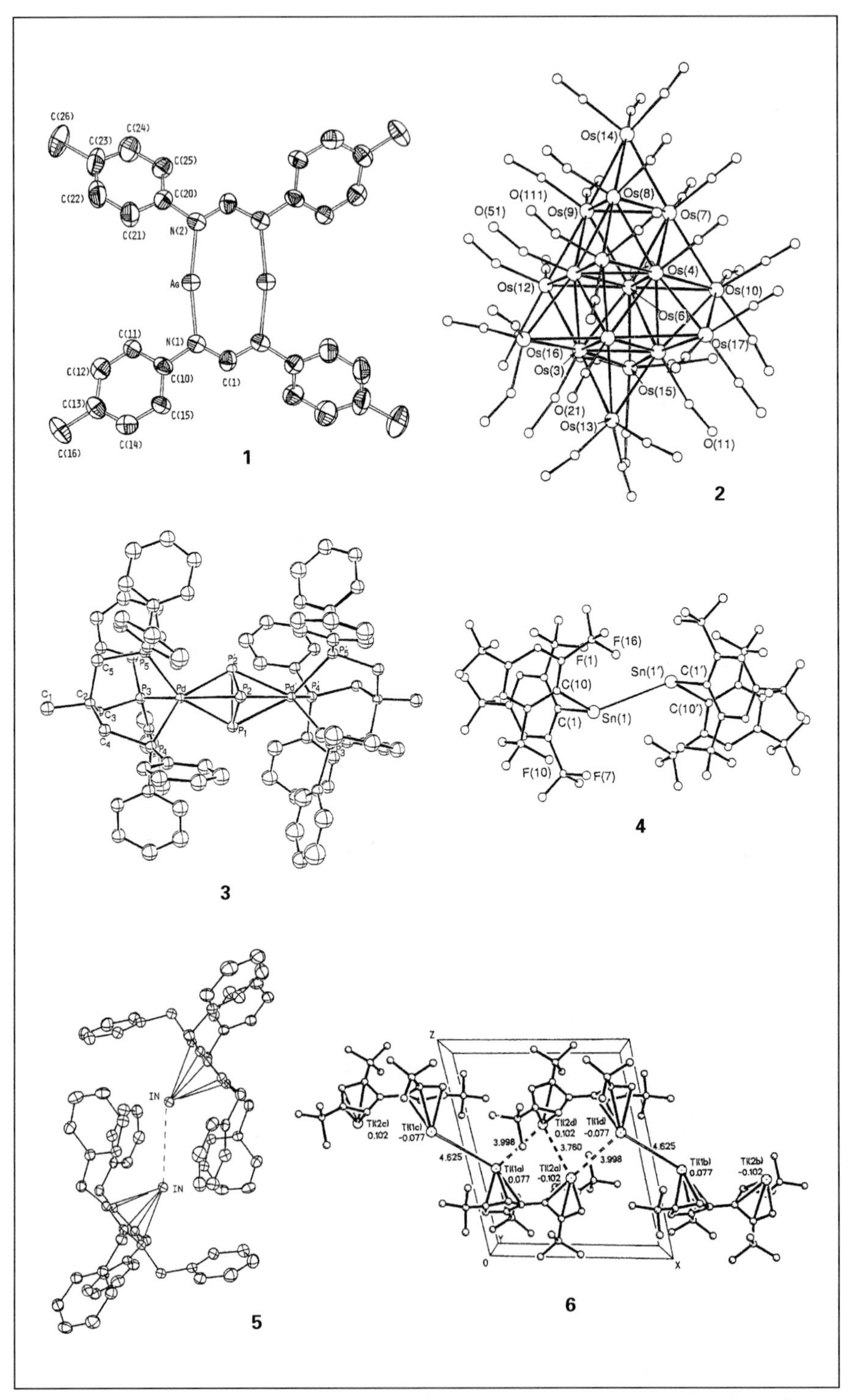

Fig. 1: Molecules challenging the record for "bond length"

There exist molecules that are monomeric in solution and in the gas phase but dimeric or polymeric in the solid phase. This suggests that the forces acting between molecules in the solid are extremely weak;[e] one thus often finds very large distances between the corresponding contact atoms.[f] The point can be illustrated by the following examples:[g] the tin–tin distance in the dimeric Sn(II) molecule **4** is 363.9(1) pm,[9] the indium–indium distance in the dimeric In(I) complex **5a** is 363.1(2) pm,[10] and the thallium–thallium distance in the analogous Tl(I) complex **5b** is 363.2 pm.[11] The question "What constitutes (or perhaps what does *not* constitute) a true bond?"[h] becomes even more insidious in the case of associations in the crystalline state of thallium compound **6**, in which three different intermolecular thallium–thallium distances are observed (376.0, 399.8, and 462.5 pm).[12]

The Shortest Bonds

Short bonds are much easier to judge. The **shortest known bond** is that in the HD molecule (74.136 pm), closely followed by those in D_2 (74.164 pm) and H_2 (74.166 pm). The **shortest metal–metal bonds** (Fig. 2) are found in dichromium(II) complexes. The chromium–chromium quadruple bonds in question have a length (better: shortness) of ca. 183 pm [182.8(2) for **7**[13] and 183.0(4) for **8**[14]]. The **shortest metal–carbon bond** occurs in the complex

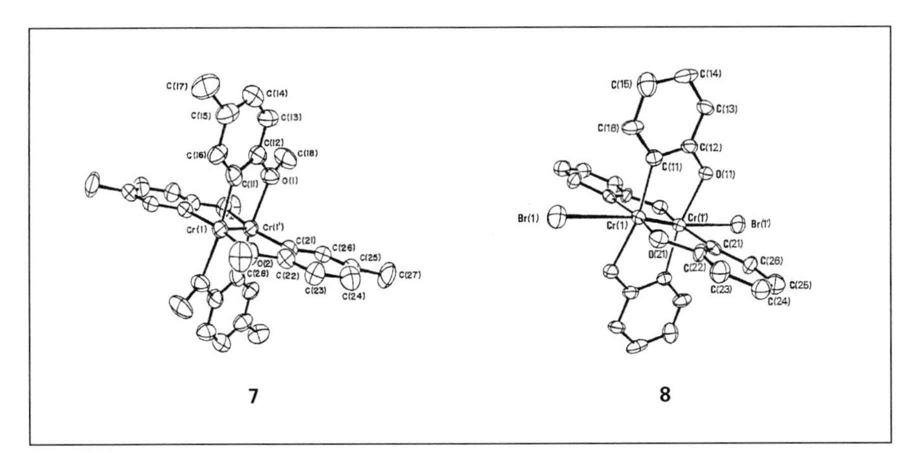

Fig. 2: The shortest metal–metal bonds

[9] U. Lay, H. Pritzkow, H. Grützmacher, *J. Chem. Soc., Chem. Commun.* **1992**, 260.
[10] H. Schumann, C. Janiak, F. Görlitz, J. Loebel, A. Dietrich, *J. Organomet. Chem.* **1989**, *363*, 243.
[11] H. Schumann, C. Janiak, J. Pickardt, U. Börner, *Angew. Chem.* **1987**, *99*, 788; *Angew. Chem. Int. Ed. Engl.* **1987**, *26*, 789.
[12] P. Jutzi, J. Schnittger, M. B. Hursthouse, *Chem. Ber.* **1991**, *124*, 1693.
[13] F. A. Cotton, S. A. Koch, M. Millar, *Inorg. Chem.* **1978**, *17*, 2084.
[14] F. A. Cotton, S. Koch, *Inorg. Chem.* **1978**, *17*, 2021.

(TPP)FeCRe$_2$(CO)$_9$ (TPP = tetraphenylporphyrinate),[15] with an iron–carbon bond only 160.5(13) pm long.[16]

The main-group elements have on occasion been the scene of a systematic search for especially long or short bonds. For this reason we present a number of potential record-breaking examples. The **shortest oxygen–oxygen bond** was established in the O_2F_2 molecule [121.7(3) pm, calculated from spectroscopic data].[17] The **longest oxygen–oxygen bond** is found in dioxirane F_2CO_2 [157.8(1) pm, determined by electron diffraction].[18] The **shortest nitrogen–nitrogen bond** is that in the N_2 molecule (109.76 pm, calculated from spectroscopic data). The **longest nitrogen–nitrogen bonds** are in ON–NO (218 pm, calculated from spectroscopic data) and ON–NO_2 (189.2 pm). Hexa-*tert*-butyldisilane (*t*Bu$_3$Si–Si*t*Bu$_3$) possesses the **longest known silicon–silicon bond** (269.7 pm),[19] and hexa-*tert*-butyldigermane (*t*Bu$_3$Ge–Ge*t*Bu$_3$) contains both the **longest germanium–germanium** (271.0 pm) and **germanium–carbon** (207.6 pm) **bonds** so far observed.[20] Diphosphene **9** very likely includes the **shortest phosphorus–phosphorus bond** [200.4(6) pm] **in a stable molecule** (Fig. 3).[21] Similarly short are the phosphorus–phosphorus bonds in a number of other diphosphenes, but also in complex **10** [201.9(9) pm].[22] The **shortest phosphorus–phosphorus bond of all** is the triple bond in the P_2 molecule (189.3 pm, calculated from spectroscopic data). The **longest phosphorus–phosphorus bond** is found in complex **11** between P(3) and P(4) [246.16(22)

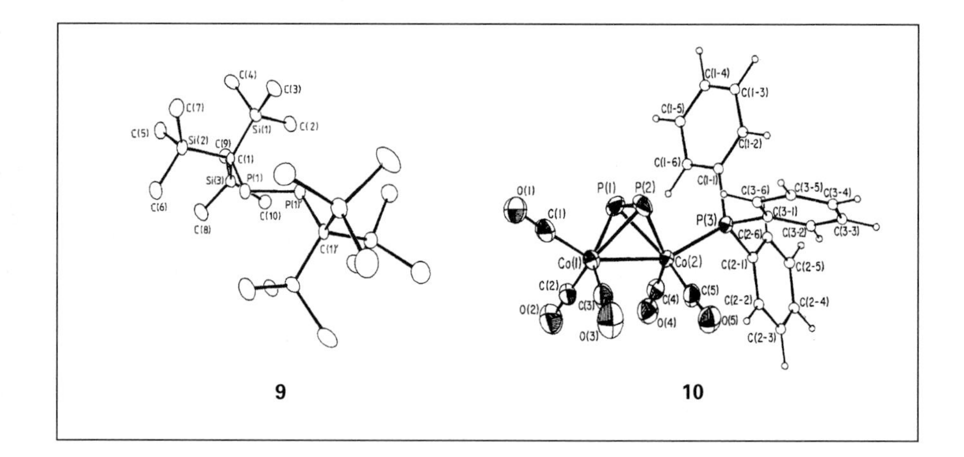

[15] W. Beck, B. Niemer, M. Wieser, *Angew. Chem.* **1993**, *105*, 969; *Angew. Chem. Int. Ed. Engl.* **1993**, *32*, 923.
[16] W. Beck, W. Knauer, C. Robl, *Angew. Chem.* **1990**, *102*, 331; *Angew. Chem. Int. Ed. Engl.* **1990**, *29*, 293.
[17] R. H. Jackson, *J. Chem. Soc.* **1962**, 4585.
[18] B. Casper, D. Christen, H.-G. Mack, H. Oberhammer, G. A. Argüello, B. Jülicher, M. Kronberg, H. Willner, *J. Phys. Chem.* **1996**, *100*, 3983.
[19] N. Wiberg, H. Schuster, A. Simon, K. Peters, *Angew. Chem.* **1986**, *98*, 100; *Angew. Chem. Int. Ed. Engl.* **1986**, *25*, 79.
[20] M. Weidenbruch, F.-T. Grimm, M. Herrndorf, A. Schäfer, K. Peters, H.-G. v. Schnering, *J. Organomet. Chem.* **1988**, *341*, 335.
[21] A. H. Cowley, *Polyhedron* **1984**, *3*, 389 and references cited.
[22] C. F. Campana, A. Vizi-Orosz, G. Palyi, L. Markó, L. F. Dahl, *Inorg. Chem.* **1979**, *18*, 3054.

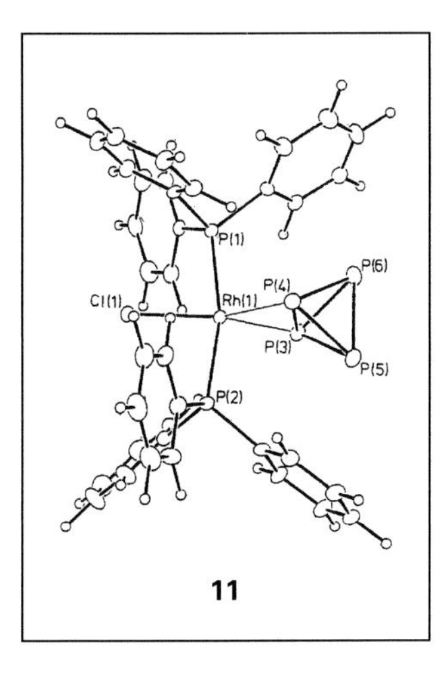

Fig. 3: Bonding extremes in phosphorus compounds

pm].[23] Because organic compounds are so important and also so numerous, candidates for record C–C bonds are treated in a separate section (→ Molecular Form, Bonding Records).

The Strongest and the Weakest Bonds

The nitrogen–nitrogen triple bond in N_2 is the **strongest of the homoatomic bonds**, with a dissociation energy of 945.3 kJ/mol. The **strongest single bond between two atoms of the same element** is that in the T_2 molecule (447.2 kJ/mol); in second place is the bond in DT (445.5), followed by those in D_2 (443.6), HT (440.9), HD (439.6), and finally H_2 (436.2).[i] The **strongest bond between two different atoms** is the carbon–oxygen bond in carbon monoxide (CO), at 1070.3 kJ/mol. **Extremely weak covalent bonds** are found in the nitrogen oxides N_2O_3 ($ON–NO_2$) and N_2O_4 ($O_2N–NO_2$). The strength of the nitrogen–nitrogen bonding is ca. 40.6 kJ/mol in the first case and 56.9 kJ/mol in the second. Similarly weak bonds are also present in many noble-gas compounds. The bond energies for the krypton–fluorine bond in KrF_2 and the xenon–oxygen bond in XeO_4 are in both cases only about 49 kJ/mol (Fig. 4). Numerous gold compounds contain **weak gold–gold bonds**, which are the result of $d^{10}d^{10}$ interactions attributed to relativistic effects (→ Atoms and Molecules, $E = mc^2$). The corresponding bond energies amount to about 30 kJ/mol, and are hence of the same order of magnitude as hydrogen bonds.[24] The **strongest hydrogen bond** has a bond energy of 150 kJ/mol and occurs in the linear, symmetrically arranged $[F–H–F]^-$ ion. The **weakest experimentally**

[23] A. P. Ginsberg, W. E. Lindsell, K. J. McCullough, C. R. Sprinkle, A. J. Welch, *J. Am. Chem. Soc.* **1986**, *108*, 403.
[24] A detailed review has been presented by B. A. Hess, *Ber. Bunsenges. Phys. Chem.* **1997**, *101*, 1; P. Pyykkö, *Chem. Rev.* **1997**, *97*, 597.

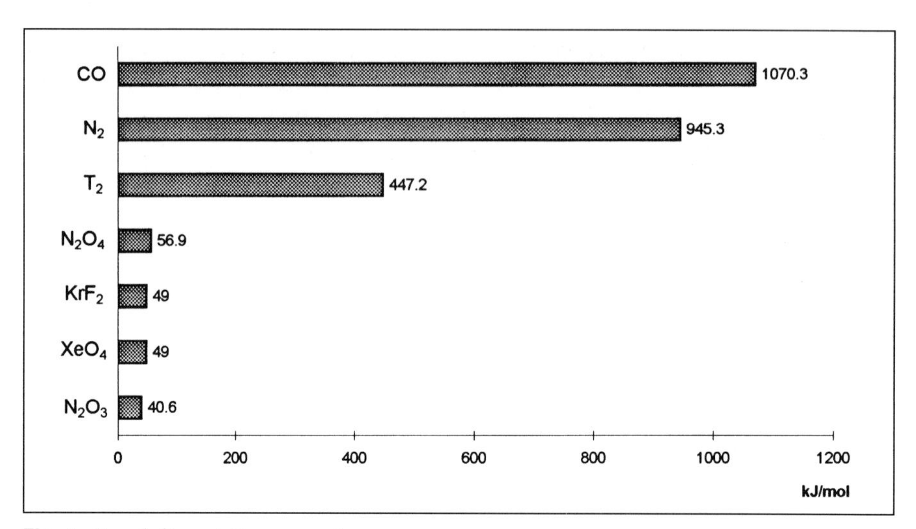

Fig. 4: Bond dissociation energies

determined bond is that in the helium dimer $^4He–^4He$.[25] This species is held together not by typical ionic or covalent forces, but rather by the relatively much weaker van der Waals forces (interactions between fleeting and induced dipoles). Moreover, such forces are weaker still in the case of the helium dimer compared to ordinary van der Waals molecules, primarily because of the extremely low polarizability of the small helium atom. The actual bond energy is only about 8×10^{-6} kJ/mol – more than ten million times less than that of a typical covalent bond! It is hardly surprising that such weakly bound particles are able to exist only at the very lowest temperatures imaginable, because thermal vibrations are sufficient to shake them apart. Dimeric helium is obtained by ultrasonic expansion of helium gas. During expansion the gas cools to about one ten-thousandth of a degree above absolute zero (0.0001 K). The two helium atoms are thus bound by a truly frosty relationship. The time-averaged length of this bond is ca. 6200 pm, making it not only the weakest, but also the **longest bond** ever established in a molecule (typical bond lengths range between 100 and 300 pm).[1] A cause of this extremely high value is the

[25] F. Luo, G. C. McBane, G. Kim, C. F. Giese, W. R. Gentry, *J. Chem. Phys.* **1993**, *98*, 3564.

[a] With respect to the problem of borderline situations between bonding and nonbonding interactions, see for example K. R. Leopold, M. Canagaratna, J. A. Phillips, *Acc. Chem. Res.* **1997**, *30*, 57.

[b] What is meant here is of course not thermodynamic but kinetic stability under normal laboratory conditions.

[c] In a host of publications one finds tables with X-ray structural data (not considered here) containing nonsensical bond lengths reaching far beyond 400 pm. The expression "bond length" should be replaced in questionable cases by "atomic distance."

[d] It should be noted that the valence-bond concept of localized two-center electron bonding cannot be applied meaningfully to clusters.

[e] In these cases, packing effects arising from the crystal lattice become more influential. See for example A. G. Orpen, *Chem. Soc. Rev.* **1993**, *22*, 191.

[f] A detailed discussion of so-called secondary bonding in molecular crystals would exceed the bounds of the discussion. See for example N. W. Alcock, *Adv. Inorg. Chem. Radiochem.* **1972**, *15*, 1.

Heisenberg uncertainty principle: it is not possible to establish precisely and simultaneously both the location and momentum of a particle. The more precisely its momentum is known, the less precisely can it be localized. Motion on the part of helium atoms near absolute zero is so slow that the uncertainty in their momentum is very small, so positional uncertainty becomes correspondingly large (→ Atoms and Molecules, Bonding Records: The Longest Bonds).

[g] The examples presented represent systems in which there is discussion of the presence of weak bonding interactions between formal valence-saturated atoms with s^2 configurations.

[h] The question whether Tl–Tl bonds even exist in crystalline cyclopentadienylthallium compounds is controversial. See C. Janiak, R. Hoffmann, *Angew. Chem.* **1989**, *101*, 1706; *Angew. Chem. Int. Ed. Engl.* **1989**, *28*, 1688 (these authors come to the conclusion: "A $Tl^I–Tl^I$ bond definitely exists in these compounds."); P. H. M. Budzelaar, J. Boersma, *Rec. Trav. Chim. Pays-Bas* **1990**, *109*, 187 (the authors interpret them as "weak donor–acceptor interactions"); P. Schwerdtfeger, *Inorg. Chem.* **1991**, *30*, 1660 (the author subjects their fundamental existence to question: "It is ... unlikely that reasonably strong Tl(I)–Tl(I) bonds exist in any of the known inorganic or organometallic compounds of the element."); P. Pyykkö, *Chem. Rev.* **1997**, *97*, 597 [the author draws the conclusion that "anything goes" (very much in the sense of P. Feyerabend, *Against Method: Outline of an Anarchistic Theory of Knowledge*, 3rd ed., Verso, London, Suhrkamp, **1993**, passim), because "all neutral closed-shell atoms and molecules are 'sticky,' anything sticks to anything, including the case of helium, whose dimer has D_e and D_0 values of 9.06×10^{-2} and 8×10^{-6} kJ/mol, respectively"].

[i] Deuterium (D) is the second isotope of the element hydrogen; the third is the radioactive form, tritium (T). In contrast to the neutronless H, the nucleus of D contains one neutron, that of T two neutrons. The resulting mass difference (D is twice and T three times as heavy as H) causes significant differences in the properties of these isotopes. A higher reduced mass and thus lower zero-point energy is the source of higher dissociation energies for the heavier isotopomers relative to H_2.

The Largest Clusters

Large molecular clusters stand on the threshold between molecular chemistry and solid-state chemistry. Their physical and chemical properties are highly dependent upon structure and size, and investigating them is of considerable interest with respect to many fundamental questions in the natural sciences. For example, clusters serve as model systems for heterogeneous catalysts, for the study of size-dependent quantum effects and cooperative characteristics, as well as for nucleation processes.[1]

The **largest cluster whose structure has been more or less firmly established** has the idealized formula $[Pd_{561}phen_{36}O_{200\pm10}]$ (phen = 1,10-phenanthroline). The 561 palladium atoms that constitute the cluster core of this giant are packed cubic-octrahedrally in five "shells" around a central Pd atom and thus constitute a section of the most densely packed cubic form of the metal.[2]

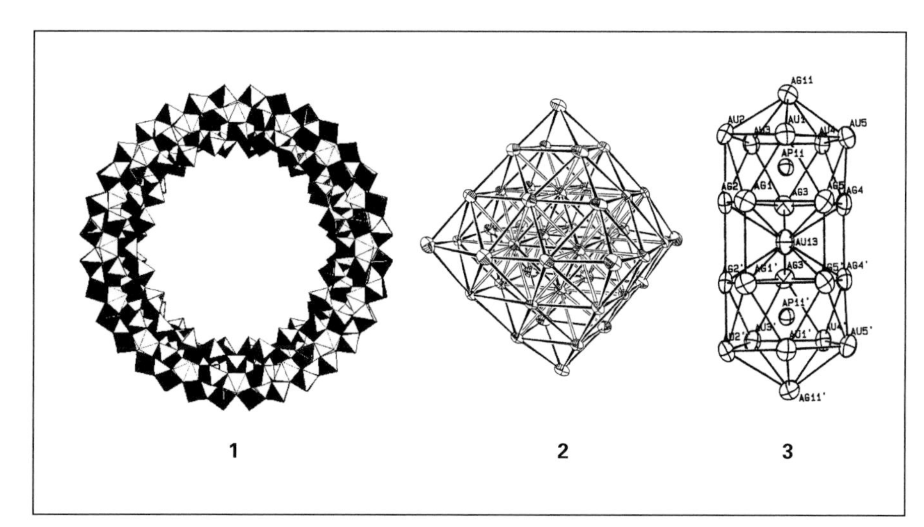

Fig. 1: Large molecular clusters

The **largest cluster-like material so far characterized by X-ray diffraction** has the composition $[(MoO_3)_{176}(H_2O)_{80}H_{32}]$ (**1**).[3] The related polyoxometallates $[As_{12}Ce_{16}(H_2O)_{36}W_{148}O_{524}]^{76-}$[4] and $[Mo_{154}(NO)_{14}O_{420}(OH)_{28}(H_2O)_{70}]^{(25\pm5)-}$[5]

---

[1] See for example G. Schmid, *Chem. Unserer Zeit* **1988**, *22*, 85.

[2] G. Schmid, *Polyhedron* **1988**, *7*, 2321. M. N. Vargaftik, V. P. Zagorodnikov, I. P. Stolyarov, I. I. Moiseev, V. A. Likholobov, D. I. Kochubcy, A. L. Chuvilin, V. I. Zaikovsky, K. I. Zamaraev, G. I. Tomofeeva, *J. Chem. Soc., Chem. Commun.* **1985**, 937.

[3] A Müller, E. Krickemeyer, H. Bögge, M. Schmidtmann, C. Beugholt, P. Kögerler, C. Lu, *Angew. Chem.* **1998**, *110*, 1278; *Angew. Chem. Int. Ed. Engl.* **1998**, *37*, 1220.

[4] K. Wassermann, M. H. Dickmann, M. T. Pope, *Angew. Chem.* **1997**, *109*, 1513; Angew. Chem. Int. Ed. Engl. **1997**, *36*, 1445.

[5] A. Müller, E. Krickemeyer, J. Meyer, H. Bögge, F. Peters, W. Plass, E. Diemann, S. Dillinger, F. Nonnenbruch, M. Randerath, C. Menke, *Angew. Chem.* **1995**, *107*, 2293; *Angew. Chem. Int. Ed. Engl.* **1995**, *34*, 2122.

are of equally impressive size. Only a little smaller is the layer-type cluster $[Cu_{146}Se_{73}(PPh_3)_{30}]$.[6]

The **largest main-group element cluster so far characterized by X-ray diffraction** is $[Al_{77}\{N(SiMe_3)_2\}_{20}]^{2-}$; a central Al atom is surrounded by three shells containing 12, 44, and 20 Al atoms.[7]

The **largest silver cluster** has the formula $[Ag_{50}(PPh)_{20}Cl_7P\textit{n}Pr_3)_{13}]$.[8] Silver five-membered rings comprise an important structural principle in this species.

The **largest dimetallic clusters** characterized by X-ray structural analysis contain 44 metal atoms. An example is anion **2**, $[Ni_{38}Pt_6(CO)_{48}H]^{5-}$, which consists of an octahedral Pt_6 core surrounded by nickel atoms. The metal atoms are arranged according to the most dense form of cubic packing.[9]

The **largest trimetallic cluster** (**3**) contains 25 metal atoms and has the composition $[(Ph_3P)_{10}Au_{12}Ag_{12}PtCl_7]^-$; the core of the cluster consists of two Au_6Ag_6 icosahedra connected by way of a common gold atom. A single gold atom is at the center of one icosahedron, a single platinum atom at the center of the other.[10]

[6] H. Krautscheid, D. Fenske, G. Baum, M. Semmelmann, *Angew. Chem.* **1993**, *105*, 1364; *Angew. Chem. Int. Ed. Engl.* **1993**, *32*, 1303.
[7] A. Ecker, E. Weckert, H. Schnöckel, *Nature (London)* **1997**, *387*, 379.
[8] D. Fenske, F. Simon, *Angew. Chem.* **1997**, *109*, 240; *Angew. Chem. Int. Ed. Engl.* **1997**, *37*, 230.
[9] A. Ceriotti, F. Demartin, G. Longoni, M. Monassero, M. Marchionna, G. Piva, M. Sansoni, *Angew. Chem.* **1985**, *97*, 708; *Angew. Chem. Int. Ed. Engl.* **1985**, *24*, 697.
[10] B. K. Teo, H. Zhang, X. Shi, *J. Am. Chem. Soc.* **1993**, *115*, 8489.

The Hardest Substances

The hardness of a solid body is a measure of the resistance it offers against penetration by another body. Hardness can be tested in a variety of ways. The well-known scratch hardness, rated in Mohs (on a scale of 1–10), is applied to minerals. Many hard materials are evaluated by the Knoop method, among others, which can also be miniaturized for application to small samples (microhardness testing). The Knoop test employs a weighted diamond pyramid of precisely defined geometry to obtain a Knoop hardness (*KH*) which, like other hardness measures, is defined in terms of the impression made in the surface as a function of imposed load. The **hardest known material** is diamond (single crystal: *KH* = 90 GPa;[a] polycrystalline: *KH* = 50 GPa); therefore, the **hardest element** is carbon in the form of diamond.[1,b] In second place is cubic boron nitride (single crystal: *KH* = 48 GPa;[a] polycrystalline: *KH* = 32 GPa), which is therefore the **hardest ceramic material**. The third hardest substance is most probably the ternary metallic cobalt tungsten boride CoWB (microhardness: 45 GPa).[c] The latter is classified as a ceramic material, but one could also regard it as the **hardest alloy**.[d] Fourth place is held by the unusual high-pressure modification of SiO_2 known as stishovite (KH = 33 GPa).[2] Stishovite is thus the **hardest oxide**. In contrast to all other modifications of SiO_2, the Si atoms in stishovite are not coordinated tetrahedrally with four oxygen atoms, but rather octahedrally with six oxygen atoms, and its density (4.387 kg/L) is the highest for any SiO_2 modification. Stishovite is found in nature only in meteorite craters. Other materials rated among the hardest of the hard (Fig. 1) include carbon-rich boron carbide (B_4C, *KH* = 30 GPa),[1] boron-rich boron oxide (B_6O, 30 GPa),[1] silicon carbide (SiC, 29 GPa),[1] and titanium carbide (TiC, 28 GPa).[3] For industrial applications the hardness of a material at high temperature is generally more interesting than its hardness at room temperature.[e] Thus, the hardnesses of both diamond and boron nitride decrease rapidly with increasing temperature, whereas that of boron carbide remains nearly constant. Above ca. 400 °C boron carbide is harder than boron nitride, and above ca. 1100 °C it is harder even than diamond.[4] There has been repeated speculation in the technical literature as to whether there might not exist some material harder than diamond.[5] Indeed, theoretical considerations suggest that this may actually be the case for one modification of carbon nitride (C_3N_4).[1]

[1] C.-M. Sung, M. Sung, *Mater. Chem. Phys.* **1996**, *43*, 1.
[2] J. M. Léger, J. Haines, M. Schmidt, J. P. Petitet, A. S. Pereira, J. A. H. da Jomada, *Nature (London)* **1996**, *383*, 401 and references cited.
[3] R. Riedel, *Adv. Mater.* **1994**, *6*, 549. In ref. 2, a *KH* value of 25 GPa is cited for TiC.
[4] R. Telle, *Chem. Unserer Zeit* **1988**, *22*, 93.
[5] See for example W. Schnick, *Angew. Chem.* **1993**, *105*, 1649; *Angew. Chem. Int. Ed. Engl.* **1993**, *32*, 1580.

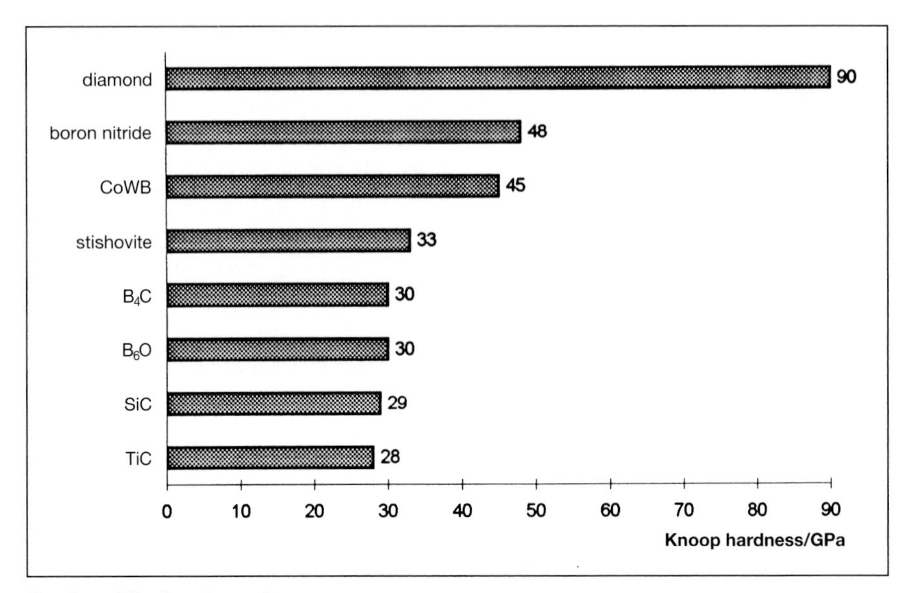

Fig. 1: The hardest substances

---

[a] One gigapascal (1 GPa = 10^6 Pa) corresponds to a pressure of 10 bar.

[b] Not taken into account were occasional pieces of nonsensical data appearing in the technical literature, such as a reported "record" hardness for GaSb of 720 GPa [I. Sh. Dadashev, G. I. Safaraliev, *Inorg. Mater. (Transl. of Neorg. Mater.)* **1990**, *26*, 975].

[c] Z. T. Zahariev, M. I. Marinov, *J. Alloys Compd.* **1993**, *201*, 1. The microhardness values reported in this paper presumably correspond to the Knoop hardness, since microhardness values listed for Al_2O_3 and TiN are very close to Knoop hardness values cited for these substances in ref. 2.

[d] An alloy is a metallic mixture with a minimum of two components, at least one of which is a metal.

[e] Apart from temperature there are numerous other parameters that play a role in the industrial application of a hard material. A detailed review of the broad field of hard-metal materials is provided by R. Menon, *Welding J.* February **1996**, 43.

The Smallest Coordination Numbers

In a chemical sense, "coordination" refers to the arrangement of atoms about a central atom as a consequence of chemical bonding.[a] The coordination number can be regarded as the number of atoms bound to a central atom (ligator formalism). The **smallest coordination number** is zero, found under ordinary conditions only with the noble gases. The next highest coordination number, one, occurs with the elements hydrogen, indium, carbon, nitrogen, phosphorus, arsenic, oxygen, sulfur, selenium, and tellurium, as well as the halogens.[1]

The Largest Coordination Numbers

The **largest coordination numbers in stable compounds** are shown by the actinoid elements. In the compound $[U(BH_4)_4]$ the coordination number of uranium is 14. From the perspective of ligator formalism the uranium atom in the compound $[U(\eta^5\text{-}C_5H_5)_4]$ actually has the coordination number 20 (Fig. 1.).[2,b]

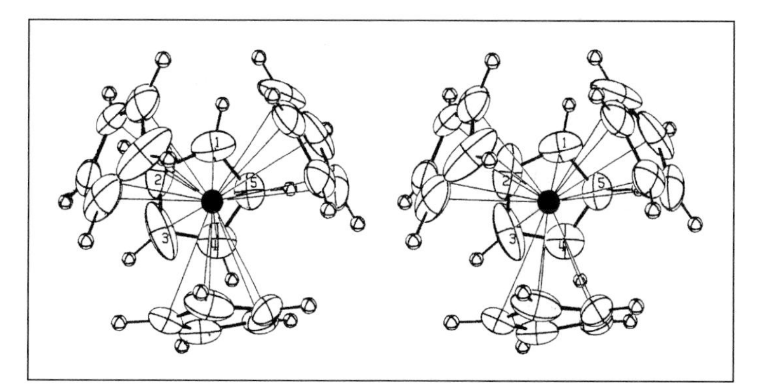

Fig. 1: Uranium atom with the coordination number 20 in $[U(\eta^5\text{-}C_5H_5)_4]$

[1] This coordination number was realized with In and As only very recently; see S. T. Haubrich, P. P. Power, *J. Am. Chem. Soc.* **1998**, *120*, 2202 (In) and R. R. Schrock, *Acc. Chem. Res.* **1997**, *30*, 9 and references cited (As).
[2] J. H. Burns, *J. Organomet. Chem.* **1974**, *69*, 225.

The Greatest Range of Coordination Number

A **probable record for the breadth of known coordination numbers** is ten, for both gold and uranium. Gold often has the coordination number two, as in the complex anion $[Au(CN)_2]^-$, but in clusters it can also display the coordination number 12. Uranium covers a span of coordination numbers from 4 (e.g., in $U(NPh_2)_4$] to 14 (20 based on ligator formalism).

---

[a] With solids a distinction must be made between chemical and geometrical coordination. A discussion of the difference between the two would exceed the bounds of this treatment (see H. Alig, M. Trömel, *Z. Kristallogr.* **1992**, *201*, 213).

[b] Geometric coordination numbers extending far above 20 (e.g., 28 for carbon in the cubic diamond lattice) are not uncommon (see for example M. Trömel, H. Alig, L. Fink, J. Lösel, *Z. Kristallogr.* **1995**, *210*, 817).

The Record Solvent Water

Most chemical reactions take place in solution, and solvents[a] constitute the blood of chemistry. A most exceptional fluid is **water** itself, and for numerous reasons. Water covers 71% of the earth's surface in the form of oceans, and it is also present in large quantities in inland waterways, in the biosphere, in the atmosphere, and even in the lithosphere. It is the most widespread and, in terms of quantity, by far the leading solvent. All the processes of life take place in aqueous medium (indeed, are perfectly adapted to this medium), and water is without question mankind's most important solvent. Water's influence on biological processes is a result largely of the intimate details of the hydrogen bonds between individual water molecules. The delicate subtlety of these details is at least suggested by the fact that heavy water, D_2O, is lethal for virtually all living organisms.[1] Photoassisted cleavage of water by sunlight, carried out along with the reduction of carbon dioxide in the context of photosynthesis by green plants and autotrophic bacteria, is one of the two largest scale chemical reactions by far conducted on earth (the other of course being the associated reduction of CO_2). One of many remarkable features of water is its density anomaly: in contrast to almost every other liquid, the density of water increases as its temperature rises from 0 to 4 °C. Only at temperatures above 4 °C does the density begin to decrease with increasing temperature. In the course of freezing, again unlike almost every other liquid, water expands – by ca. 9%. In other words, ice floats. For these reasons, deep bodies of water rarely freeze all the way to the bottom, a very convenient "coincidence" with respect to living systems present in these environs. Nevertheless, a few living species have the extra protection of an emergency antifreeze supply.[1] Thus, arctic fish, which inhabit seawater with an average temperature of –1.5 °C, have blood that would normally freeze around –0.5 °C, but they are also equipped with cleverly designed "antifreeze" peptides and glycopeptides that delay the lethal formation of ice nuclei until the temperature reaches –2 °C. In analogy to the principle of adding antifreeze to an automobile radiator, certain insects can survive temperatures below –10 °C because their bodily fluids contain high concentrations of materials that lower the freezing point (glycerin, glucose, etc.). Apart from the density anomaly, water also displays an interesting viscosity anomaly:[b] its viscosity passes through a local minimum slightly below 40 °C – that is, in the vicinity of the body temperature of many warm-blooded creatures. Because of the fact that ice is less dense than cold water, the pressure dependence of the melting point of water is most unusual. Again unlike almost all liquids, the melting point of water decreases under pressure, reaching –22 °C, for example, at 2000 bar. This behavior makes ice skating far easier than it would otherwise be.

---

[1] F. Franks, *Chem. Unserer Zeit* **1986**, *20*, 146.

Water is characterized by an extremely high dielectric constant (80.20 at 20 °C), and it is a polar solvent with a great deal of solvation power. Water's enthalpy of melting is much higher than is the case with other low-melting compounds. The values for vaporization enthalpy, surface tension, specific heat capacity, sound absorption, etc., all lie significantly above those of comparable compounds. Water is the only homoleptic nonmetallic compound of hydrogen that is not a gas under normal conditions but is instead a liquid.[c] Given all these remarkable facts it is not surprising that the scientific literature on the subject of water is more extensive than on almost any other topic. Water is the only solvent that has its own scientific journals (in Germany, for example, the series "Vom Wasser", published by the Water Chemistry Interest Group of the German Chemical Society).

The "Most Critical" Solvent

An unusual, but industrially very significant solvent is **supercritical CO_2**.[2] When liquids or gases are heated under pressure, they are transformed at temperatures above their corresponding critical temperatures (T_c) and pressures above their critical pressures (p_c) into a state referred to as "supercritical", in which no distinction can be made between liquid and vapor (gas): the corresponding liquid and gaseous phases would be identical from the standpoint of density and all other characteristics. The fact that one sometimes refers to such materials as "supercritical liquids" is essentially a matter of semantics. A supercritical fluid has a lower density and a much lower viscosity than a comparable "normal" fluid, and its solvation power is much higher in the supercritical state. For industrial applications, especially extraction, it is very advantageous that a supercritical solvent can be removed simply by reducing the pressure. Supercritical CO_2 (T_c = 31 °C, p_c = 73 bar) is typically used at 40 °C and 80–200 bar for gentle extraction of natural products, such as removing caffeine from coffee.[d] Inclusion of additional components makes it possible to extract even very hydrophilic biomolecules such as proteins.[3] Supercritical CO_2 is also an interesting solvent for homogeneous catalysis, one that in the future might be substituted industrially for ordinary nonpolar and slightly polar reaction media.[4]

---

[2] A broad-based overview is provided by the papers in issue 10 of *Angew. Chem.* **1978**, *90*; *Angew. Chem. Int. Ed. Engl.* **1978**, *17*.

[3] E. J. Beckman, *Science* **1996**, *271*, 613. K. P. Johnston, K. L. Harrison, M. J. Clarke, S. M. Howdle, M. P. Heitz, F. V. Bright, C. Carlier, T. W. Randolph, *Science* **1996**, *271*, 624.

[4] Review: P. G. Jessop, T. Ikariya, R. Noyori, *Science* **1995**, *269*, 1065.

The Most Exotic Solvent

One rather more exotic solvent is **liquid lead**. Hittorf used this solvent to isolate by recrystallization the phosphorus modification that bears his name.

An equally unusual solvent is **liquid xenon**, which finds applications in research, especially for investigating solutions of exceedingly reactive species that would react with other solvents.[5] Its solvent properties are comparable to those of pentane. Liquid xenon is especially well suited to the study of vibrational spectra, since it is transparent in the infrared.

[5] See for example B. A. Arndtsen, R. G. Bergman, T. A. Mobley, T. H. Peterson, *Acc. Chem. Res.* **1995**, *28*, 154 and references cited. M. Tacke, *Chem. Ber.* **1995**, *128*, 1051. P. A. Hamley, S. G. Kazarian, M. Poliakoff, *Organometallics* **1994**, *13*, 1767.

[a] A distinction should be made between solvents and solution agents. Ethanol is a popular "solution agent" in many cultures.

[b] The data that follow apply to measurements on thin layers of water enclosed between plates of quartz glass; see G. Peschel, K. H. Adlfinger, *Naturwissenschaften* **1969**, *56*, 558.

[c] Even HF boils as low as 19.51 °C.

[d] This separation method was invented in Germany [K. Zosel, US-Pat. 3969196 (Priority: April 16, 1963), Studiengesellschaft Kohle].

The Strongest Oxidizing and Reducing Agents

The **strongest chemical oxidizing agent** is OF_2 [standard reduction potential E^0 = +3.294 V (acidic), +3.197 V (neutral)], closely followed by F_2 [E^0 = +3.07 V (acid), +2.89 V (neutral)]. Compounds containing xenon in high formal oxidation states come next on the list [H_4XeO_6: E^0 = +2.38 V (acid); XeO_3: +2.10 V (acid)]. A less exotic very strong oxidizing agent is ozone [E^0 = +2.075 V (acid)]. The **strongest chemical reducing agent** is N_3^- [E^0 = –3.608 V (neutral), –3.334 (acid)]. The alkali and alkaline earth metals follow (Li: E^0 = –3.040 V; K: –2.936 V; Ba: –2.906 V; Sr: –2.899 V; Ca: –2.868 V; Na: –2.714 V). Figure 1 presents this information graphically.

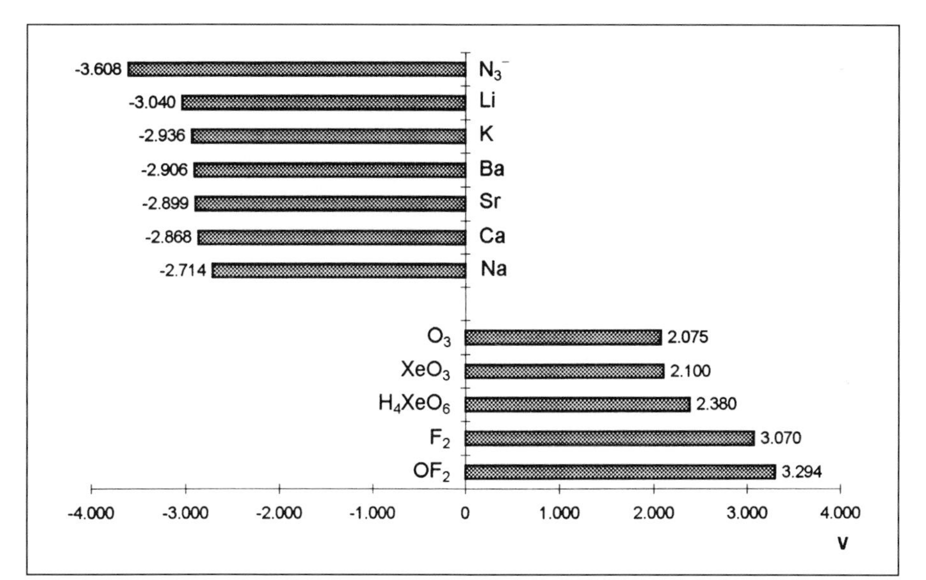

Fig. 1: Redox potentials

The Strongest Relativistic Effects

The special relativity theory states that, relative to a stationary observer, the mass of an object increases as it approaches the speed of light. This has important consequences for the chemistry of heavy elements with massive and highly charged nuclei. Electrons in s orbitals, which unlike p, d, and f electrons have a non-zero probability density at the nucleus and therefore are most sensitive to nuclear attraction, travel at velocities that approach the speed of light when subject to a high nuclear charge. They are thus susceptible to detectable relativistic effects. The mass of these electrons increases, and the sizes and energies of the corresponding orbitals decrease, as a function of gravity (direct relativistic effects). As a result, the sizes and energies of the corresponding d and f orbitals increase due to increasingly effective nuclear shielding by the s orbitals (indirect relativistic effects).[1] The **strongest relativistic effects** associated with stable elements are encountered with gold, platinum, and mercury. Thus, the atomic radius of gold is smaller than that of silver, for example. For doubly coordinated metal(I) compounds the covalent radius of silver is 133 pm, that of gold only 125 pm.[2] Relativistic effects are also responsible for the fact that "gold-plated" compounds of carbon and nitrogen appear in very unusual – namely hypercoordinated – coordination relationships. Thus, in the dication **1**, $[(Ph_3PAu)_6C]^{2+}$,[3] carbon has the coordination number 6, and in the dication **2**, $[(Ph_3PAu)_5N]^{2+}$,[4] nitrogen has the coordination number five (Fig. 1). Moreover, relativistic effects are ultimately responsible for $d^{10}d^{10}$ interaction between the two valence-saturated metal centers constituting the platinum–platinum bond in platinum(0) compound **3**,[5] which is dimeric in the gas phase as well as in the solution and crystalline states.[6] Even more spectacular than the binuclear platinum(0) complex **3** is compound **4**, with three such valence-saturated d^{10} centers in a nearly linear unbridged arrangement.[7] Relativistic contributions to the bond energy (→ Atoms and Molecules, Bonding Records: The Strongest and the Weakest Bonds) can in the case of gold and platinum amount to more than 50% of the total![8] As a final and especially striking example of the consequences of relativistic effects we note the exceptionally weak metallic bonding in elemental

[1] A detailed discussion can be found for example in B. A. Hess, *Ber. Bunsenges. Phys. Chem.* **1997**, *101*, 1. P. Pyykkö, *Chem. Rev.* **1997**, *97*, 597.
[2] A. Bayler, A. Schier, G. A. Bowmaker, H. Schmidbaur, *J. Am. Chem. Soc.* **1996**, *118*, 7006.
[3] F. Scherbaum, A. Grohmann, B. Huber, C. Krüger, H. Schmidbaur, *Angew. Chem.* **1988**, *100*, 1602; *Angew. Chem. Int. Ed. Engl.* **1988**, *27*, 1544.
[4] A. Grohmann, J. Riede, H. Schmidbaur, *Nature (London)* **1990**, *345*, 140.
[5] T. Yoshida, T. Yamagata, T. H. Tulip, J. A. Ibers, S. Otsuka, *J. Am. Chem. Soc.* **1978**, *100*, 2063.
[6] J. Strähle in *Unkonventionelle Wechselwirkungen in der Chemie metallischer Elemente* (Ed. B. Krebs), VCH, Weinheim, **1992**, pp. 357–372.
[7] T. Tanase, Y. Kudo, M. Ohno, K. Kobayashi, Y. Yamamoto, *Nature (London)* **1990**, 344, 526.
[8] Cf. for example D. Schröder, J. Hrušák, I. C. Tornipoth-Oetting, T. M. Klapötke, H. Schwarz, *Angew. Chem.* **1994**, *106*, 223; *Angew. Chem. Int. Ed. Engl.* **1994**, *33*, 212. J. Hrušák, R. H. Hertwig, D. Schröder, P. Schwerdtfeger, W. Koch, H. Schwarz, *Organometallics* **1995**, *14*, 1284. P. Schwerdtfeger, J. S. McFeaters, M. J. Liddell, J. Hrušák, H. Schwarz, *J. Chem. Phys.* **1995**, *103*, 245. C. Heinemann, H. Schwarz, W. Koch, K. G. Dyall, *J. Chem. Phys.* **1996**, *104*, 4642.

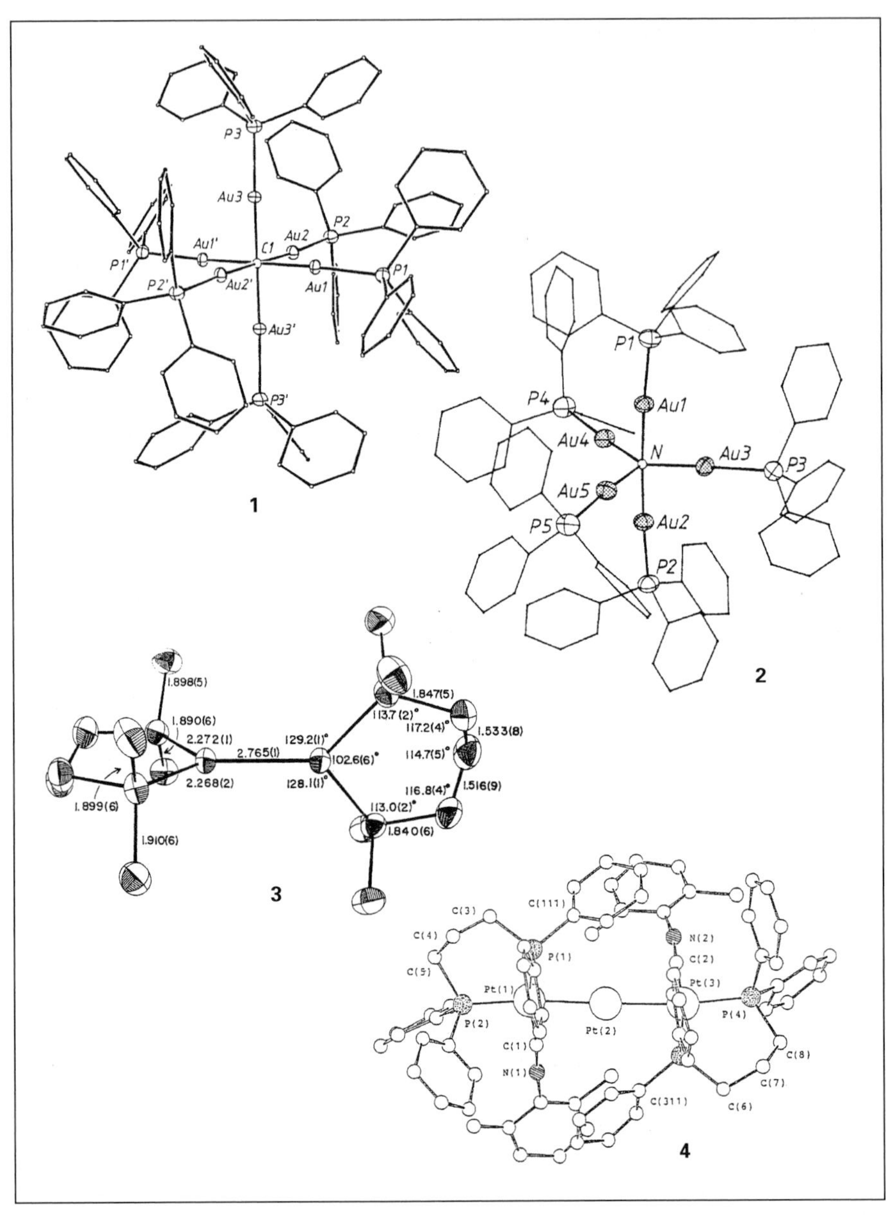

Fig. 1: Unusual coordinations and bonds attributable to relativistic effects

mercury, which leads to the fact that in contrast to all the other metals, mercury under normal conditions is a liquid. The cause of this weak bonding is strong contraction of the fully occupied 6s orbital, whose electron pair thereby becomes so "inert" that mercury behaves like the noble gases and has even been referred to as "pseudohelium."

The Strongest Acids

Along with poisons (→ Poisons), acids and bases are usually viewed as the archetypal playground of chemistry. This alone is sufficient reason to include the present subject in our account of records. An assortment of common (Brønsted) acids is given in Table 1. A useful measure of acid strength is the pK_a value,[a] which is the negative logarithm to the base ten of the equilibrium constant K_a. The strongest acids have the *lowest* pK_a values, a relationship that is apparent from the data in Table 1. Whereas water itself is seen to dissociate into ions to only a miniscule extent (the reason why pure water is a relatively poor conductor of electricity), the remaining pK_a values decrease as one moves from hydrocyanic acid (present in trace amounts in bitter almonds) through acetic acid and phosphoric acid. The **strongest of the common acids** are the so-called mineral acids: hydrochloric, nitric, and sulfuric acids. All three have pK_a values that are negative. With acids as strong as these, the measurement of pK_a becomes increasingly difficult, so the data in Table 1 should be regarded as approximate.

Table 1: The pK_a values for selected common acids at 25 °C.[a]

Acid	Formula	pK_a
water	H_2O	15.74
hydrocyanic acid	HCN	9.22
acetic acid	CH_3COOH	4.76
phosphoric acid	H_3PO_4	2.15
nitric acid	HNO_3	–1.3
hydrochloric acid	HCl	–2.2
sulfuric acid	H_2SO_4	–5.2

[a] A. Streitwieser, C. H. Heathcock, E. M. Kosower, *Introduction to Organic Chemistry*, 4th ed., Macmillan New York, **1992**. Only the first dissociation is considered in the case of polyprotic acids.

Nevertheless, even the strong mineral acids are exceeded in acidity by what have come to be called "**superacids.**" This term is applied to compounds whose powers of protonation exceed those of sulfuric acid. In order to quantify superacid behavior, recourse is taken to the acidity function H_0 suggested by Hammett, which is defined as the difference between the pK_a value of a weak indicator base (e.g., an aromatic nitro compound) and the negative logarithm to the base ten of the concentration ratio of the protonated and nonprotonated forms of the base.

$$H_0 = \mathrm{p}K(\mathrm{BH}^+) - \log([\mathrm{BH}^+]/[\mathrm{B}])$$

The H_0 values for several superacids are listed in Table 2.[1] Again, the more negative H_0 the more powerful the acid. It has long been known that simply

[1] D. Lenoir, H.-U. Siehl, *Houben-Weyl: Methoden der Organischen Chemie*, Vol. E19c, Thieme, Stuttgart, **1990**, 18.

Table 2: Hammett H_0 values for selected superacids.

Acid	H_0	Ref.
H_2SO_4	–11.93	[a]
H_2SO_4 + 10% SO_3	–13.93	[a]
HSO_3F	–15.07	[a]
HSO_3F + 10% SbF_5	–18.94	[a]
HSO_3F + 90% SbF_5	–26.5	[b]

[a] D. Lenoir, H.-U. Siehl, *Houben-Weyl: Methoden der Organischen Chemie*, Vol. E19c, Thieme, Stuttgart, **1990**, 18.
[b] V. Gold, K. Laali, K. D. Morris, L. Z. Zdunik, *J. Chem. Soc., Chem. Commun.* **1981**, 769–771.

introducing SO_3 into sulfuric acid with the resulting formation of oleum (fuming sulfuric acid) leads to an increase in protonation capacity by two orders of magnitude. The SO_3 reacts with H_2SO_4 to give $H_2S_2O_7$, which as a stronger acid is capable of protonating sulfuric acid itself to the superacidic cation $H_3SO_4^+$. An even stronger acid is fluorosulfonic acid HSO_3F, with an H_0 value of –15.07. The superacidic species in this compound is the $H_2SO_3F^+$ cation, generated by autoprotolysis. Its concentration can be further increased by trapping the other autoprotolysis product, the SO_3F^- anion, through introduction of the strong Lewis acid SbF_5. The resulting mixture has been called "**magic acid**",[b] with an H_0 value as low as –26.5 depending on the amount of added SbF_5.[2] Such a mixture is thus 10^{15} times as acidic as concentrated sulfuric acid! This enormous protonation capacity is useful in protonating such unreactive compounds as alkanes, transforming them into reactive electrophilic species. Superacids serve also as nonnucleophilic media for the experimental investigation of carbocations (→ Reactive Intermediates, Carbocations).

Astonishing acidic character is not restricted exclusively to inorganic compounds, however. Certain organic compounds also contain readily deprotonatable CH bonds, so that they too can legitimately be described as acids. Special structural and electronic prerequisites must be met by a CH acid for the resulting base, a carbanion, to be stabilized. When this is not the case, the pK_a value of the hydrocarbon is very high, as illustrated by methane with $pK_a = 50$. On the other hand, molecules of the 1,2-difullerene type **1a–c** (Fig. 1) are among the **strongest CH acids** known. Even the alkylated (and thus electron-rich) fullerene derivative **1a** has a pK_a of 5.7.[3] Less electron-rich substituents in the neighboring position, as in the parent hydrocarbon **1b**, reduce the pK_a to 4.7.[4] With a pK_a value of 2.5,[5] α-cyano derivative **1c** is the

[2] V. Gold, K. Laali, K. D. Morris, L. Z. Zdunik, *J. Chem. Soc., Chem. Commun.* **1981**, 769–771.
[3] P. J. Fagan, P. J. Krusic, D. H. Evans, S. A. Lerke, E. Johnson, *J. Am. Chem. Soc.* **1992**, *114*, 9697–9699.
[4] M. E. Niyazymbetov, D. H. Evans, S. A. Lerke, P. A. Cahill, C. C. Henderson, *J. Phys. Chem.* **1994**, *98*, 13 093–13 098.
[5] M. Keshavarz-K, B. Knight, G. Srdanov, F. Wudl, *J. Am. Chem. Soc.* **1995**, *117*, 11 371–11 372.

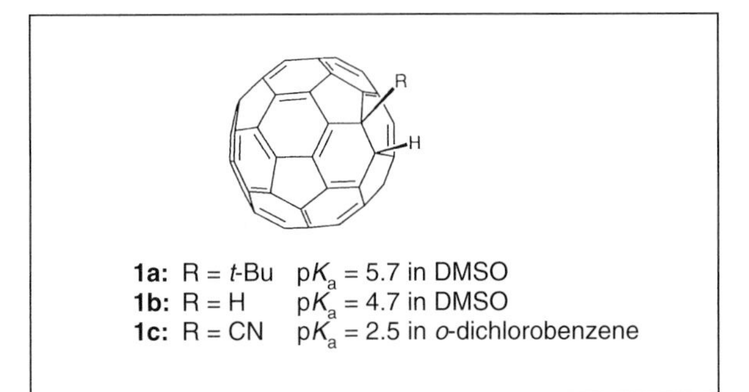

Fig. 1: The strongest known CH acids: dihydrofullerenes

strongest CH acid so far reported. A comparison with Table 1 shows that these compounds are roughly one hundred times as acidic as acetic acid.

Apart from the mineral acids, all of which can be found in nature, the question might be raised as to what is the **most acidic natural product**. A good candidate is the mycotoxin (→ Poisons) moniliformin **2** (Fig. 2), also known as semisquaric acid and discovered along with its sodium and potassium salts in the mold species *Fusarium moniliforme*.[6] Experiments have shown that **2** has a pK_a value of 0.88,[7] making it stronger than any other organic acid. This high acidity is attributed to the great resonance stabilization of the resulting anion.

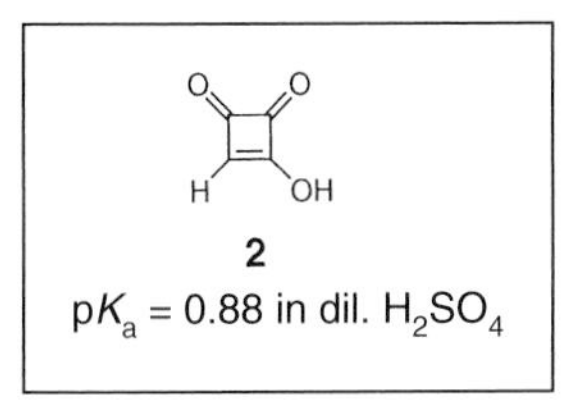

Fig. 2: Semisquaric acid, the most acidic natural product

The Strongest Bases

How does it look at the other end of the scale, where the strong bases are located? From the discussion above it follows that the anions of the weakest acids (e.g., the alkanes) would certainly be the **strongest bases**, since their tendency to acquire protons should be especially pronounced. Indeed, organometallic compounds such as butyl lithium, in which, put very simply, the structure is that of a salt consisting of a butyl carbanion and a Li^+ cation, are

[6] J. P. Springer, J. Clardy, R. J. Cole, J. W. Kirksey, R. K. Hill, R. M. Carlson, J. I. Isidor, *J. Am. Chem. Soc.* **1974**, *96*, 2267–2268.
[7] H.-D. Scharf, H. Frauenrath, P. Pinske, *Chem. Ber.* **1978**, *111*, 168–182.

among the **strongest bases employed in the laboratory**. In analogy to pK_a one can also define for a base a pK_{BH+} value associated with the equilibrium

$$BH^+ \rightleftharpoons B + H^+$$

The methanide ion thus has a pK_{BH+} of 50, whereas the corresponding value is only 15.74 for hydroxide ion in water (as in common "caustic" solutions of potassium or sodium hydroxide). Figure 3 is instructive for comparisons with the so-called **neutral bases**. Tertiary aliphatic amines such as trimethylamine **3** have pK_{BH+} values roughly equivalent to the hydroxide ion. On the other hand, if one examines more complex nitrogen bases such as diazabicyclo[5.4.0]undec-7-ene (DBU) **4**, values are encountered that lie well above 20. The phosphazene bases developed by Schwesinger[8] add a new dimension to the picture. The most basic of these substances, compound **5**, has a pK_{BH+} value never before approached by a neutral base: 46.9. Indeed, this suggests that chemical deprotonation of nonactivated alkanes may someday prove feasible.

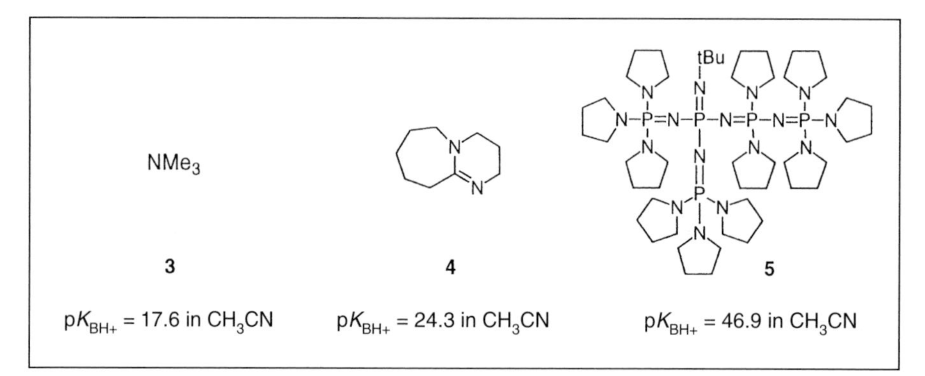

Fig. 3: The strongest neutral bases

[8] R. Schwesinger, H. Schlemper, C. Hasenfratz, J. Willaredt, T. Dambacher, T. Breuer, C. Ottaway, M. Fletschinger, J. Boele, H. Fritz, D. Putzas, H. W. Rotter, F. G. Bordwell, A. V. Satish, G.-Z. Ji, E.-M. Peters, K. Peters, H. G. von Schnering, L. Walz, *Liebigs Ann.* **1996**, 1055–1081.

[a] The term "acid" is often used qualitatively as well in comparison to water. In this case the corresponding reference quantity is the pH; that is, the negative logarithm to the base ten of the hydrogen-ion concentration, [H^+]. Pure water has a pH of 7; the pH value for an acidic solution is lower, that for a basic solution higher.

[b] The designation "magic acid" derives from an astonished outcry by the German postdoc J. Lukas, who in the course of a Christmas celebration in the research group of G. A. Olah (→ Appendix, Nobel Prize Winners: in Chemistry) so expressed himself when he witnessed a wax candle dissolving in this acid. "Magic Acid" is now a registered trademark. See G. A. Olah, *Angew. Chem.* **1995**, *107*, 1519–1532, *Angew. Chem. Int. Ed. Engl.* **1995**, *34*, 1393–1405.

The Highest Transition Temperatures

In 1911, Kamerlingh-Onnes discovered that the electrical resistance of solid mercury disappears at 4.15 K (–269 °C), and below this specific transition temperature (T_c) electric current passes freely without any losses (superconductivity). It is now known that the phenomenon of superconductivity is displayed by most metals and alloys as well as a few semiconductors, ceramics, and other materials, including some molecular substances. Meissner and Ochsenfeld demonstrated in 1933 the especially spectacular effect of magnetic levitation: a magnet can be caused to "float" freely above a superconductor. (A superconductor is perfectly diamagnetic, so an external magnetic field generates an opposing field within the superconductor that precisely compensates for the external field.) Apart from the loss-free transmission of electric current – obviously very interesting from the standpoint of power generating companies – there are a number of potentially attractive applications for superconductors, including magnetically levitated trains (Meissner–Ochsenfeld effect). But superconducting materials would be truly interesting for such applications only if their transition temperatures fell within a readily achievable range, one farther removed from absolute zero. The simplest and best material would of course be one with a transition temperature near room temperature, but the search for such a material has so far proven elusive.

The **metal with the highest transition temperature** is niobium (T_c = 9.2 K = –263.9 °C); among **alloys** the winner is Nb_3Ge (23.2 K = –249.9 °C). Much attention was directed to the observation that certain **alkali metal fullerides** become superconducting at surprisingly high transition temperatures; the record in this case is held by a species with the stoichiometry $Rb_{2.7}Tl_{22}C_{60}$ (T_c = 45 K = –228 °C).[1] Even more spectacular are the transition temperatures of a great many **ceramic superconductors**, the first of which was discovered in 1986 by Bednorz and Müller (→ Appendix, Nobel Prize Winners: in Physics), initiating a veritable race to identify the "hottest" superconductor. This race is still in progress, and has led to many premature reports of records. At the moment, a ceramic with the approximate composition $HgBa_2Ca_2Cu_3O_8$ is in the lead (T_c = 135 K = –138 °C;[2] indeed, a transition temperature of 164 K = –109 °C has been achieved under pressure).[3] This material thus possesses the **highest reproducible transition temperature for a superconductor** (Fig. 1). Superconductors of this type have transition temperatures significantly above the boiling point of liquid nitrogen (77.3 K = –195.8 °C), which is commercially available in enormous quantities from the liquefaction of air and is a

[1] Z. Iqbal, R. H. Baughman, B. L. Ramakrishna, S. Khare, N. S. Murthy, H. J. Bornemann, D. E. Morris, *Science* **1991**, *254*, 826.

[2] A. Schilling, M. Cantoni, J. D. Guo, H. R. Ott, *Nature (London)* **1993**, *363*, 56. C. W. Chu, L. Gao, F. Chen, Z. J. Huang, R. L. Huang, R. L. Meng, Y. Y. Xue, *Nature (London)* **1993**, *365*, 323.

[3] L. Gao, Y. Y. Xue, F. Chen, Q. Xiong, R. L. Meng, D. Ramirez, C. W. Chu, J. H. Eggert, H. K. Mao, *Phys Rev. B* **1994**, *50*, 4260.

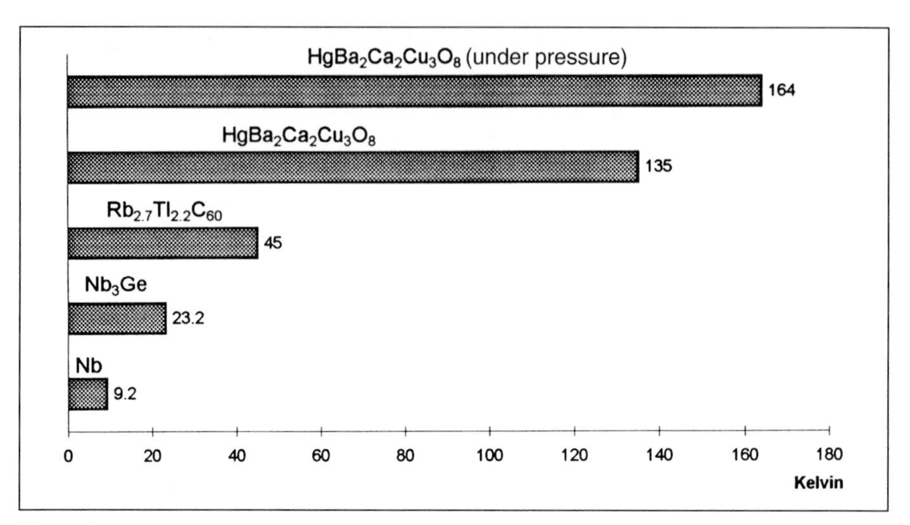

Fig. 1: Transition temperatures

convenient and readily accessible cooling agent. Nevertheless, industrial application of these high-temperature superconducting ceramics is limited by problems with their processing.[a] The **first report of high-temperature superconductivity** appeared as early as 1946 when Ogg reported that if a solution of sodium in ammonia is rapidly cooled to ca. 93 K (–180 °C) it can be shown subsequently to be superconducting up to 190 K (–83 °C). Ogg's efforts attracted little attention, however, because of problems with reproducibility. He suggested electron-pair formation as an explanation for the superconducting behavior of these quenched solutions of sodium in ammonia. At first glance the idea of such pair formation seems counterintuitive due to the electronic repulsion that prevails between two electrons. Nevertheless, it is a possibility, since coulombic repulsion under certain conditions can be more than compensated by lattice forces. Electrons with opposite spins and coupled in pairs through quantized vibrations of the crystal lattice (phonons) have an overall spin of zero and thus class as bosons. The Pauli exclusion principle applies only to particles with spins of one-half (fermions, including electrons), so phonons can in principle exist all together in one and the same quantum state of lowest energy. This has as a consequence that there would be no energy exchange with the environment, so the phonons could move about through the lattice without resistance. To paint a greatly simplified picture of the situation, the second electron of the pair swims through the lattice in the wake of the first electron, and vice versa – each essentially pulling the other along without any scattering from components constituting the lattice. A quantification of this explanation originally proposed by Ogg was provided in 1957 by Bardeen,

[4] R. A. Ogg, Jr., *Phys. Rev.* **1946**, *69*, 243. *Ibid.* **1946**, *69*, 668. *Ibid.* **1946**, *70*, 93.

Cooper, and Schrieffer (BCS theory) (→ Appendix, Nobel Prize Winners: in Physics). Since then the electron pairs responsible for superconductivity have been referred to as Cooper pairs. Unfortunately, satisfactory application of the BCS theory fails precisely in the case of the high-temperature superconducting ceramics.

[a] Superconducters are currently utilized especially in two fields: for the measurement of extremely weak magnetic fields with so-called SQUIDs (superconducting quantum interference devices) and in achieving extremely high magnetic fields. (In 1996 it proved possible to produce stable flux densities of 11 tesla with an electromagnet based on a 40-ton coil made from a superconducting niobium–tantalum alloy. This corresponds to 220 000 times the mean magnetic field of the earth; see *Phys. Unserer Zeit* **1997**, *28*, 43.) The superconductors employed in these applications are not high-temperature ceramic materials, but rather the much more easily machined metallic superconductors. The requisite high investment in cooling – in this case with liquid helium – is simply tolerated in such special situations. For information regarding the latest developments in materials and processing for ceramic high-temperature superconductors (for example, the "powder-in-a tube" method) in the quest to prepare flexible wires, see for example H.-J. Kalz, H. Eckhardt, *Spektrum d. Wiss. Digest* January **1996**, *3*, 38.

World Record in the High-Jumping Crystal Contest

Molecules of a crystalline solid prove to be remarkably mobile despite the apparently placid exterior of the average crystal. Changes in temperature can cause such molecules to reorient themselves not only intramolecularly but also relative to one another, which on the macroscopic level can lead to significant demonstrable changes in crystalline form. Crystalline phase transitions of this type are especially striking with compounds in which energy changes accompanying a molecular repositioning manifest themselves as tiny "crystalline leaps." Thus, crystals of the *myo*-inositol derivative **1**,[1] for example, jump several centimeters high when warmed from room temperature to about 70 °C. This odd behavior is even reversible: upon cooling to ca. 40 °C the crystals proceed to jump again. Investigations of single crystals of the compound[2] have shown that during the phase transformation the needle-shaped crystals become bent and ca. 10% shorter. Reorientation of the molecules spreads from one end of the crystal ($30 \times 0.5 \times 0.3$ mm) to the other within one one-hundredth of a second. Mechanical tensions that arise between individual crystal layers are presumably responsible for the observed "hopping" effect.

Although crystal-hopping is a relatively underrepresented discipline in the realm of solids, **1** is not the only compound to display this sporting behavior. Crystals of the perhydropyrene **2**[3] and of the palladium complex **3**[4] show a tendency to execute more or less large leaps (6 cm in the case of **2**, for example). However, the **unofficial champions** are crystals of the alloy MnCoGe.[5] Upon heating they have been known to spring to heights of as much as 30 cm!

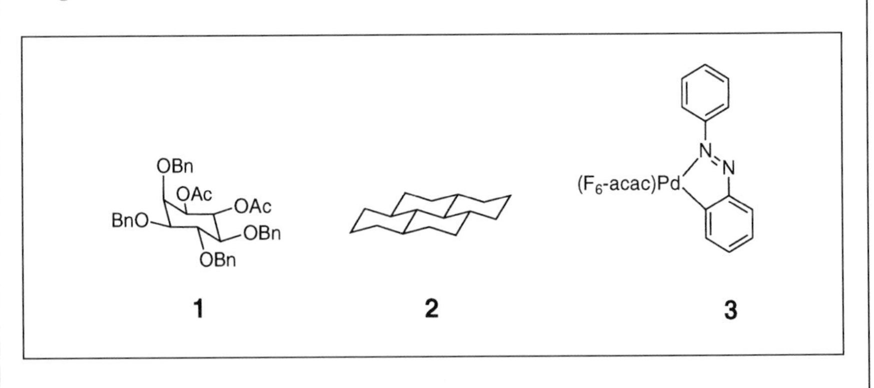

[1] J. Gigg, R. Gigg, S. Payne, R. Conant, *J. Chem. Soc., Perkin Trans. 1* **1987**, 2411–2414.
[2] T. Steiner, W. Hinrichs, W. Saenger, R. Gigg, *Acta Crystallogr.* **1993**, *B49*, 708–718.
[3] B. Kohne, K. Praefcke, G. Mann, *Chimia* **1988**, *42*, 139–141.
[4] M. C. Etter, A. R. Siedle, *J. Am. Chem. Soc.* **1983**, *105*, 641–643.
[5] W. Jeitschko, *Acta Crystallogr.* **1975**, *B31*, 1187–1190.
[6] Cited according to ref. [1].

General

The **most robust ceramic**[1] **with respect to all its characteristics** has the composition $SiBN_3C$.[2] It is so far the nonoxidic ceramic that is most stable toward oxidation, and it can tolerate thermal loads under inert conditions up to 1900 °C while remaining amorphous. Other advantageous properties include a low density (1.8 kg/L), low thermal coefficient of expansion (2×10^{-6} K^{-1}), low thermal conductivity (0.4 W/m K at 1500 °C), extremely high resistance to thermal shock, impressive hardness (about the same as corundum, with a Mohs hardness of 9), and high mechanical loading capacity (→ Atoms and Molecules, Hardness Records). This ceramic can be processed into hair-like fibers that can even be interwoven. Similarly robust is a related ceramic of the composition $Si_{3.0}B_{1.0}C_{4.3}N_{2.0}$.[3]

The **lowest-melting alloy** is the mercury–thallium eutectic (8.5% Tl) with a melting point of –60 °C.

The **lowest-melting robust glasses** consist of thallium oxide, selenium oxide, and arsenic oxide; some of them melt as low as 125 °C.

The **most dense colorless material** is lutetium tantalate (9.75 kg/L).

[1] Brief introductions to various aspects of the extensive field of high-performance ceramics are provided by G. Petzow, F. Aldinger, *Spektrum d. Wiss. Digest* January **1996**, 3, 44; H. Prielipp, J. Rödel, N. Claussen, *ibid.*, 49; R. Hamminger, J. Heinrich, *ibid.*, 54; R. Riedel, *ibid.*, 56.

[2] H.-P. Baldus, M. Jansen, *Angew. Chem.* **1997**, 109, 338; *Angew. Chem. Int. Ed. Engl.* **1997**, *36*, 328.

[3] R. Riedel, A. Kienzle, W. Dressler, L. Ruwisch, J. Bill, F. Aldinger, *Nature* (London) **1996**, *382*, 796.

The Most Dynamic Growth[1]

Individual genes were isolated and incorporated into other organisms for the first time about 25 years ago. Since then there has been a downright explosion in biotechnological knowledge. It should also be possible to recast this technology in the form of commercial success. The OECD therefore assumes that biotechnology will develop into one of the most economically powerful branches of learning. At this point it is difficult to present even a rough estimate of its economic importance, however, because biotechnology is increasingly becoming an ancillary industry, the products and knowledge from which also have their effects in more established sectors ("interdisciplinary technology"). Included, among others, are the pharmaceutical (→ Pharmaceuticals) and food industries, seed production, crop protection (→ Crop Protection), chemistry, and environmental technology. This results from the fact that today many biotech companies are pursuing a technology-oriented strategy; that is, they are less interested in bringing their own products to market than in becoming specialists in particular methods and techniques that can in turn be sold and/or licensed to other well-established firms.

Depending upon how biotechnological products are defined, current annual sales within Europe are estimated to be $15–35 billion, with predicted annual growth rates for the coming years of 15–20%. Based on data from the EU commission, approximately 9% of the net European industrial product is being generated by industries heavily influenced by genetic engineering and biotechnology.

Using a narrower definition of the biotech industry, sales in biotechnology in 1995 amounted to only $1.1 billion.

World Champions in the Licensing and Utilization of Drugs Prepared by Genetic Engineering

Given the impassioned and prolonged discussion in Germany regarding the advantages and disadvantages of genetic engineering it may come as a surprise to learn that more genetically engineered drugs have been licensed there than in either the United States or Japan (Fig. 1).It is *not* surprising, however, that of the 22 genetically engineered active substances permitted in Germany, only 4 are also produced in that country (Table 1).

[1] data from: *Handelsblatt*, July 22, 1996.

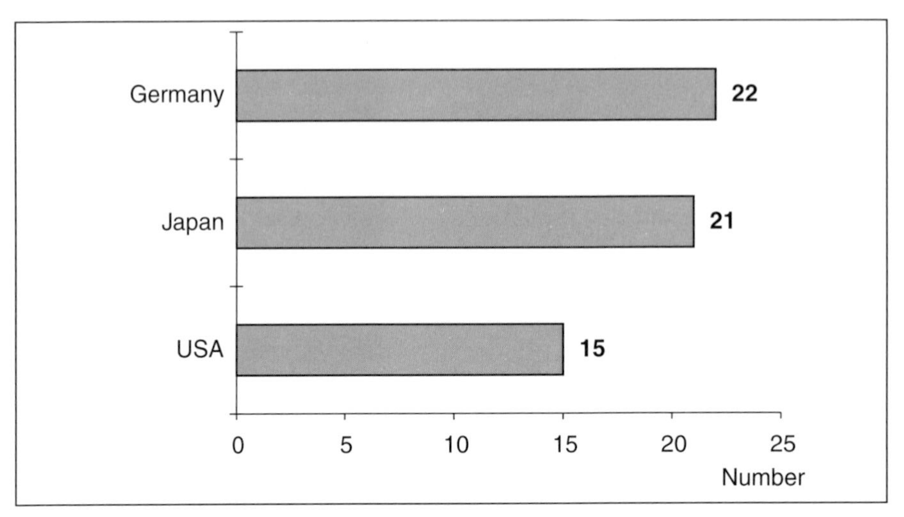

Fig. 1: Number of genetically engineered active ingredients licensed in various countries (end of 1995)[2]

Table 1: Genetically engineered active ingredients manufactured in Germany

Substance	Company	Indication
tissue-plasminogen activator	Thomae	heart attack
tissue-plasminogen activator	Boehringer Mannheim	heart attack
pro-urokinase	Grünenthal	heart attack
erythropoietin	Boehringer Mannheim	renal anemia

Recognition that genetic engineering can or soon will play an important role in the treatment of disease (e.g., diabetes, hemophilia, cancer) has so far

The First Genetically Engineered Drug

Blood-clotting factor VIII, produced by Bayer, is the **largest protein** (→ Molecular Form, Giants) **so far prepared by genetic engineering**. Assembled from 2332 amino acids, it has a molecular weight of 300 000. (For comparison: human insulin consists of only 51 amino acid residues.) Factor VIII, which is highly glycosylated, is also the **first genetically engineered product** from an animal cell culture that has been **accepted worldwide** for long-term treatment. As a coagulating agent it plays a critical life-sustaining role. In hemophiliacs, this protein is missing or does not function properly. As a result, even a minor incident of untreated bleeding can lead to death.

[2] data from: *CHEManager*, September, 1996.

apparently not stimulated the insight that industrial application of similar methods for producing drugs is a vital consideration for Germany as a site for industry (Table 2).

Table 2: Genetically engineered drugs

Active ingredient	Indication	Company
Aldesleukin (interleukin 2)	hypernephrom	Chiron
Alglucerase (modified glucocerebrosidase)	Gaucher's disease	Genzyme
Dornase alfa	cystic fibrosis	Roche
Epoetin alpha and beta (erythropoietin)	anemia	Janssen-Cilag, Boehringer Mannheim
factor VIII	hemophilia	Baxter, Bayer
Filgastrim (G-CSF, cytokine)	neutropenia	Amgen/Hoffmann-La Roche
follitropin alpha	sterility	Ares Serono
follitropin beta	sterility	Organon
hepatitis A/B vaccine	combination vaccine	SmithKline Beecham Pharma
hepatitis B/DPT vaccine	combination vaccine	SmithKline Beecham Pharma
hepatitis B vaccine	hepatitis B vaccine	SmithKline Beecham Pharma, Pasteur Mérieux, Behringwerke
insulin (Lispro)	diabetes	Lilly
insulin (human)	diabetes	Lilly, Novo Nordisk
interferon alpha 2a and alpha 2b	hair cell leukemia	Essex, Roche
interferon beta 1b	multiple sclerosis	Schering
interferon gamma 1b	chronic granulomatosis	Thomae
lenograstim	neutropenia (white blood cell deficiency; e.g., after chemotherapy)	Rhône-Poulenc Rorer
molgramostim	neutropenia (rhu-GM-CSF, cytokine)	Sandoz/Essex
MSD/Beh-Alteplas (tPA)	heart attack	Thomae
rec. factor VIIa	special hemophilia	Novo Nordisk
reteplase	heart attack	Boehringer Mannheim
somatotropin	pituitary hyposomia	Upjohn, Pharmacia, Lilly, Novo Nordisk, Serono, Ferring

(no guarantee of completeness; status on December 10, 1996)

Leading Biotech Regions

The biotechnological industry is expected to be one of the key industries in the 21st century. It already encompasses more than 3000 companies worldwide. About 1300 are in the United States, and somewhat more than 1000 are in Europe. Although Europe has succeeded in significantly closing the gap to the U.S. in recent years, the latter remains clearly in the lead in biotechnology. Thus, U.S. concerns recorded $17.4 billion in **sales** during 1997, nearly six times as much as their European counterparts, at the same time expending about 4.3 times as much for R&D (Fig. 2).

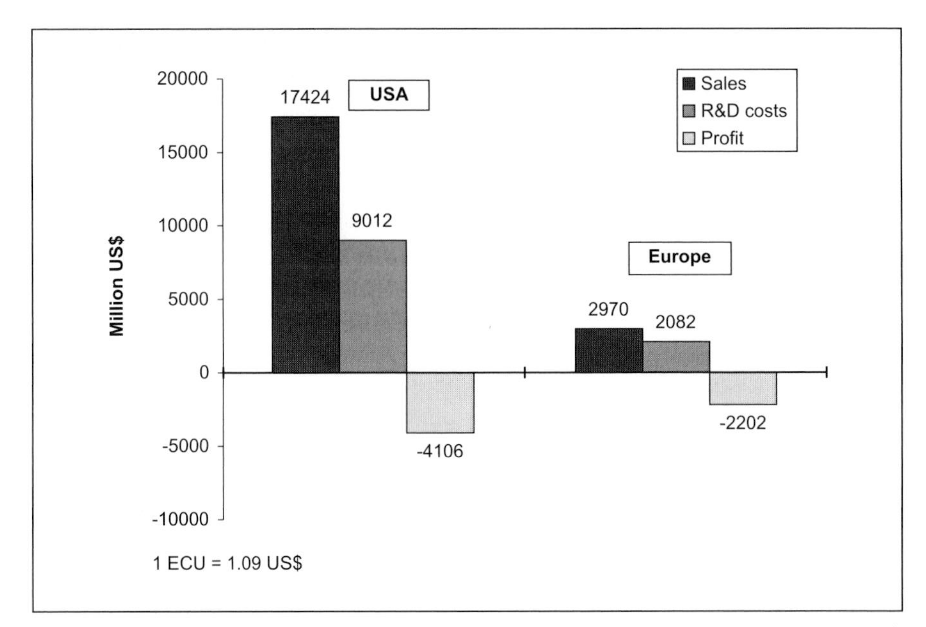

Fig. 2: Comparisons between the European and U.S. biotechnology industries, 1997 (in $ million) [3]

With respect to **personnel**, too, American companies (140 000 employees) far outstripped the Europeans (39 000 employees). The same applies to corporate losses, however, which in the U.S. amounted to $4.1 billion in 1997.

The heavy losses can be ascribed to the fact that they were recorded mainly by young companies attempting to use special biotechnological know-how to discover and develop new products and processes. This strategy demands high R&D expenditures, which are difficult to cover during a company's early years when there is little or no income available. Many companies simply fail to survive the inevitable dry spell.

---

[3] data from: Ernst & Young, *European Life Sciences '98: Continental Shift*, p. 11.

Leading Firms

The total market value (current price per share × number of shares outstanding) for the five leading biotechnology firms (according to market capitalization) in the United States at the end of 1997 amounted to $27 billion, 7.4 times that of the corresponding European firms ($3.64 billion).

The leader of the U.S. list was Amgen, with a market capitalization of $15.4 billion, more than the market value of the next four companies combined (Fig. 3a). Among these, Chiron with a market value of $3.9 billion was in clear possession of second place, having lifted itself from its third-place position at the end of 1995.

By comparison, the spectrum of the top five European biotech companies at the end of 1997 was much more homogeneous. The company in fifth place, Shire Pharmaceuticals, could still claim a market value 52% that of the leader, British Biotech, which was forced to abandon its lead in the first quarter of 1998 (first place: Innogenetics; second place, Qiagen; Fig. 3b).

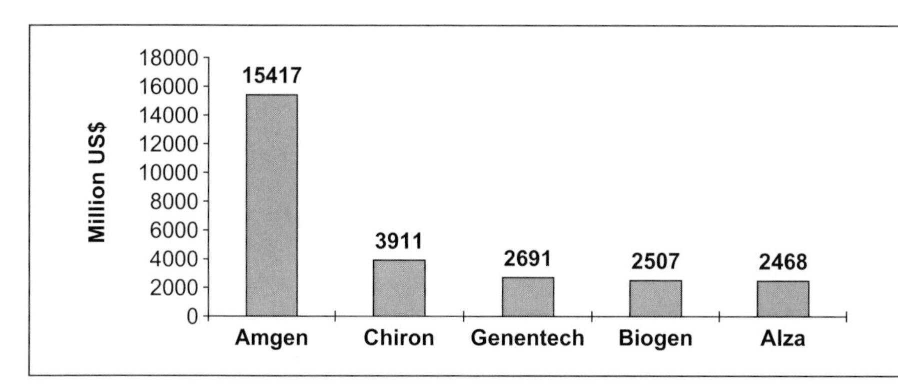

Fig. 3: a) Ranking of U.S. biotechnology companies (based on market value, end of 1997)[4]

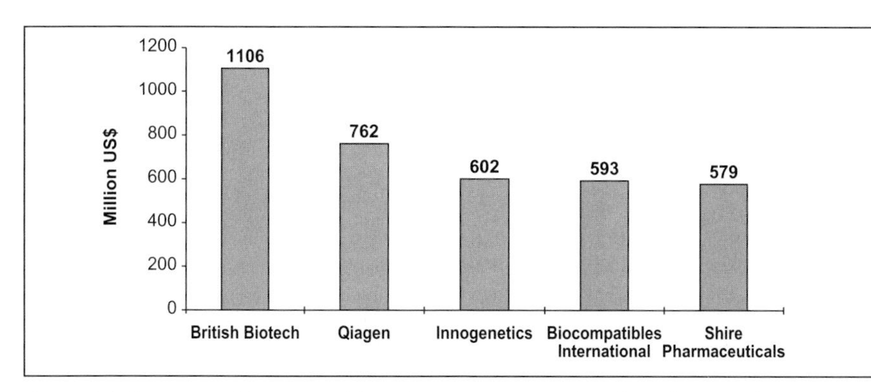

Fig. 3: b) Ranking of European biotechnology companies (based on market value, end of 1997)[4]

[4] data from: Ernst & Young, *European Life Sciences '98: Continental Shift*, p. 12.

No single European firm could point to a market capitalization comparable to those of the top five in the U.S. Indeed, their combined market values amounted to only one-fourth that of Amgen.

If the highly capitalized biotech companies are arranged according to sales, a different order emerges. In Europe in 1997 Qiagen occupied first place (with sales of $74.3 million) ahead of Scotia Holdings, where the latter is not even present in the list of the top five based on market capitalization. It is noteworthy that, with the exception of the leader, R&D expenditures relative to sales were still exceedingly high. It is also apparent that sales for the European firms were quite modest compared with their U.S. counterparts (Fig. 4). In this case the leader alone managed to report sales in 1997 of a respectable $2.7 billion (→ Top Ten Biotech Products).

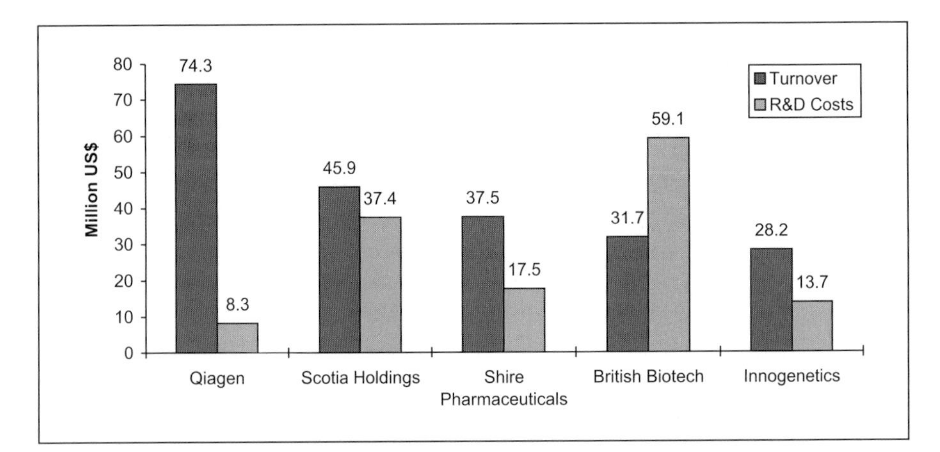

Fig. 4: a) Sales of leading European biotechnology companies, 1997[4]

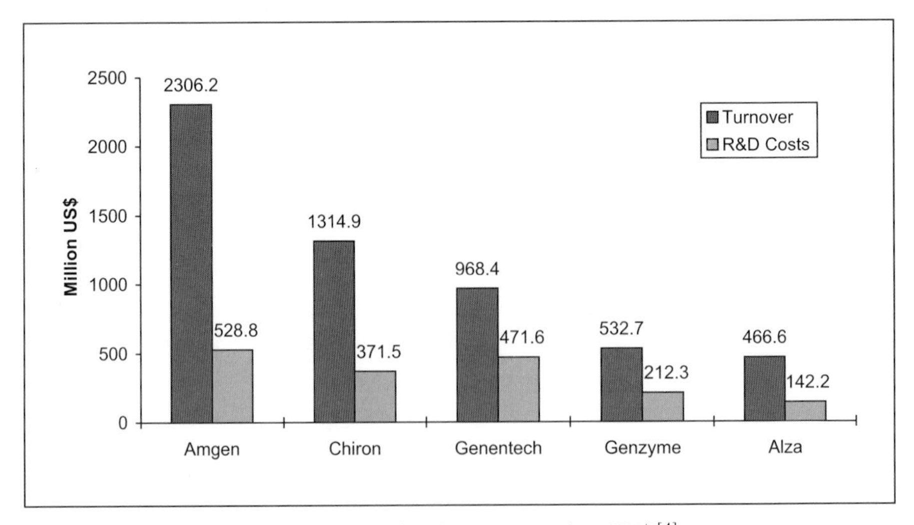

Fig. 4: b) Sales of leading U.S. biotechnology companies, 1997[4]

High Expenditures for Research and Development

Biotech companies are noted for relatively high research budgets. The 1997 **R&D costs per employee** for the five most research-intensive European "biotech flagships" (the companies with the highest market capitalization) ranged from $155 000 for Celltech Group to $89 000 for Scotia Holdings. The corresponding figures for American companies ranged from $196 000 (Biogen) to $86 000 (Alza), substantially above those of their European counterparts (Fig. 5).

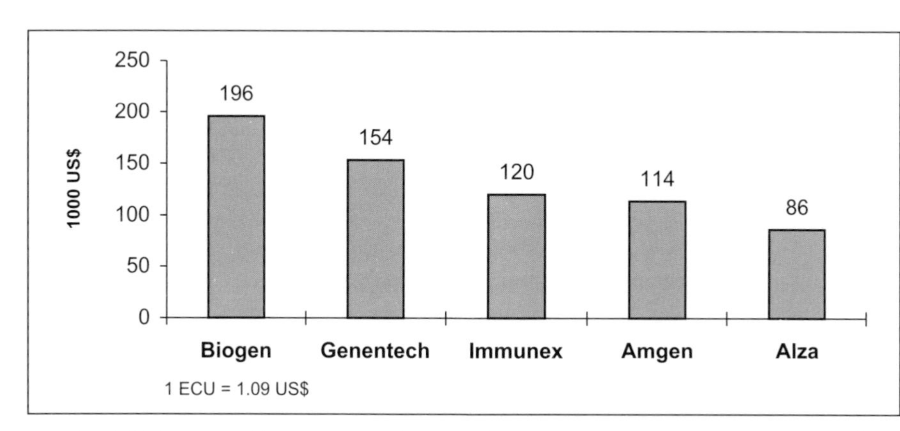

Fig. 5: a) R&D expenditure per employee of large U.S. biotechnology companies[4]

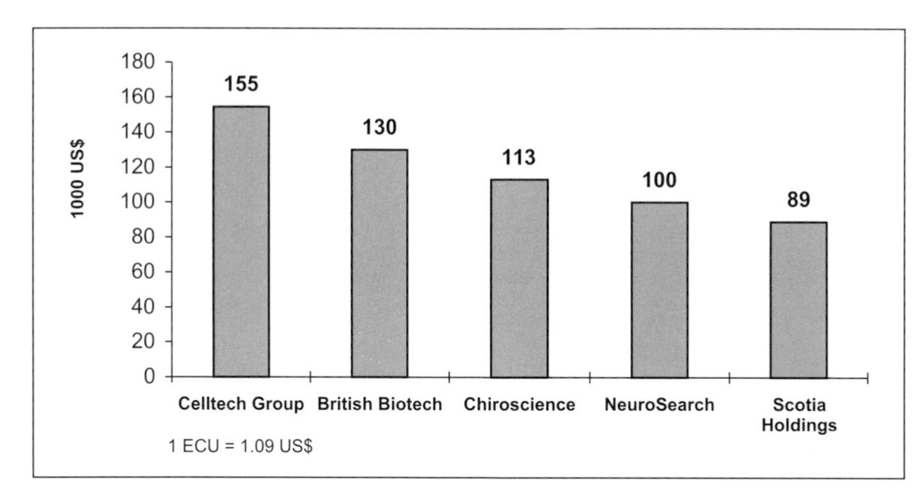

Fig. 5: b) R&D expenditure per employee of large European biotechnology companies[4]

A comparison with R&D expenditures for the most research-intensive pharmaceutical companies – a group well-known for high research costs – underscores again the towering significance of research and development in

Biotechnology Facts and Figures

the biotech field. In 1997 the same European biotechnology firms cited above spent 2.5–4.3 times as much per employee on R&D as the pharmaceutical giant Glaxo Wellcome.

Top Ten Biotech Products

Worldwide sales of the **top ten biotech products** combined amounted in 1996 to $6.35 billion. Nearly half of this total represented products developed by Amgen, the source of Epogen® – a growth factor for red blood cells for the treatment of kidney disease – which was the leader in 1996, and also the other two contenders for the lead: Neupogen® (a growth factor for white blood cells) and Procrit® (Fig. 6).

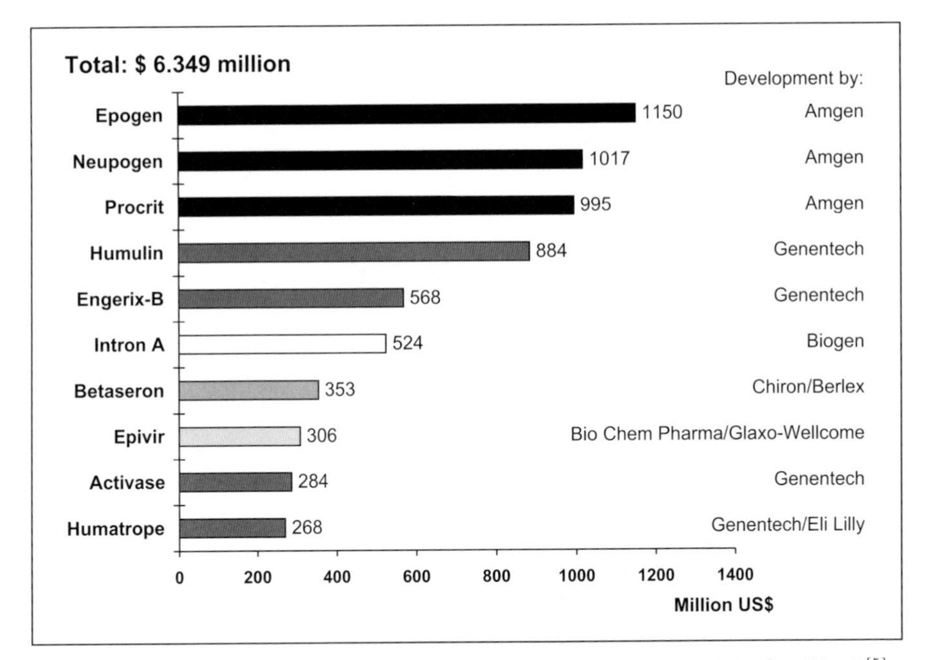

Fig. 6: Worldwide sales for the top ten biotechnology products, 1996 (in $ million) [5]

With respect to number of products for the top ten derived from in-house research, the frontrunner is Genentech, whose researchers managed to steer four products to the hit list, although one of these (Humatrope®) was developed jointly with Eli Lilly.

[5] data from: Ernst & Young, *Erster Deutscher Biotechnologie Report*, p. 21.

Agricultural Biotechnology: Transgenic Crops

Genetic engineering is gaining increased attention also in the field of agriculture. More and more of the crops that are planted display improved properties as a result of genetic modification (at the present time, especially soybeans, corn, cotton, and summer rape). It has thus become possible to grow corn that is resistant to the effects of European corn-borer larvae, which can otherwise result in crop losses of 7–20%. Moreover, certain plants have also been made resistant to the effects of herbicides.

Such genetically engineered plants in 1997 accounted already for nearly 32 million acres of cropland, most of which could be attributed to herbicide-resistant species (Fig. 7).

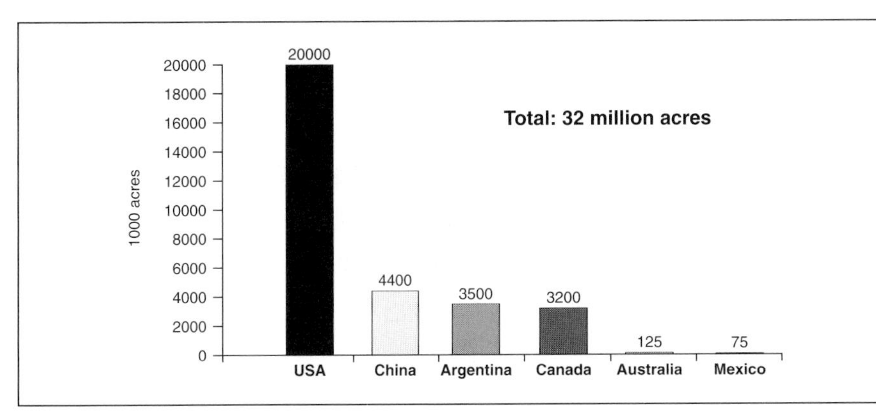

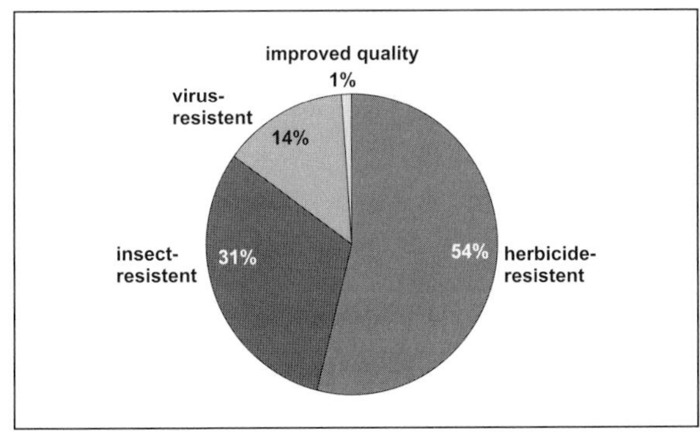

Fig. 7: Land devoted in 1997 to transgenic crops, depicted by country and by the characteristics modified[6]

The **primary region for the use of transgenic crops** in 1997 was North America, which accounted for nearly three-quarters of the total acreage worldwide. Indeed, 64% of the total was in the United States alone, where in 1997 14% of the soybean (only 2% in 1996) and 20% of the cotton acreage was devoted to genetically modified plants.[6]

[6] Information from: Dechema biotechnology secretariat, August, 1998.

The Smallest and the Largest Catalysts

The **smallest catalyst** is the **proton**, and the **largest (homogeneous) catalysts** are **enzymes**. Enzymes have molar masses between ca. 10 000 and 400 000 g/mol.[a] In other words, they can be nearly four million times heavier than a proton. Their molecular structures are correspondingly complex. Looked at simplistically, countless enzymes amount to nothing more than protons embedded in incredibly clever architectural settings: The proton-catalyzed reactions they are intended to mediate occur only with substrates that fit well into the corresponding environment.

The Most Active Catalysts

The most active catalyst for the polymerization of ethene is described as shown in **1** (Fig. 1), activated with methylaluminoxan (MAO); its activity corresponds to 60 kg of polyethylene per milligram of zirconium and hour (or 5473 kg per millimole of zirconium and hour) at 100 °C in toluene under 14 bar of ethene.[1] It is very likely that **2**, also activated with MAO, is the **most active catalyst for the polymerization of propene**, with an activity of 9.6 kg of isotactic polypropylene per milligram of zirconium and hour (or 875 kg per millimole of zirconium and hour) at 70 °C in liquid propene.[2]

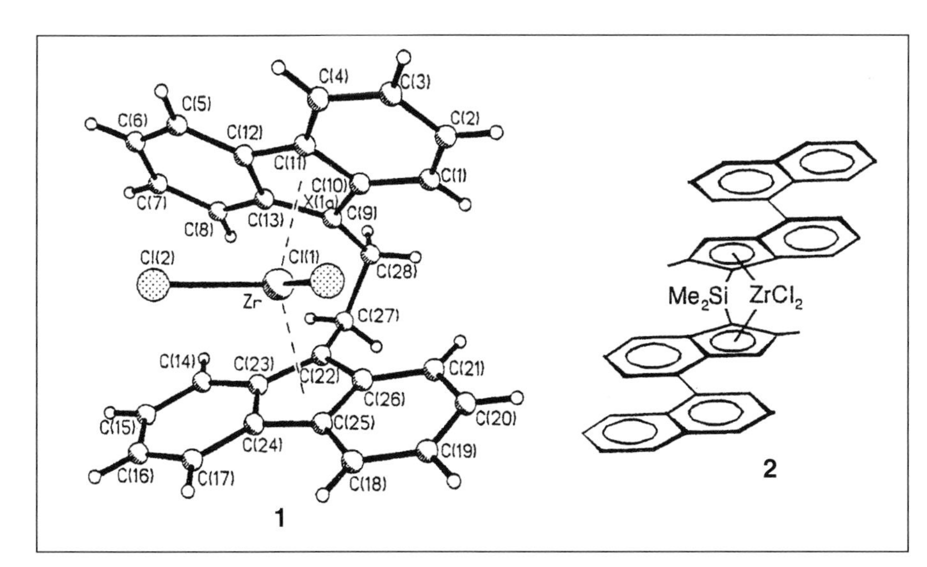

Fig. 1: The most active catalysts (metallocenes) for the polymerization of olefins

[1] H. G. Alt, W. Milius, S. J. Palackal, *J. Organomet. Chem.* **1994**, *472*, 113.
[2] M. Aulbach, F. Küber, *Chem. Unserer Zeit* **1994**, *28*, 197.
[a] Desulfovibriohydrogenase, with a molar mass of ca. 9000 g/mol, is an unusually small enzyme. Multienzyme complexes such as pyruvate dehydrogenase often have molar masses on the order of 10^6 g/mol.

The World Market for Catalysts

According to information from The Catalyst Group, a marketing research institute, the world market for catalysts in 1995 amounted to $8.6 billion.

The largest share of this market consisted of catalysts for emission control ($3.1 billion), including those for both automobiles and the industrial sector (Fig. 2).

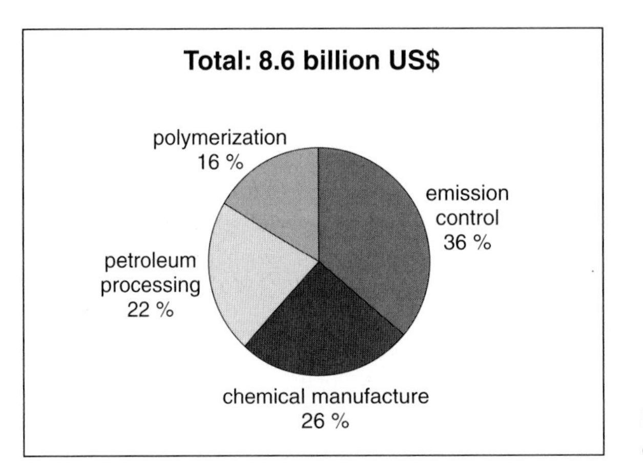

Fig. 2: World market for catalysts, 1995[3]

The **most important market** for catalysts in 1995 was North America, with a volume of nearly $3 billion, ahead of Western Europe ($2.2 billion) and Japan ($0.9 billion). The markets in this triad shared one characteristic not prominent elsewhere in the world: They were dominated by emission-control catalysts. Catalysts of this type are expected to show the highest growth rate worldwide in the years ahead.

The catalyst market as a whole is regarded as a growth market; its volume is expected to expand by the year 2001 to $10.7 billion.

[3] data from: *ECN Chemscope* (06), June 1, 1996, pp. 22–23; *Chem. Marketing Reporter* (18), April 29, 1996.

Top Ten Employers

Of the 500 companies worldwide with the highest sales figures in 1997, just as in 1996 only three (1995: four) were chemical corporations with six-figure employee rosters. Among these, Bayer (144 600) was by a wide margin the **leading chemical employer**. Hoechst and BASF, two other German firms, followed in second and third place ahead of the largest US-based chemical employer, Du Pont, with ca. 98 400 employees (Fig. 1).

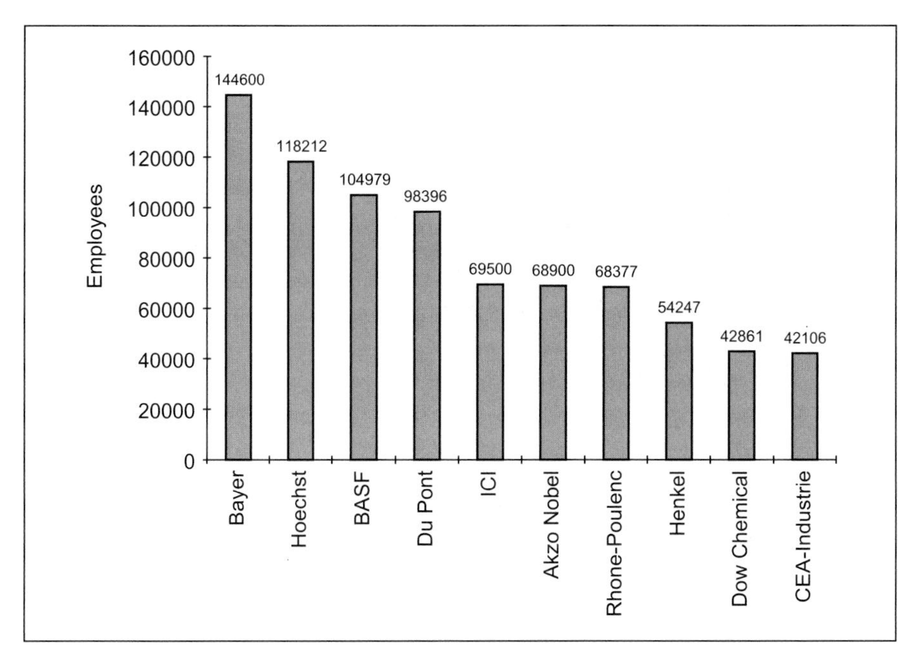

Fig. 1: Top ten chemical employers worldwide, 1997[1]

The top ten chemical employers together accounted for a total of ca. 812 200 positions (1995: 900 120). This roughly corresponds to the number employed by Wal-Mart Stores, the second largest employer after the U.S. Postal Service among the 500 corporations with the largest sales worldwide.

Chemical Jobs in Western Europe

In 1995 the European chemical industry, still the **largest collection of chemical manufacturers in the world**, offered **employment opportunities** for

[1] data from: *Fortune*, "The Global 500," Aug 3, **1998**, pp. 74f.

roughly 1.8 million workers. Approximately 30% of these were in Germany, placing it in the lead position among the top five, with more employees in this branch than second- and third-place contenders France and the United Kingdom combined. Italy and Spain rounded out the list of Western European countries whose chemical industries employed more than 100 000 workers (Fig. 2).

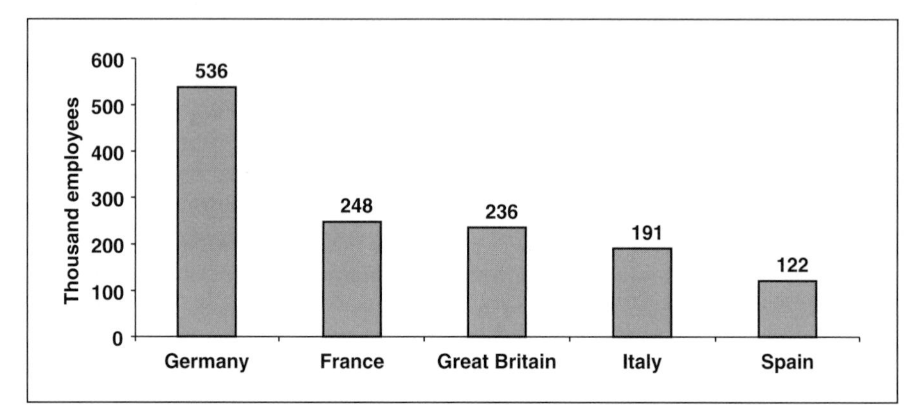

Fig. 2: Top five chemical countries in Western Europe, 1995 (based on number of employees) [2]

The number of employees in Western Europe's chemical industry has shrunk in recent years due to restructuring efforts set in motion by increasing international competition.

Trends in Employment Figures in Germany

From a level of 550 000 in 1984, the number of employees in the West German chemical industry increased steadily for several years at a rate of 0.5–1.8%, reaching a total of 592 000 in 1990. As a result of German reunification, a **record level of 717 000 employees** was posted in 1991 (Fig. 3).

However, it proved impossible to come even close to maintaining this new mark. Within only three years the employee rolls had contracted to 570 000, roughly equivalent to the 1986 level prior to reunification.

The downward trend has slowed significantly in more recent years. In 1996 the chemical industry constituted Germany's fourth largest employer, with 518 000 workers, exceeded only by the mechanical engineering, automotive, and metal-products industries.

[2] data from: *CHEManager* 10/96, p. 2.

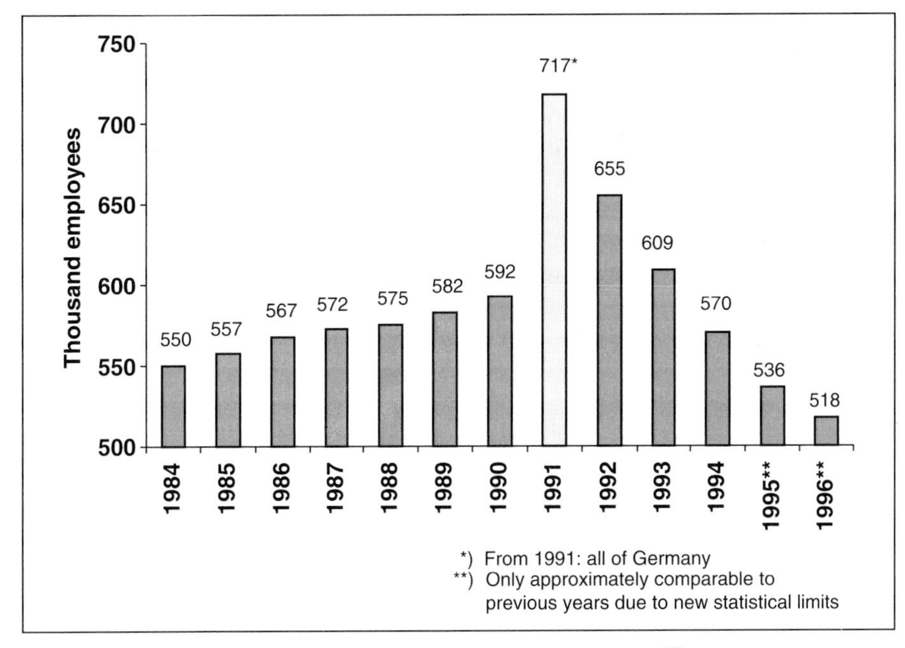

Fig. 3: Number of employees in the German chemical industry[3]

Nevertheless, current data suggest that the decline in employment is not yet over. Thus, the number of chemical employees in Germany had fallen by the end of the first quarter of 1997 to 507 140.

The Country with the Highest Labor Costs

Labor costs are an important factor in the competitiveness of German chemical products in the world market. In this respect, just as in the preceding year, the situation for West Germany was not particularly bright in 1997. Compared with its 11 European competitors, Japan, and the United States, the West German chemical industry continued to show the **highest labor costs per employee-hour**. Germany can thank among other things the high cost of fringe benefits (currently 98.5% of wages) for its leading position.

[3] data from: Verband der Chemischen Industrie VCI, *Chemiewirtschaft in Zahlen 1996*, supplemented by more recent communications.

Germany's lead over its competitors has somewhat diminished as a result of both currency-exchange rates and modest wage increases. Still, an hour of labor in Japan (fourth place) remained \$9.50 (23.4%) less expensive than in Western Germany, and the corresponding advantage for industries in the United States was \$17.90 (43.9% less) (Fig. 4).

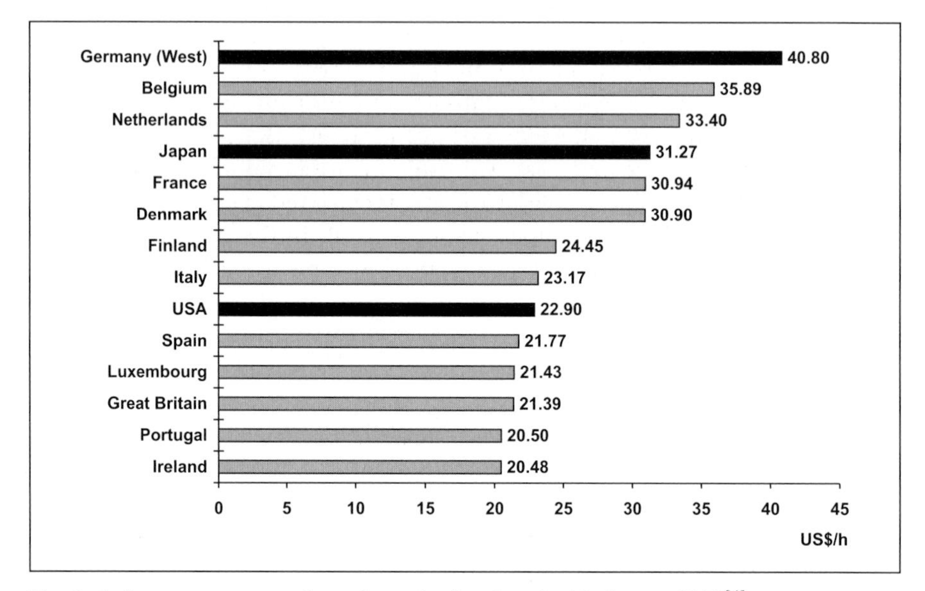

Fig. 4: Labor costs per employee hour in the chemical industry, 1997[4]

[4] data from: *Europa Chemie 22–23*, **1998**, p. 6.

Chemistry and the Automobile

The **role of chemistry in the automobile** has increased significantly in recent years. A modern middle-class car may today contain as much as 255 kg of chemical material, representing about 25% of its total weight. These materials correspond to a value of $800–1100 and make a critical contribution both to automobile safety (→ Molecular Energy, Explosives: Airbags) and to travel comfort. Moreover, a reduction in weight and the associated decrease in fuel consumption results in less threat to the environment.[1]

The future of the automobile is closely coupled with the development of novel materials that have their genesis primarily in the chemical laboratory. One can therefore anticipate that chemistry's share in the automotive market will continue to grow. The automotive sector is expected to expand in the years ahead in virtually all parts of the world – especially in the Far East – (Fig. 1), and it can thereby be regarded as a promising growth market for chemical manufacturers.

The number of automobiles on the road worldwide is estimated now to be roughly 650 million.

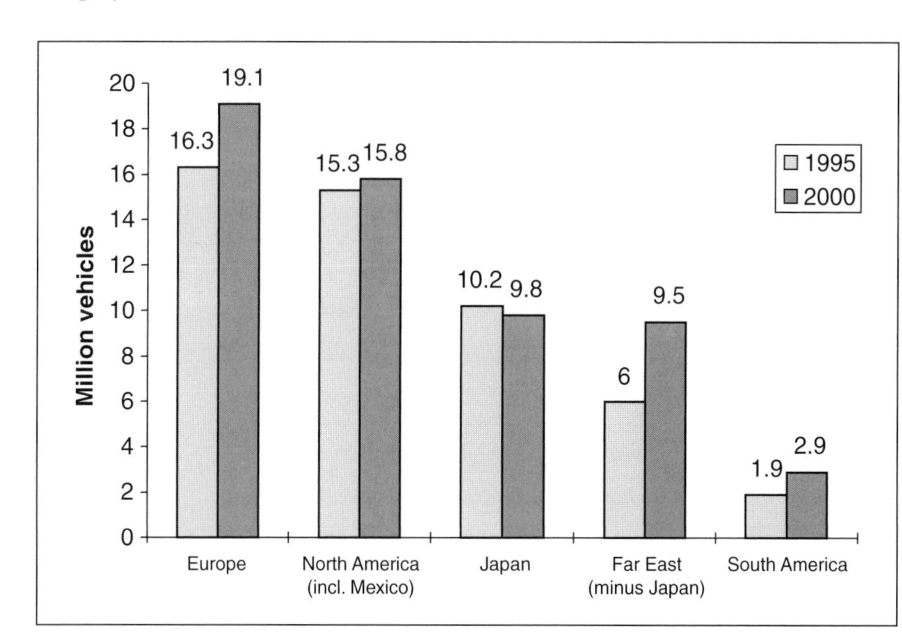

Fig. 1: Automobile manufacture by region (millions of passenger and commercial vehicles)

As noted above, as much as 255 kg of chemical material may be present in a **modern middle-class automobile** (Fig. 2). Of this total, plastics alone account

[1] data from: Bayer press release, Sept. 12, 1996.

for 100–125 kg (dashboards, body parts, lights). The second most important material in this category is rubber, found primarily in the tires but also present in engine mounts, gaskets, drive belts, and the like. Polyurethane consumption worldwide by the automotive industry (ca. 900 000 tons per year) is directed chiefly toward seats, upholstered safety features, fenders, and side panels. Interior comfort is enhanced by textiles and tapestries, and an appealing exterior and corrosion protection for one's costly vehicle are assured by paints and underbody corrosion protection. Finally, chemicals are essential as fuels and operating fluids, including brake fluid and antifreeze.[1]

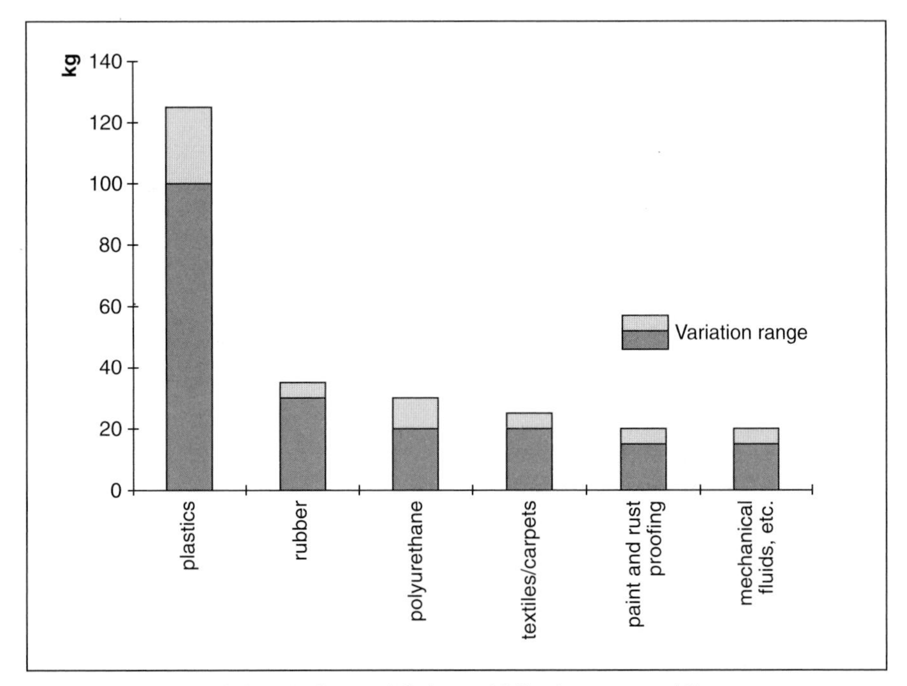

Fig. 2: Utilization of chemical materials in a middle-class automobile

Chemistry and Construction

The construction industry is **one of the most important clients** for the chemical industry. Its share of national sales for (West) German chemical manufacturers was estimated by the VCI to have been 10.2% in 1994 ($5.9 billion).

Among the countless products offered to this market, with applications ranging from energy conservation through construction itself, to pure esthetics, the following might be regarded as representative:

- Insulating materials
- Fire-retardant panels
- Adhesives and impregnating resins
- Caulking compounds
- Plastics for pipes
- Dispersions for protective paints
- Concrete additives
- Fibers for floor coverings and decorator fabrics
- Resins for façade panels

Against the background that ca. 32% of all the energy consumed in Germany is used for heating buildings, effective insulation has a central role to play in energy conservation.[2]

The chemical industry offers a variety of products to meet this challenge, including polyurethanes, melamine resins, extruded polystyrene (XPS), and expandable polystyrene (EPS), widely known under the trade name Styropor®. The latter is recognized by the public more for its use as a packing material, but its main application is as insulation in the construction industry. About 60% of the worldwide output of EPS is consumed in this way, and in Germany the figure is as high as 85%. From the European perspective, EPS is the **second most important insulating material** after mineral fibers (fiberglass and rock wool), as shown in Figure 3.

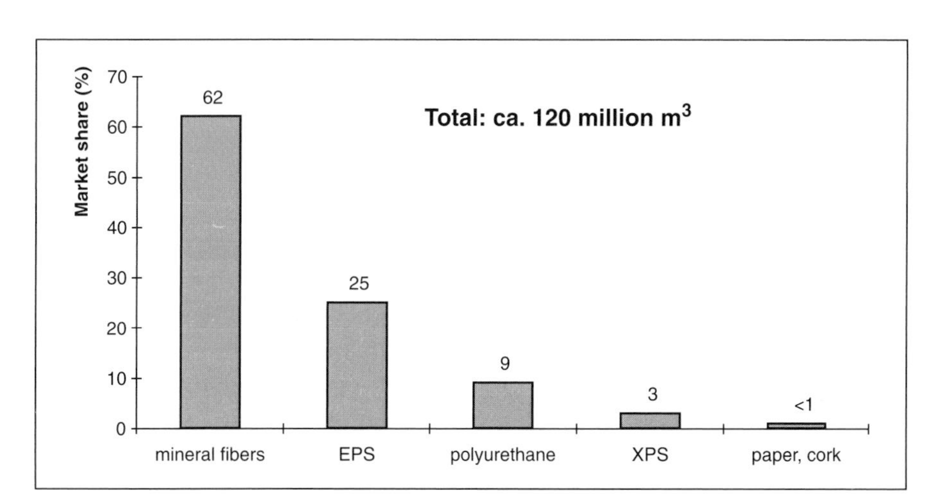

Fig. 3: Western European insulation market, 1995

There exists a huge backlog of demand for insulating materials, not only in Germany, where 70% of the dwellings still class as underinsulated, but especially in regions of the world other than Western Europe, as suggested by per capita consumption of EPS (Fig. 4).

[2] BASF press conference on thermal insulation and the environment, June 17, 1996.

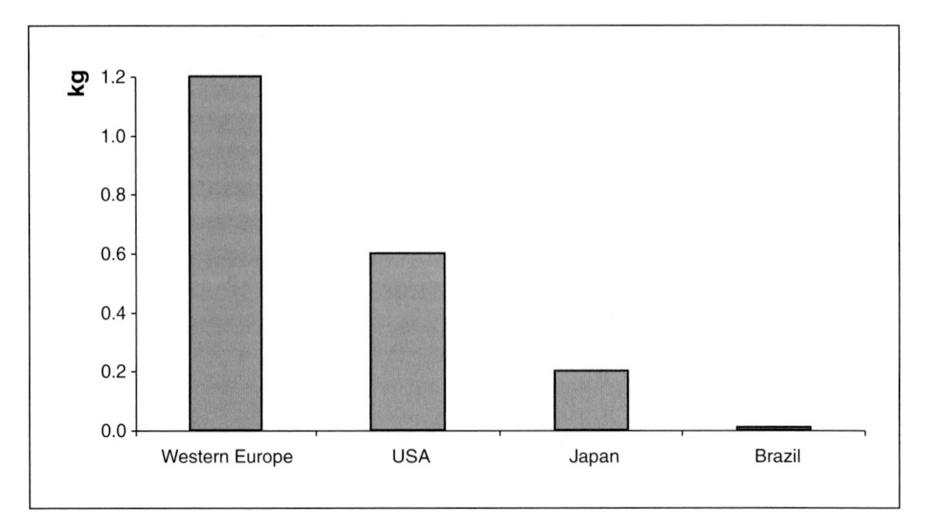

Fig. 4: Per capita consumption of expandable polystyrene (EPS) as insulation

It is often overlooked that more effective insulation in warmer zones can lead to far less energy consumption by electrical air-conditioning equipment. In these regions as well, effective insulation of buildings contributes significantly to the conservation of natural resources and ultimately to protection of the environment. Energy conservation also decreases the emission from power plants of such exhaust gases as carbon dioxide, nitrogen oxides, and sulfur dioxide.

Thermal Insulation

Insulation of a single-family house with EPS or XPS (extruded polystyrene) over a 50-year period has the potential to save 80 metric tons of heating oil. This in turn corresponds to the fuel consumption of a fully loaded jumbo jet during a flight from Frankfurt to New York (→ Raw Materials and Energy).

Plastics and Packaging

Plastics are very much in the foreground when it comes to packaging. Nevertheless, a closer look shows that they constitute only 12.5% by weight of the **packaging materials consumed in Germany**, thus ranking behind paper, cardboard, and glass (Fig. 5).

The total value of products generated in this area was estimated at $8.7 billion in 1995, having risen only slightly since 1992. On the other hand, the

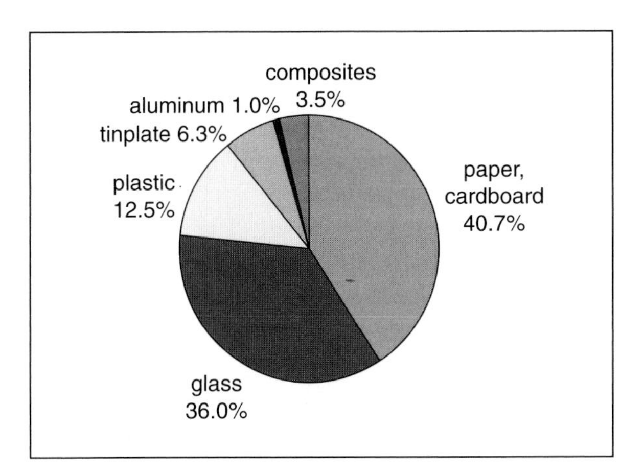

Fig. 5: Packaging materials in Germany

amount of packaging material manufactured showed a modest increase: 2.7% per year since 1992, to ca. 2.25 million metric tons per year in 1995 (Fig. 6).

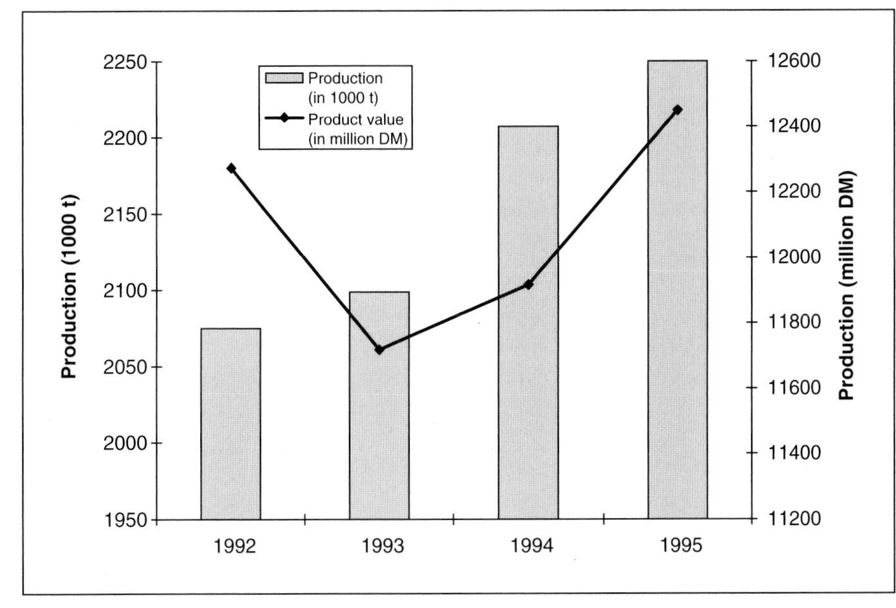

Fig. 6: Trends in plastic packaging material usage in Germany[3]

Plastics are not the only contribution of the chemical industry to packaging materials; chemical products are also central to the manufacture, coating, and labeling of paper and glass.

[3] data from: Society for Packaging Market Research (Germany); Federal Statistics Office (Germany).

The Most Spectacular Packaging Action

An especially spectacular packaging event took place in 1995 in Berlin when the world-famous Bulgarian "packaging artist" Christo enclosed the Reichstag building in roughly 100 000 m^2 of polypropylene fabric. This ca. 70-metric-ton fire-resistant "shroud," worth ca. $350 000, was vapor-blasted with aluminum as a way of giving it the desired silvery appearance.

The Most Important Plastic Packaging Materials

Polyolefins (polyethylene and polypropylene) were the **most important plastics** in the European packaging-material market in 1995, capturing a 74% share. The emphasis was clearly on polyethylene, with a preference for the LD (low-density) variety over the HD form. Poly(vinyl chloride) (PVC), often a subject of public debate, in 1995 still occupied fourth place behind polypropylene (Fig. 7).

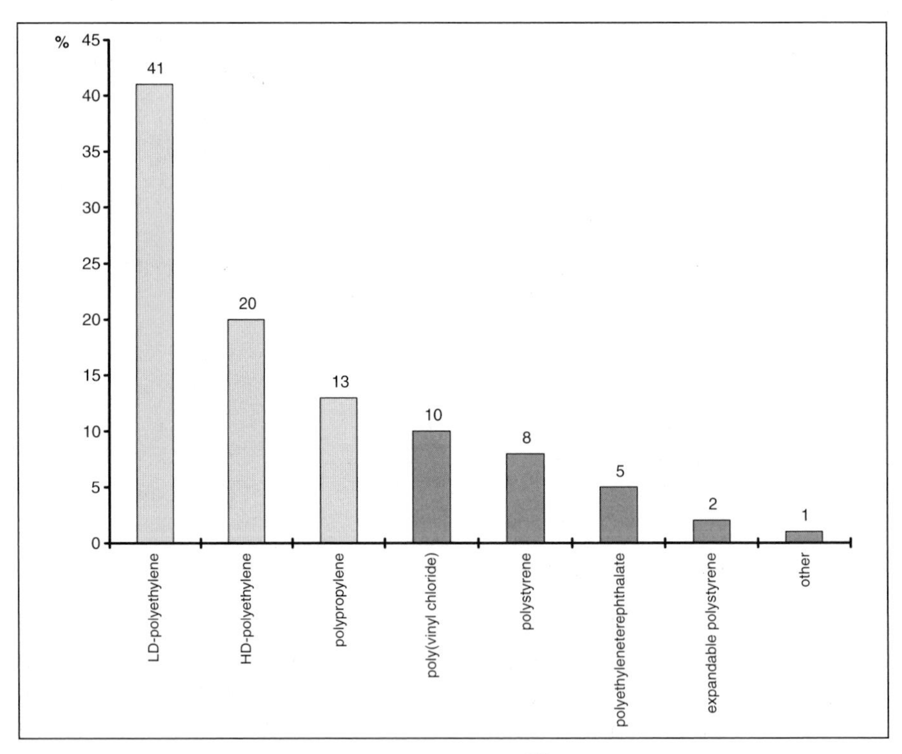

Fig. 7: Plastic packaging materials in Europe, 1995[4]

[4] data from: *CHEManager*, 9/96.

The Largest Chemical Companies

A list of the 500 **leading companies in the world based on sales** in 1997 included 16 chemical companies (firms in which the largest share of the revenue was derived from chemical activities). Of these, the clear leader was Du Pont (including Conoco) with net sales of $41.3 billion. Du Pont was followed at some distance by a purely German trio, after which came fifth-place Dow Chemical, separated by a gap of ca. $10 billion.

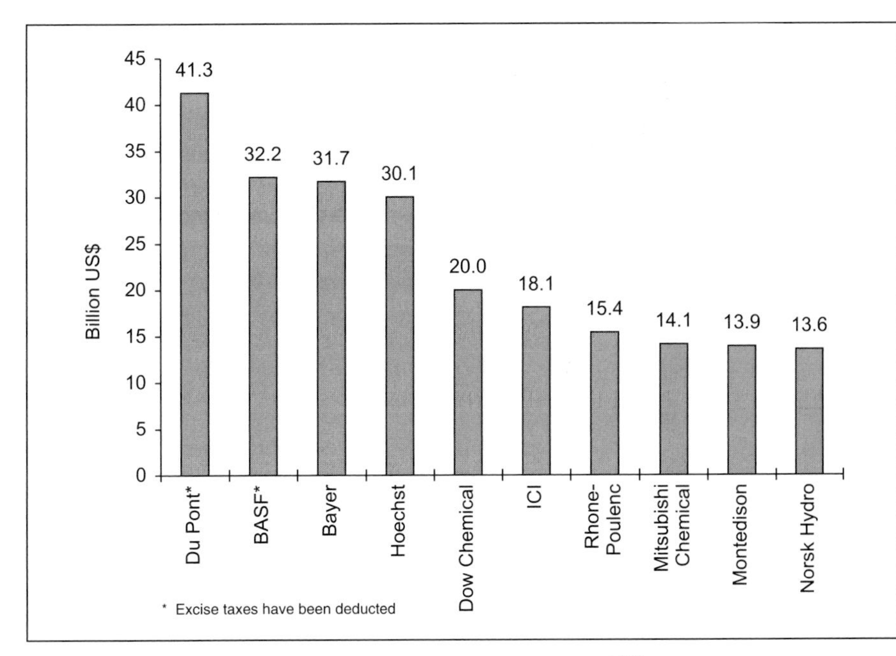

Fig. 1: Total revenues for major chemical companies, 1997[1]

Taken together, the ten largest chemical companies had net sales in 1997 of $230.4 billion. It is interesting to note by way of comparison that the three largest electronics firms (General Electric, Hitachi, and Matsushita Electric) had total sales of $224 billion. Another reference point is the fact that the largest chemical company from the standpoint of sales, Du Pont, occupied 49th place in the rank-ordered list of 500 largest companies in the world (1995: 58th; 1996: 55th).

The powerful position of European corporations in the chemical world is shown by the fact that seven of the ten largest companies on the list are based in Europe, representing somewhat more than 2/3 of the total sales of the top ten. Two of the others are located in the United States, and only one is in Japan (Fig. 2).

[1] data from: *Fortune*, "The Global 500," August 3, **1998**, pp. 74f.

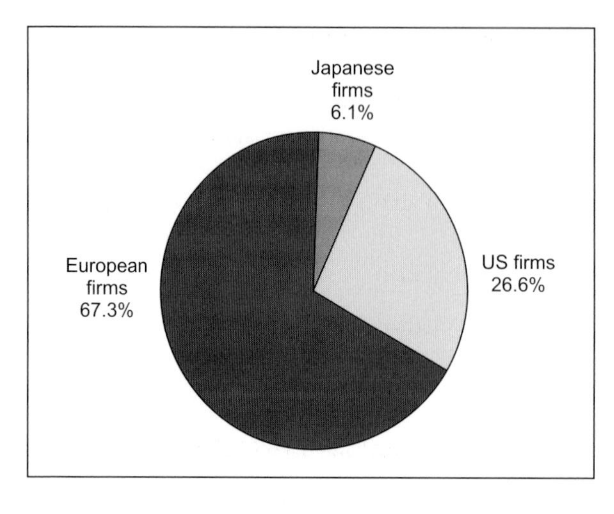

Fig. 2: Revenue shares as a function of corporate national origin[1]

Most Profitable Chemical Companies

Du Pont, winner of the title as sales leader in 1997, was also the chemical company with the **highest after-tax profits** in the "Global 500." Second place went to BASF, the leader of a trio of firms, all three of which were well ahead of the rest of the field (Fig. 3).[1]

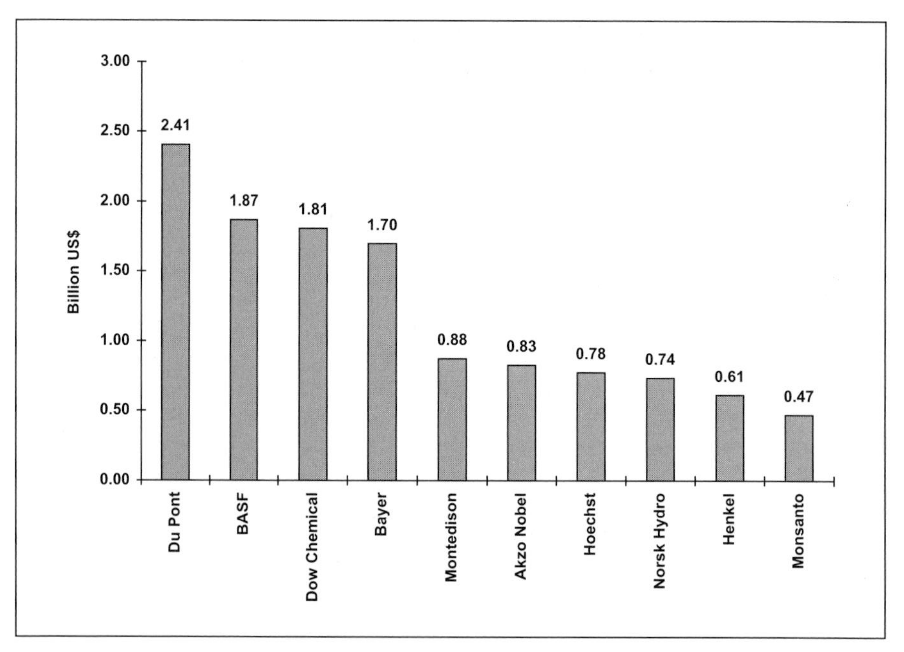

Fig. 3: Top ten chemical companies based on profits, 1997[1] (profits after taxes, after extraordinary credits or changes if any appear on the income statement, and after cumulative effects of accounting changes)

Chemical Companies with the Highest Net Yields

The **net corporate after-tax yields** (i.e., net profits as a percentage of sales) for the 16 world chemical corporations with the highest sales in 1997 ranged from 9.0% (Dow Chemical) to –4.3% (Rhône-Poulenc). If one examines the top ten companies in this list more closely one is struck by the fact that not a single Japanese firm is represented. These first appear in places 13 and 14. Instead, the top ten list consists of seven European and three United States corporations (Fig. 4).

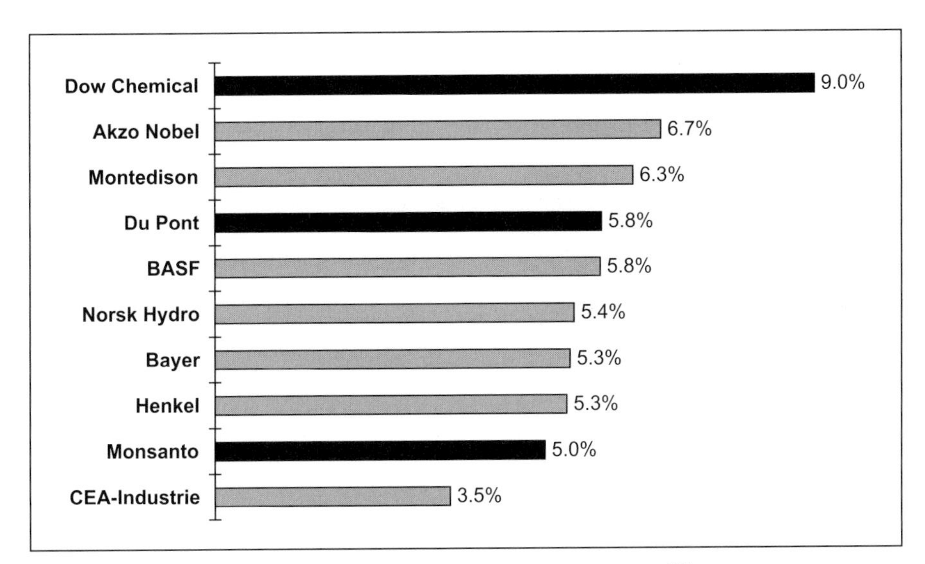

Fig. 4: After-tax yields of the top ten chemical companies, 1997[1]

Average net yields for the Global 500 (*Fortune*) chemical companies, roughly 4.1% in 1997, could be classed as "good" relative to 2.9% in the automotive industry (motor vehicles and parts), despite the fact that (as expected) they fail even to approach yields achieved in the pharmaceutical industry (15.1%).

Leading Producers of Chemicals

Considering the 50 **leading producers** of chemicals (ranked according to chemical sales; excluded from the data are formulated products such as pharmaceuticals and cosmetics, as well as energy and other "nonchemical activities"), total profits of \$381 billion were reported in 1997 (Fig. 5). Approximately 55% of that was registered by European firms. The dominance of the latter was again reflected in the list of the top ten (Fig. 6).

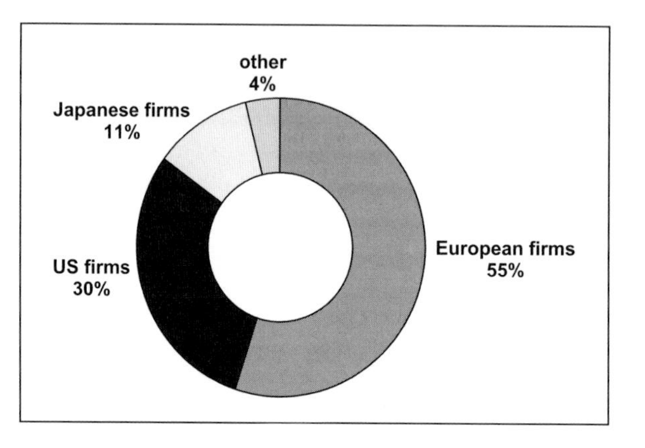

Fig. 5: Revenue as a function of corporate home[2]

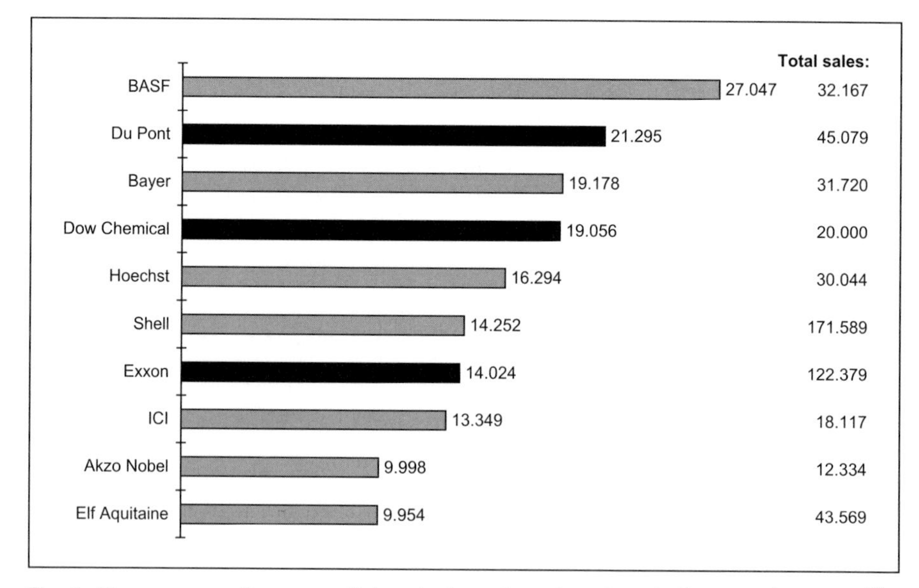

Fig. 6: Top ten manufacturers of chemical products based on 1997 sales (\$ billion)[2]

[2] data from: *Chemical & Engineering News*, July **1998**, 76, (issue 29), pp. 37–39.

The list showed a comfortable lead for BASF, which had displaced Hoechst in 1995. Hoechst moved down from second to fifth place as a result of restructuring.

Asian firms recorded 11% of the sales of the top 50, but none placed among the top ten.

Research and Development

Expenditures for research and development (Table 1) are investments in the future. In this respect, Germany has been laying a certain amount of groundwork for tomorrow ever since 1985. Thus, expenditures for research and development in the economy as a whole climbed to a new record of ca. 60 billion German marks in 1995, representing an annual average increase of 4.3%. Nevertheless, the days of clear growth are relegated to the first half of this period (6% growth per year during 1985–1990), which might provide some pause for thought.

Table 1: Research and development expenditures in Germany by operational and financial sector

	1995 (million German marks)	%	1993 (million German marks)	%
domestic expenditures, total	78 820		76 721	
higher education sector	14 900	18.9	13 838	18.0
public sector	11 800	15.0	11 647	15.2
unclassified	450	0.6	515	0.7
internal expenditures, business sector	51 670	65.6	50 721	66.1
total expenditures, business sector	59 400		58 394	
mechanical engineering	5 490	9.2	5 370	9.2
automotive industry	12 900	21.7	11 890	20.4
electrical industry	13 800	23.2	15 320	26.2
chemical industry	10 439	17.5	10 547	18.1
pharmaceuticals	4 322	7.3		

R&D expenditures in the chemical industry increased over the period 1985–1995 from 10.9 billion German marks. With an average growth of 2.9% per year the chemical industry thus lagged behind the economy as a whole. Expenditures in 1996 rose to about 7.8 billion German marks to 10.4 billion German marks (Fig. 7).[3]

[3] data from: VCI, *Chemiewirtschaft in Zahlen 1996* together with current reports.

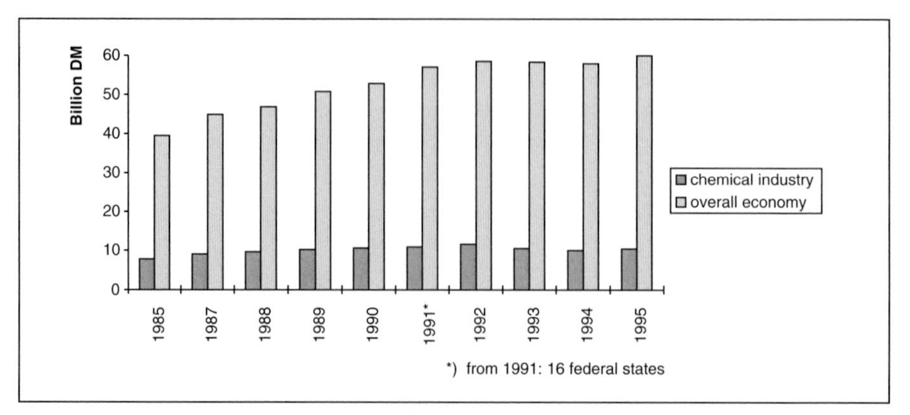

Fig. 7: a) German R&D expenditures in the chemical industry and in the overall economy

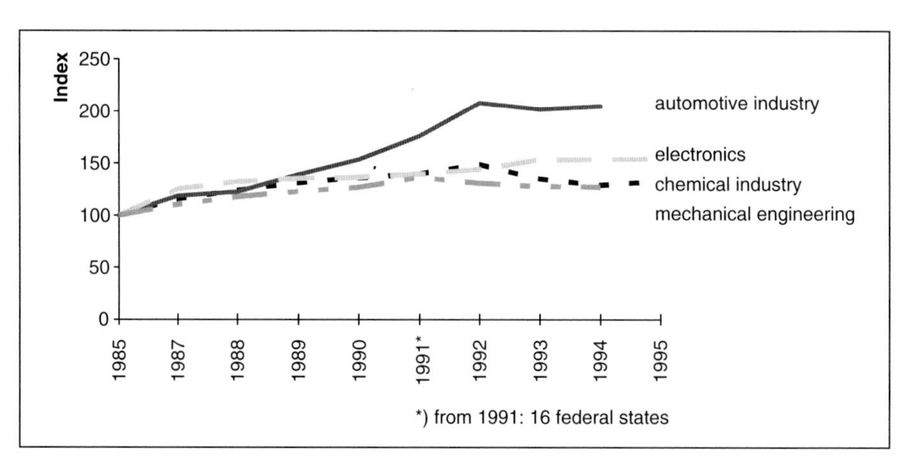

Abb. 7: b) Trends in R&D expenditures in selected industrial sectors (Germany)

Scientists: Busy Bees!

Ninety percent of all scientists who have ever carried out research in the course of human history are actively researching and inventing now, in the last decade of the 20th century. More than 5000 scientific papers are being published every day.

Foreign Investment

Foreign investment by the German chemical industry set a new record in 1994 of 54 billion German marks (Table 2).[4]

Table 2: Trends in German foreign investment (in million German marks)

Location	1992	1993	1994
EU countries	18 617	20 275	21 672
USA	14 837	16 883	17 301
Japan	2 620	3 056	3 133
Hong Kong	195	175	224
India	101	123	97
Malaysia	199	221	234
Singapore	53	65	116
South Korea	250	371	393
other	9 515	9 934	10 865
overall	46 387	51 103	54 035

Just as in 1992 and 1993, again in 1994 the heaviest investments were in countries within the European Union (40%) as well as in the United States (32%), as shown in Fig. 8. The relative share of investments in Asia outside Japan remained very limited despite an impressive rate of increase, with an **investment emphasis** on South Korea, Malaysia, and Hong Kong.

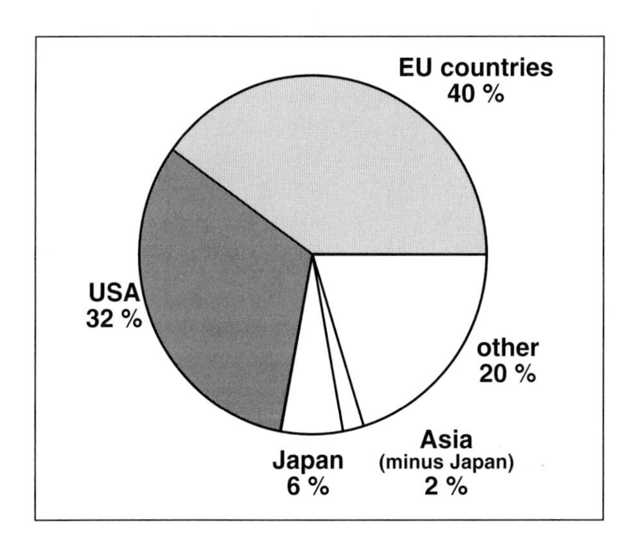

Fig. 8: Regional distribution of German chemical investments, 1994

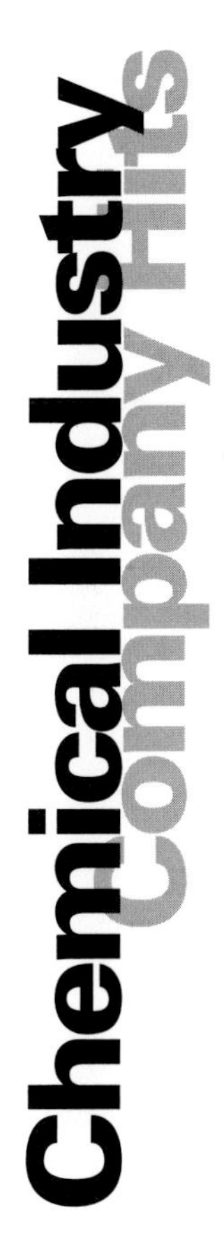

[4] data from: *Chem. Industrie*, 7–8/96.

"Megadeals" Since January, 1996

The chemical industry has in the recent past been the scene of a great many mergers, acquisitions, and spin-offs (Table 3). The driving force behind these activities is a strategy of globalization coupled with more concentration on core activities.

The intent has been to improve competitive position and at the same time increase the profitability of invested capital.

Table 3: Mega-deals since January 1996 (status on October 1996)

a) Mergers

Merging partners	New corporation	Value of the transaction
Ciba (Switzerland) and Sandoz (Switzerland)	Novartis (Switzerland)	$28.2 billion
Pharmacia (Sweden) and Upjohn (USA)*	Pharmacia & Upjohn (USA)	$13.0 billion
Degussa (Germany) and Hüls (Germany)	Degussa-Hüls (Germany)	$12.5 billion

* Late 1995

b) Spin-offs

Corporate segment	New corporation	Value of the transaction
specialty chemicals from Novartis (Switzerland)	Ciba Specialty Chemicals (Switzerland)	$5.2 billion
pharmaceuticals from Hafslund Nycomed (Norway)	Nycomed (Norway)	$2.2 billion
chemicals from Hanson (UK)	Millenium Chemicals (USA)	$1.7 billion
chemicals from Monsanto (USA)	Solutia (USA)	
carbon from Hoechst (Germany)	SGL Carbon (Germany)	$1.0 billion
chemicals from ICI (UK) in Australia	Orica (Australia)	$1.6 billion

[5] Press reports, corporate annual reports.

Table 3: c) Acquisitions

Corporate segment/company	Acquirer	Value of the transaction
Corange (including Boehringer Mannheim)	Roche (Switzerland)	$11.0 billion
chemicals from Unilever (Netherlands)	ICI (UK)	$8.0 billion
Arco Chemical (USA)	Lyondell	$6.5 billion
Courtaulds (UK)	Akzo Nobel (Netherlands)	$3.8 billion
specialty chemicals from Hoechst (Germany)	Clariant (Switzerland)	$3.7 billion
Betz Dearborn (USA)	Hercules (USA)	$3.1 billion
industrial chemicals from ICI (UK)	Du Pont (USA)	$3.0 billion
Herberts (Germany)	Du Pont (USA)	$3.0 billion
polyester in North America from Hoechst (Germany)	Kosa (USA)	$2.5 billion
Allied Colloids (UK)	Ciba Specialty Chemicals (Switzerland)	$2.4 billion
Dekalb (USA)	Monsanto (USA)	$2.4 billion
dialysis from Grace (USA)	Fresenius (Germany)	$2.3 billion
petrochemicals from Occidental (USA)	Equistar (USA)	$2.0 billion
Delta & Pine (USA)	Monsanto (USA)	$1.9 billion
Arcadian (USA)	Potash Corp. (Canada)	$1.7 billion
Protein Technologies (USA)	Du Pont (USA)	$1.5 billion
seeds from Cargill (USA)	Monsanto (USA)	$1.4 billion
Gist-Brocades (Netherlands)	DSM (Netherlands)	$1.3 billion
Loctite (USA)	Henkel (Germany)	$1.3 billion
diagnostics from Chiron (USA)	Bayer (Germany)	$1.1 billion
Tastemaker (USA)	Roche (Switzerland)	$1.1 billion
construction chemicals from Sandoz (Switzerland)	SKW (Germany)	$1.1 billion
Inspec (UK)	Laporte (UK)	$1.0 billion
Holden's (USA)	Monsanto (USA)	$1.0 billion

Table 3: d) Increased holdings

Company	Acquirer	Value of the transaction
31.7% of Rhône-Poulenc Rorer (USA)	Rhône-Poulenc (France)	$4.6 billion
43% of Roussel-Uclaf (France)	Hoechst (Germany)	$3.2 billion
50% of Du Pont Merck (USA)	Du Pont (USA)	$2.6 billion
50% of Montell (USA)	Shell (Netherlands)	$2.0 billion
50% of Dow Elanco (USA)	Dow (USA)	$1.2 billion
50% of Foamex (Canada)	Trace (USA)	$1.1 billion

The Leading Chemical Markets

The world market for chemicals in 1997 had a gross value of \$1.428 billion. By 2010 it is expected to grow at an annual rate of 3.5% to \$2.245 billion.

Western Europe and North America, with 32.3% and 30.1% of world consumption, were the **most important chemical markets** in 1997. Japan (14.2%) followed next after a wide gap, 2% ahead of South and East Asia. Nevertheless, demand in the latter region in 1997 was greater than that in South America, Eastern Europe, Oceania, Africa, and West Asia combined.[1]

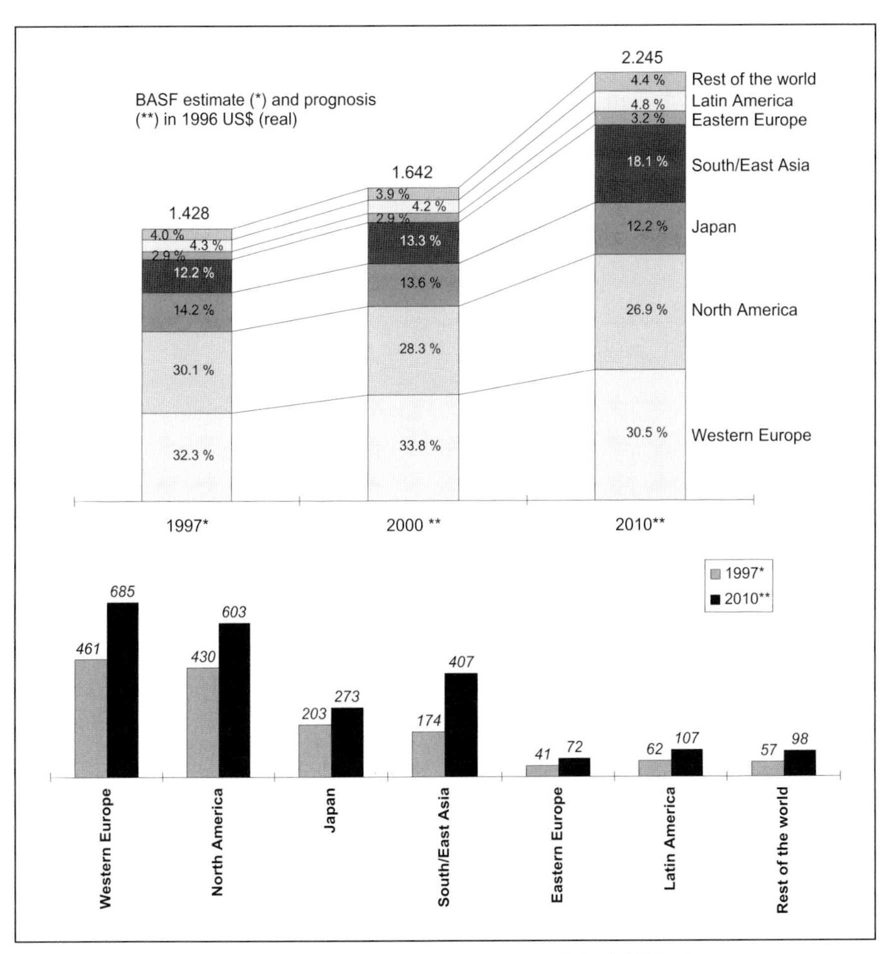

Fig. 1: Consumption of chemical products by region 1997 (\$ billion)

[1] BASF.

Changes are to be anticipated in this regional distribution in the years ahead. Each of the markets is expected to grow, but at significantly different rates. The projected real annual growth rates for the time period 1997–2010 range from 2.3% for Japan through 2.6% and 3.1% for North America and Western Europe to 6.8% in South and East Asia. This highest-growth region will thus surpass Japan as a consumer of chemicals in the near future, although not reaching by the year 2010 an importance comparable to North America or Western Europe.

The Contribution of Chemistry to the Trade Surplus

Processing industries in Western Europe were responsible in 1995 for the **largest trade surplus** (143.9 billion ECU) since 1986. The contribution of the chemical industry toward this record was 34.1 billion ECU.

During the decade in question, the share of the surplus attributable to chemicals varied between 19% and 36%. Absolute values ranged from 20.5 billion ECU to 34.1 billion ECU, and thus showed significantly less fluctuation than was the case with other products (Fig. 2).

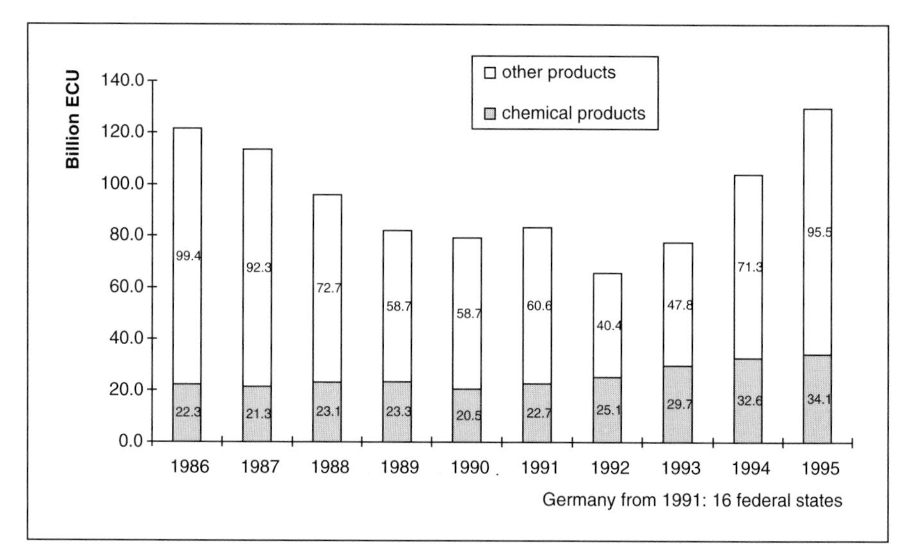

Fig. 2: Trends in trade surpluses for Western Europe, 1986–1995[2]

[2] data from: CEFIC, *Facts & Figures*, November **1996**.

The Largest Chemical Exporting Countries

In 1997 the United States led the hit list of most important exporting countries for chemicals, slightly ahead of Germany. France followed at some distance in third place, leading a group of nations all of which (with the exception of Japan in seventh place) were located in Western Europe (Fig. 3).

Taken together, the ten top countries exported chemical products with a value of $367 billion, of which the United States and Germany alone accounted for 37.5%.

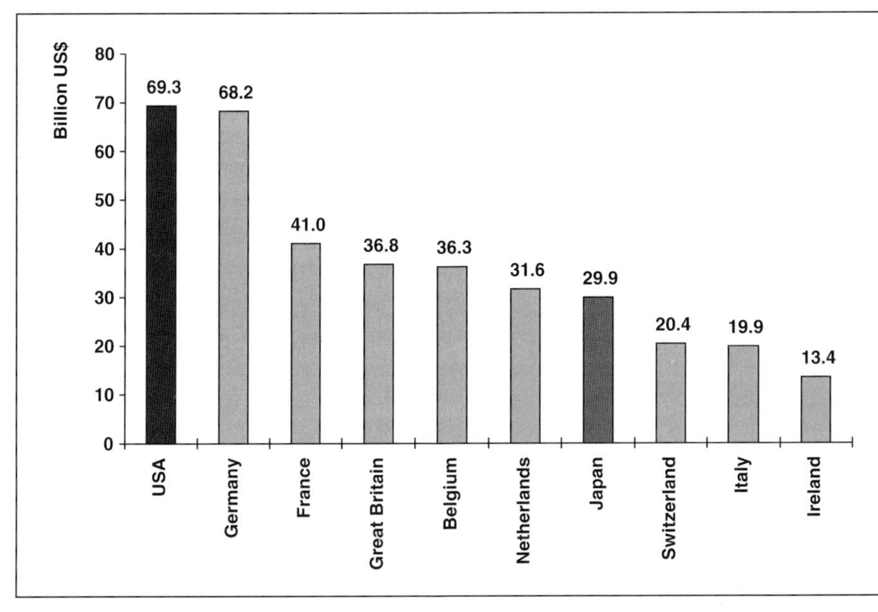

Fig. 3: The principal chemical exporting countries in the world, 1997[3]

European Union with the Best Chemical Trade Balance

For years the EU has recorded the **largest chemical trade surpluses** in the leading triad (Fig. 4). In 1996 this surplus amounted to $41.8 billion and was thus more than twice that of the United States ($18 billion). The chemical trade activity for Japan was nearly balanced in the second half of the 1980s, but then gradually moved into a positive range (1996: $5.6 billion).[4]

[3] data from: *Europa Chemie 22–23*, **1998**, p. 4.

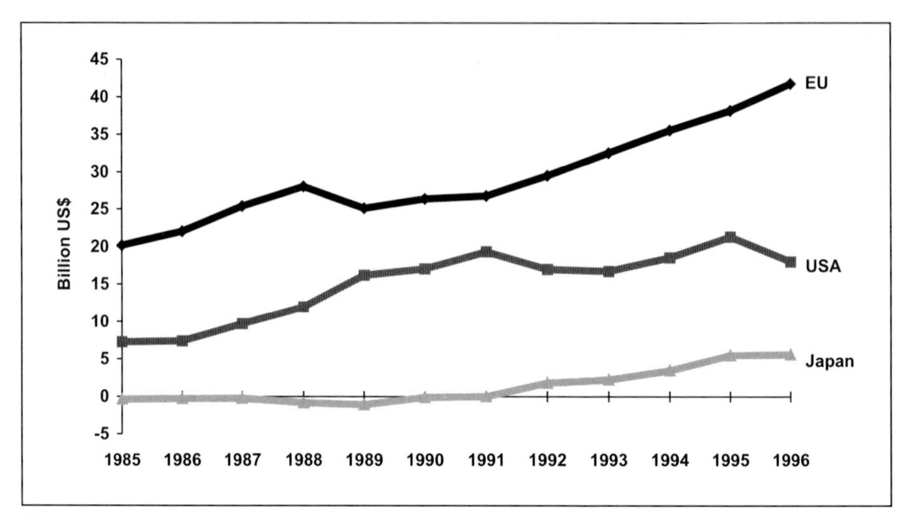

Fig. 4: Historical chemical trade balances for the EU, the United States, and Japan, 1985–1996[4]

The chemical trade surplus of the EU was a result mainly of trade with countries outside the triad – especially trade with Asia (Fig. 5).

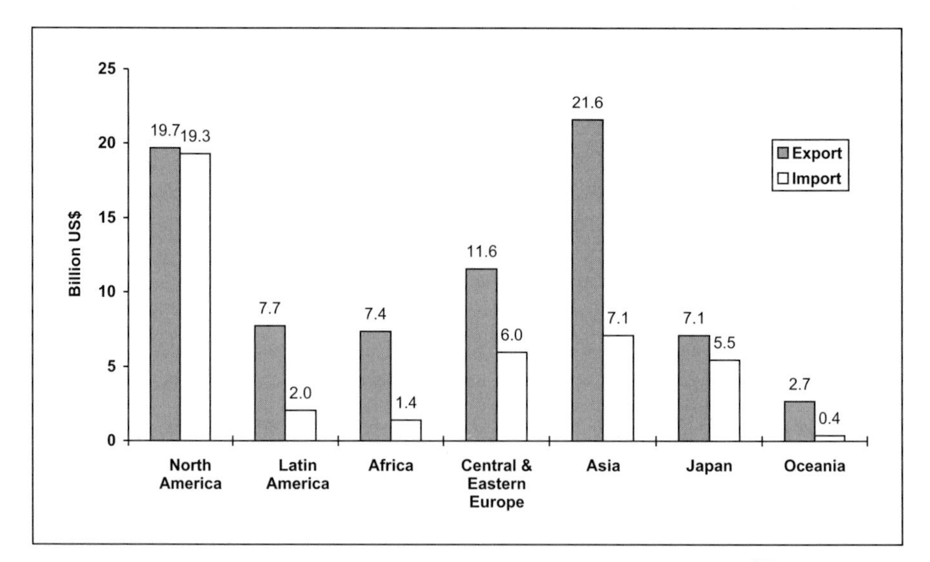

Fig. 5: Chemical trade between the EU and various world regions, 1996[4]

[4] data from: CEFIC, *Facts & Figures 97*, pp. 27, 33.

The Most Important Domestic Consumers of Chemicals (Germany)

Based on estimates by the VCI, the chemical industry itself was the **most important customer** in the German domestic chemical market, responsible for about 16.9% of the domestic trade of $57.6 billion. The healthcare industry was in second place, followed by three sectors of comparable importance: construction, private consumption, and automotive industry.

After a wide gap, agriculture came in sixth with a share of only 5%, though still far ahead of the electrical and paper industries (Fig. 6).

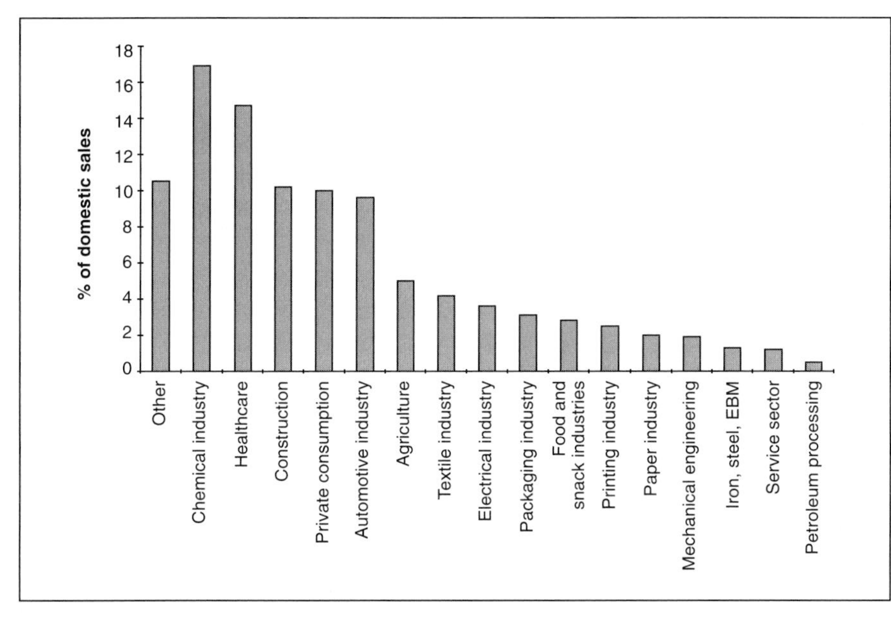

Fig. 6: Domestic market structure of the German chemical industry, 1994 (estimate)[5]

Production Figures by Branch

The largest shares of chemical production in Germany (based on value) were attributable in 1994 to pharmaceuticals (19.9%), organics (15.8%), and plastics (15.7%). These together constituted ca. 50% of total production value.

[5] data from: Verband der Chemischen Industrie VCI, *Chemiewirtschaft in Zahlen 1996.*

If recent data are compared with those from 1985 (Fig. 7), it is striking that the same branches were important a decade ago as well, but their relative shares have changed. Pharmaceuticals have grown considerably in importance, whereas organics have declined. The role of plastics has increased slightly (→ Plastics).

Marked changes also occurred with respect to crop protection agents (2.6% → 1.7%) (→ Crop Protection) and fertilizers (2.4% → 0.6%) (→ Fertilizers).[6]

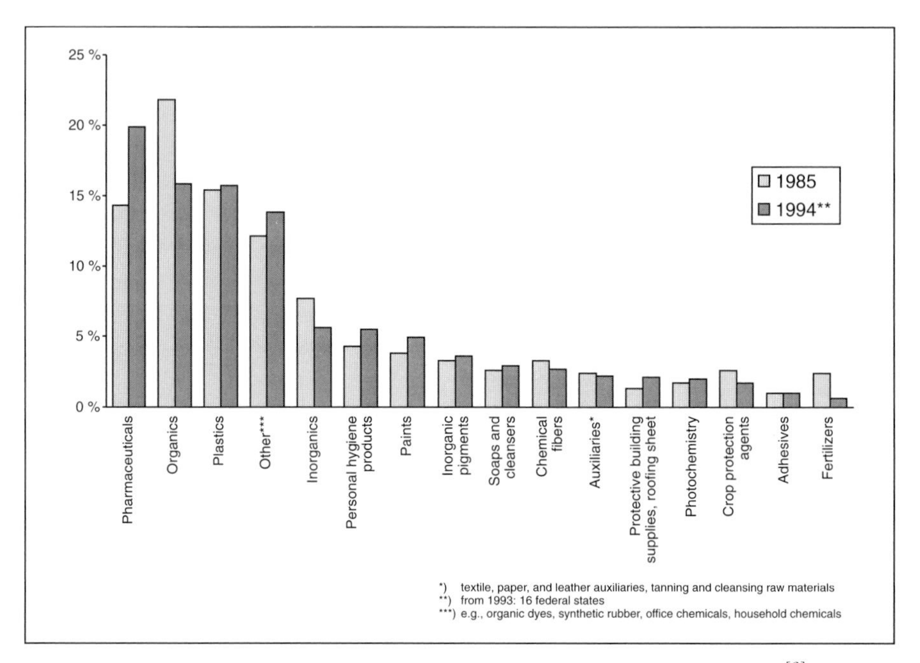

Fig. 7: Shares of chemical production in Germany in a ten-year comparison[6]

[6] data from: Verband der Chemischen Industrie VCI, *Chemiewirtschaft in Zahlen 1995*, p. 33.

Ammonia

Ammonia is one of the most important basic products of the chemical industry. Relative to production capacity it ranks second behind sulfuric acid.

Roughly 87% of the production is designated for further processing to fertilizers (→ Fertilizers), with the rest used for the synthesis of a multitude of chemical products. As would be anticipated from its major application, the demand for ammonia has increased steadily as a consequence of the undiminished growth in population in recent decades. Only in the early 1990s were there declines caused by the crisis in Eastern Europe. In the meantime demand has risen again, with the 100-million-ton-per-year level being exceeded for the first time in 1998 (Fig. 1).

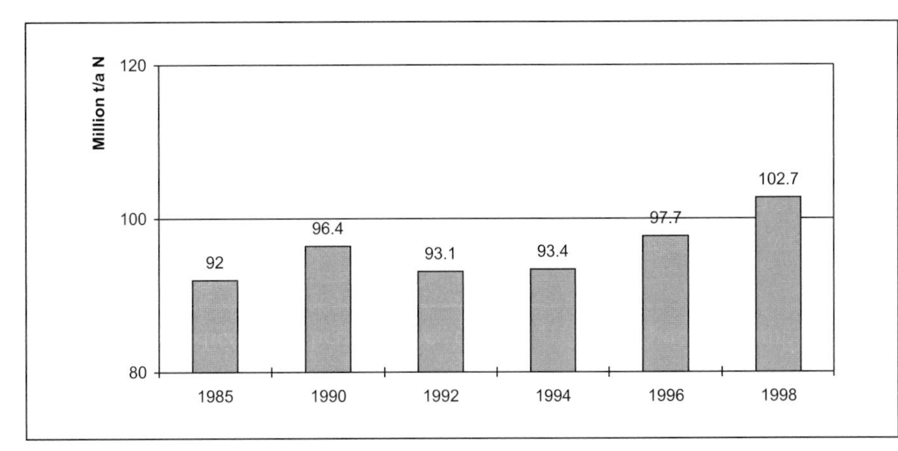

Fig. 1: a) Historical development of demand for ammonia (in millions of tons of nitrogen per year)

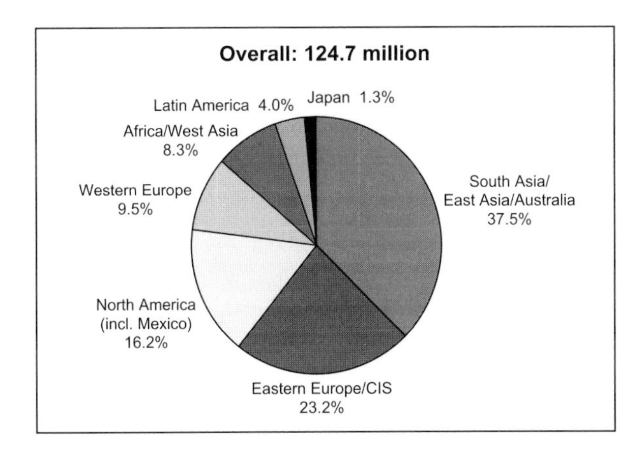

Fig. 1: b) Ammonia production capacities, 1998[1]

Chemical Products Hit List

[1] BASF.

The worldwide production capacity for ammonia in 1998 amounted to 124.70 million metric tons (based on nitrogen). In contrast to the late 1960s when Western Europe and the United States still accounted for more than half of the world's capacity, the emphasis is now on Asia. China plays a special role, alone controlling ca. 21% of world capacity (albeit mainly in the form of small facilities based on old technology). Nevertheless, the list of top ten producers was led by a convincing margin by a Western European firm, Norsk Hydro, the largest manufacturer of mineral fertilizers in Western Europe. Overall, the ten leading producers accounted for 22% (27.4 million tons of nitrogen per year) of the world's ammonia-producing capacity. (Fig. 2).

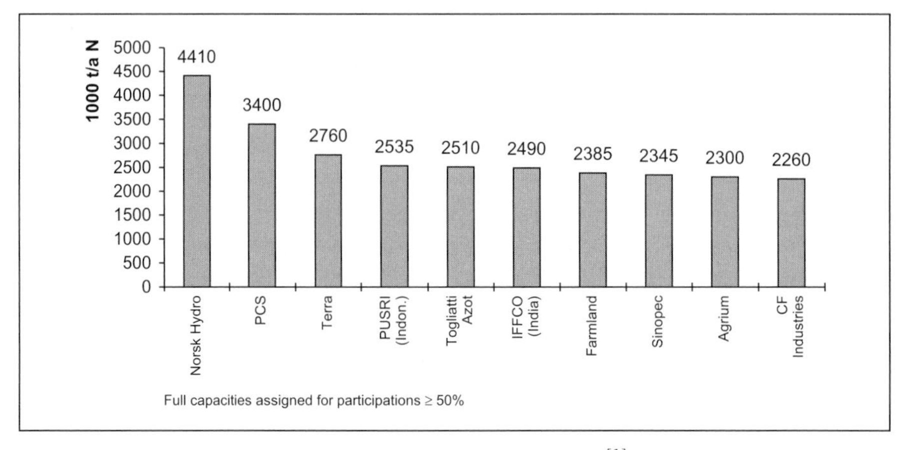

Fig. 2: Ammonia capacities of the top ten producers, 1998[1]

Benzene

Benzene, most familiar to many as an additive in fuels for internal combustion engines, represents the prototype for aromatic compounds. It is the starting material for a host of derivatives, of which the styrene precursor ethylbenzene is the most important. Benzene is obtained primarily through refinement of crude oil.

The **world capacity** for benzene in 1997 was 36.78 million metric tons, nearly half of it located in North America (including Mexico) and Western Europe (Fig. 3).

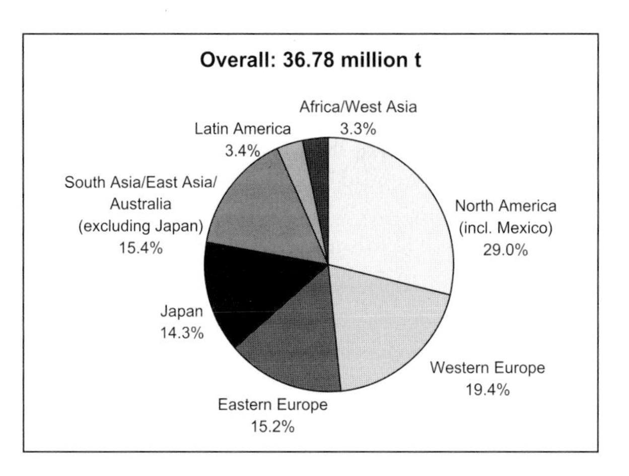

Fig. 3: Benzene production capacities by region, 1997[1]

Among the most important producers of benzene are of course several petrochemical companies. One of these, Shell, set the production record in 1997 with 2.41 million tons per year, 6.6% of the world output (Fig. 4). Taken together, the eight leading companies had access to a production capacity of 10.19 million tons per year, 27.7% of the world capacity.

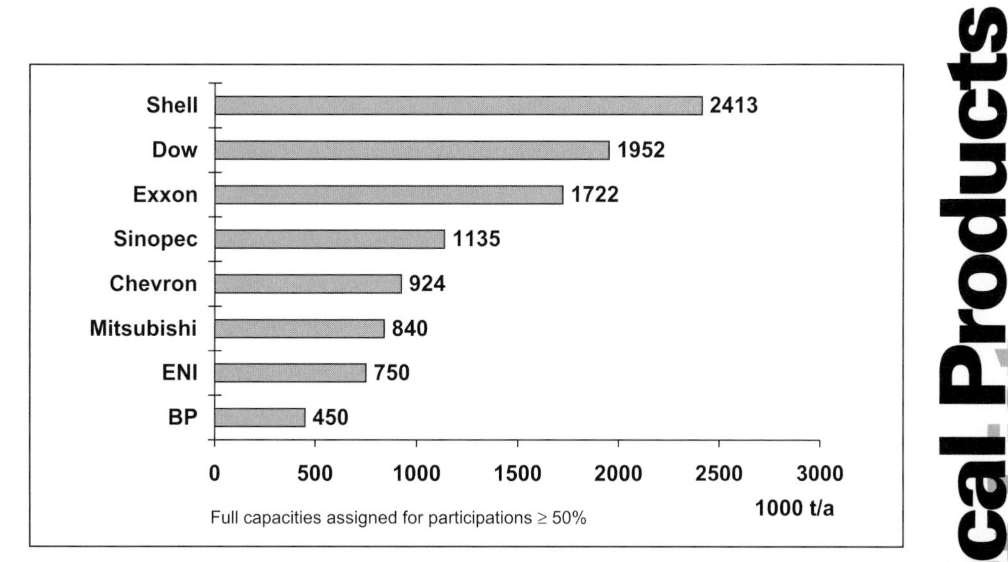

Fig. 4: Capacities of the top ten benzene producers, 1997[1]

Butadiene

Butadiene is obtained as a byproduct during ethylene recovery after the cracking of naphtha. It is an important building block in the manufacture of plastics, including polystyrene/polybutadiene and synthetic rubber.

The **world capacity** for butadiene in 1998 was 9.15 million metric tons (Fig. 5). Almost half of this was located in the two leading producer regions, Western Europe (2.38 million tons per year) and North America (2.16 million tons per year).

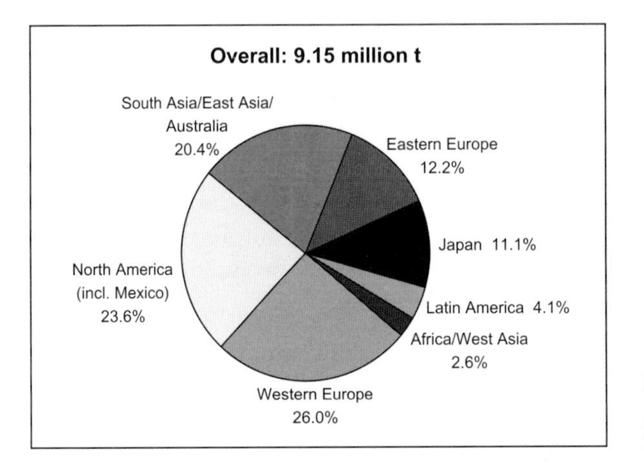

Fig. 5: Butadiene production capacities by region, 1998[1]

The leading producer by a wide margin was Shell, with a capacity of 783 000 tons per year (8.6% of world capacity), as shown in Figure 6.

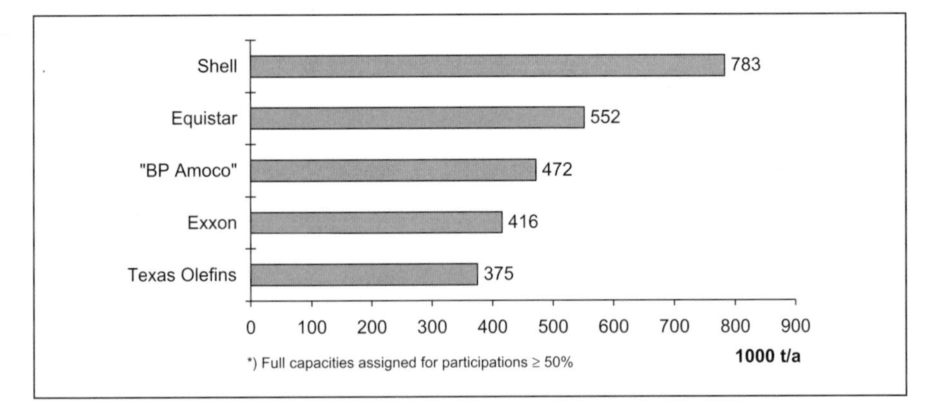

Fig. 6: Butadiene capacities* of the five leading producers, 1998

Chlorine

Chlorine is another key basic chemical, particularly because of its use in organic chemical chlorination reactions. Its significance with respect to a highly refined chemical economy will be clear from the fact that, based on VCI data for 1995, 60% of all sales in the German chemical industry involved products prepared with the aid of chlorine.

Chlorine is obtained mainly by the electrolysis of aqueous sodium chloride solutions. **Worldwide production capacity** in 1997 was 50.34 million metric tons per year, roughly half of which was accounted for by facilities in North America and Western Europe. By far the most important producing nation was the United States, followed by China and Japan with roughly equal capacities (Fig. 7). **Germany, clearly the leading chlorine producer in Western Europe**, possessed the world's fourth largest production capacity.

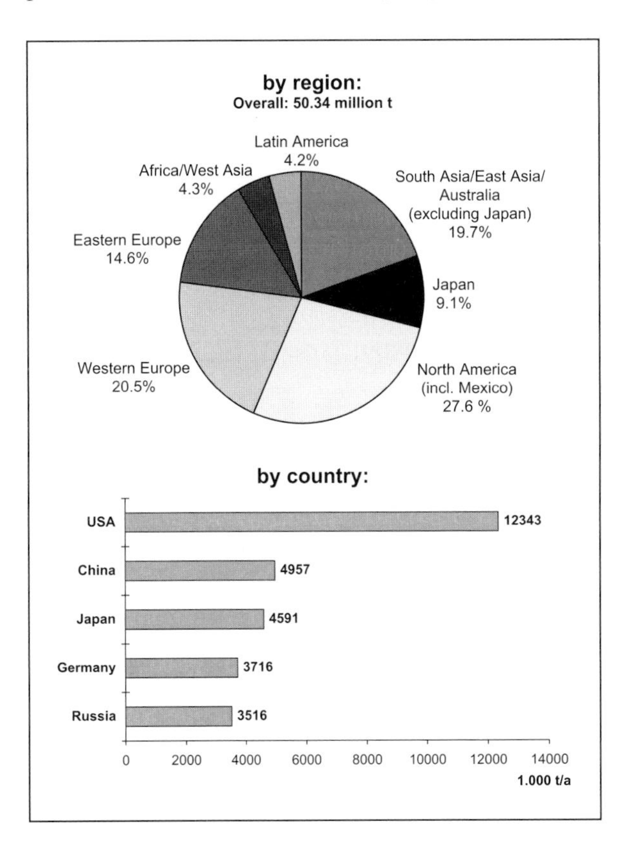

Fig. 7: Production capacities for chlorine, 1997[1]

Chemical Products Hit List

The dominant companies in chlorine production were Dow and Occidental, with 11.2% and 6.7% of world capacity. The remaining producers in the top ten were limited to shares of less than 4% each (Fig. 8).

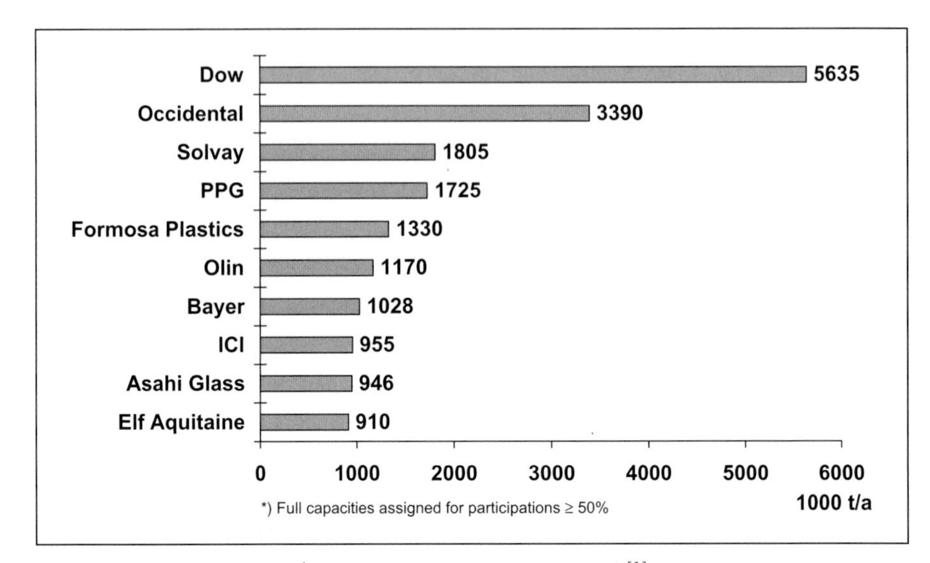

Fig. 8: Chlorine capacities* of the leading producers, 1997[1]

Demand for chlorine is increasing slowly, with a significant increase for PVC production being offset by less use in the manufacture of paper as well as for chlorofluorocarbons and chlorinated solvents.

Ethylene (Ethene)

Ethylene, **the simplest of the alkenes** (hydrocarbons containing a double bond), is one of the most important intermediates in chemistry derived from oil and natural gas. It is obtained from these raw materials by thermal cracking, and is then converted to a host of products. Plastics are at the forefront of this product spectrum from a quantitative perspective, and roughly half the world's ethylene supply is consumed in the manufacture of polyethylene.

In 1997 the worldwide production capacity for ethylene stood at 88.22 million tons per year, about one-third of it in the NAFTA region (Fig. 9).

The **ten leading ethylene producers** – Dow topped the list with 6.0% of world capacity – again represented somewhat more than one-third of total world capacity in 1997 (Fig. 10).

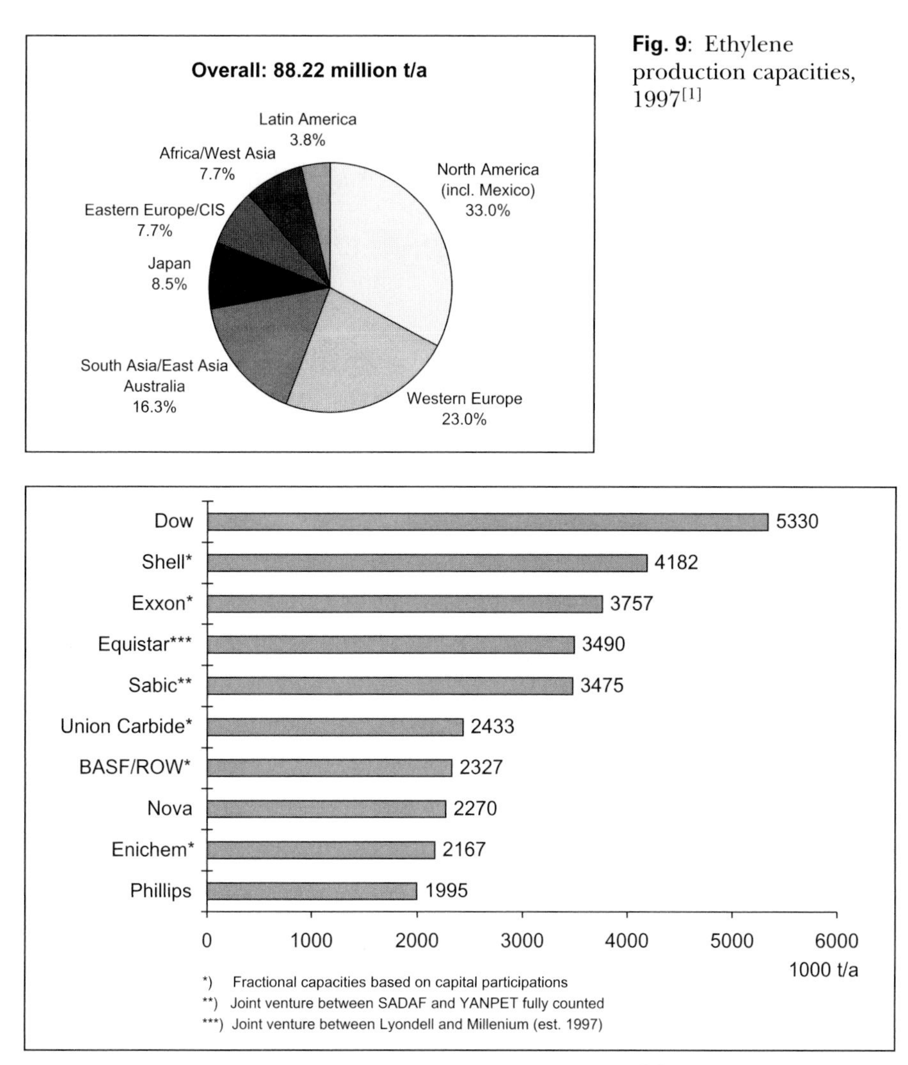

Fig. 9: Ethylene production capacities, 1997[1]

Fig. 10: Capacities of the top ten producers of ethylene, 1997[1]

Incidentally, ethylene is important not only for the chemical industry, but also in the fruit trade. Ethylene gas, which is classed among the plant hormones (→ Sensors, Signaling Agents), is used to accelerate the ripening process in stored green bananas.

Ethylene Oxide (Oxirane)

Ethylene oxide is a colorless gas now prepared mainly by the direct oxidation of ethylene over a silver catalyst. It is a very reactive compound that can be transformed into a multitude of chemicals, including ethylene glycol, monoethanolamine, and glycol monoalkyl ethers.

Worldwide **production capacity** in 1997 was 13.41 million metric tons per year, with North America being the chief producing region by a wide margin (Fig. 11).

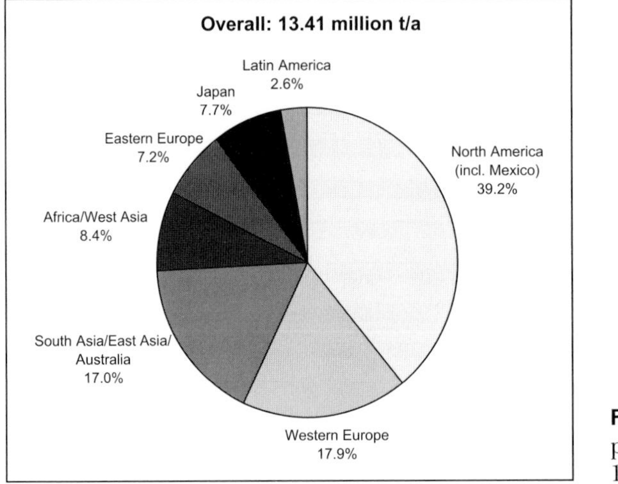

Fig. 11: Ethylene oxide production capacities, 1997[1]

Union Carbide was the leading company among the world's top ten producers, which together accounted in 1997 for 63.6% of worldwide capacity (Fig. 12).

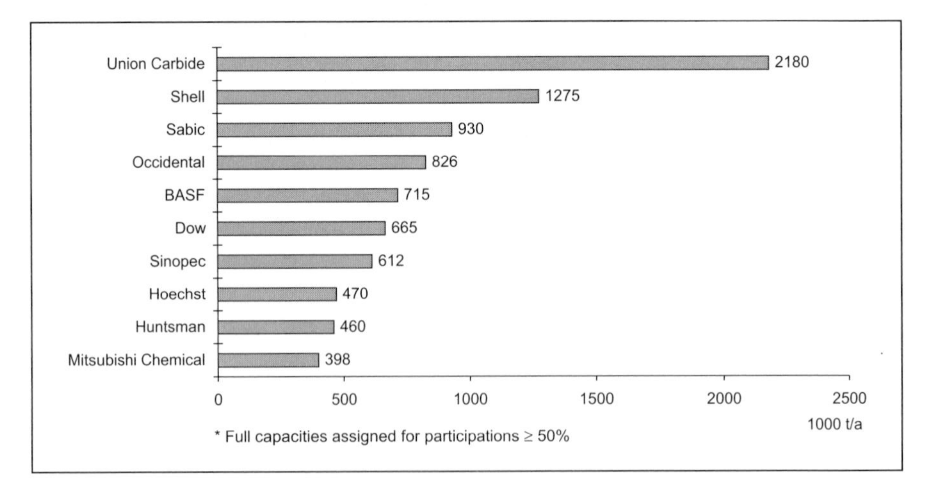

Fig. 12: Capacities* of the top ten producers of ethylene oxide, 1997[1]

Propylene (Propene)

Propylene is familiar to many as LPG (liquid petroleum gas). However, its main application is in the preparation of gasoline and as a starting material for the synthesis of chemical products. In this respect the most important examples are acrylonitrile and propylene oxide, together with the increasingly popular plastic polypropylene, which are responsible for 55% of world consumption.

Propylene is obtained during the cracking of crude oil. The world production capacity in 1997 was rated at 54.37 million metric tons per year. The rank-order of producer regions for propylene corresponded to that for ethylene (propylene is often obtained as a byproduct in the preparation of ethylene), albeit with somewhat less dominance by North America (Fig. 13).

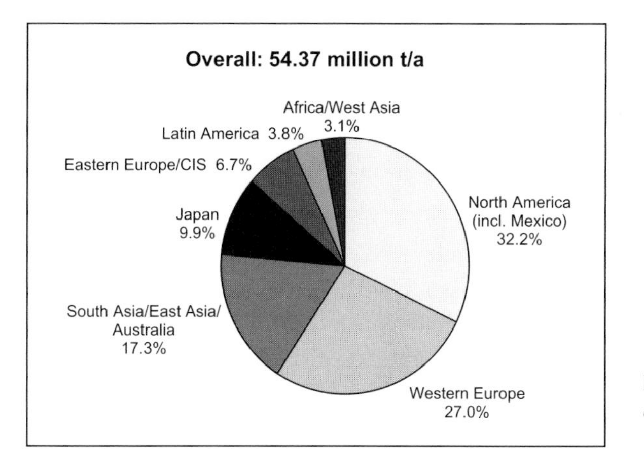

Fig. 13: Propylene capacities by region, 1997[1]

Analogies are also apparent in a comparison of the lists of top ten producers of the two gases. Indeed, the first four entries are the same, albeit in a different sequence. Shell and Exxon led with propylene production capacities of 8.0 and 6.4%, ahead of Equistar (formed by a 1997 merger of the olefin and polymer divisions of Lyondell and Millenium), which dislodged Dow from third place. Taken together, the top ten accounted for 36% of world production capacity (Fig. 14).

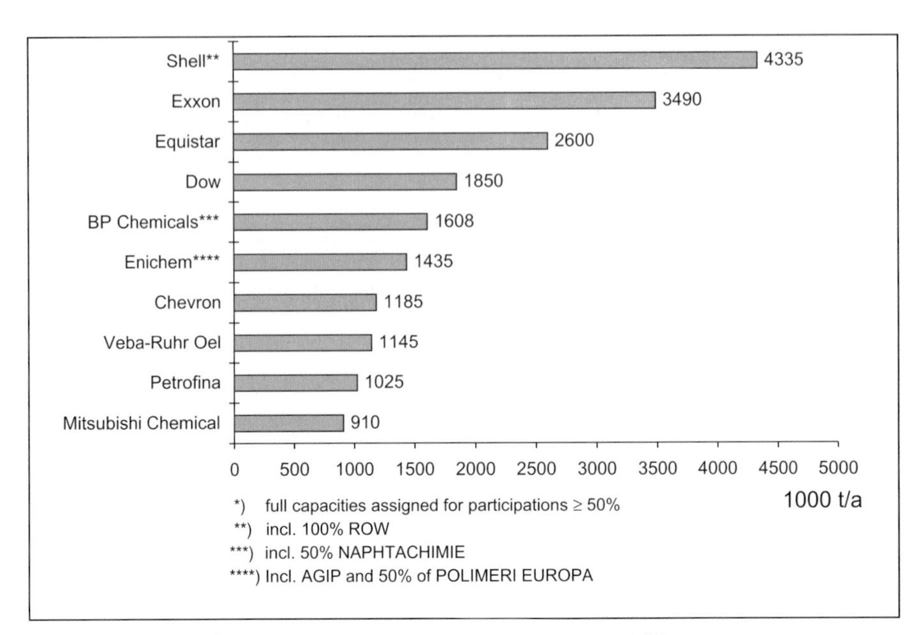

Fig. 14: Capacities* of the top ten propylene producers, 1997[1]

Propylene Oxide

Propylene oxide is the epoxide of propene and is a significant intermediate in the manufacture of polyalcohols and various other compounds.

Roughly three-fourths of the 1997 world capacity of 5.08 million metric tons resided in North America (the most important region) and Western Europe. The Asiatic sector, apart from Japan, has so far played only a very subordinate role (Fig. 15).

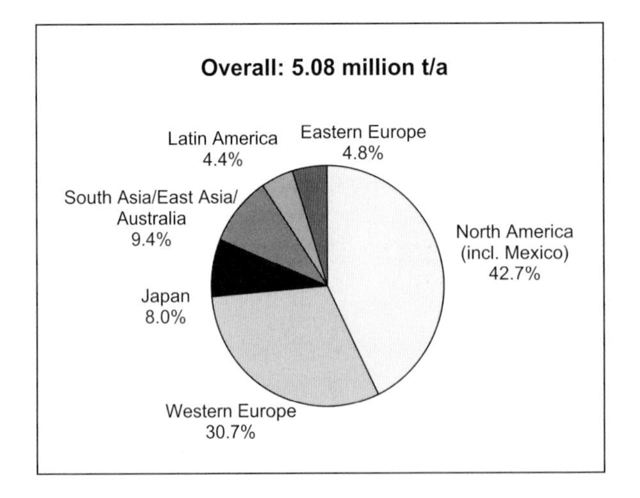

Fig. 15: Propylene oxide capacities by region, 1997[1]

Events on the producing side were clearly dominated by Arco and Dow, with equal shares of world capacity (32.9%). None of the other producers accounted for more than 7%. Taken together, the seven leading firms in 1997 controlled 84.1% of the world production capacity for propylene oxide (Fig. 16).

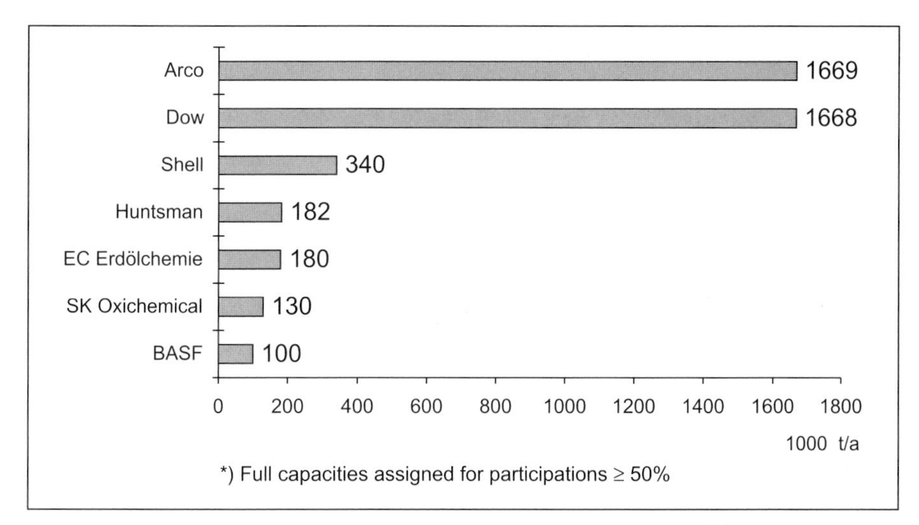

Fig. 16: Capacities* of the leading propylene oxide producers, 1997[1]

Styrene

Styrene is one of the most important intermediates in the production of synthetic thermoplastics. It is most widely known to the general public through polystyrene, but serves also as a component of styrene copolymers. World production capacity for styrene in 1997 was 22.10 million metric tons per year. As in the case of benzene, North America classes as the chief producing region (Fig. 17).

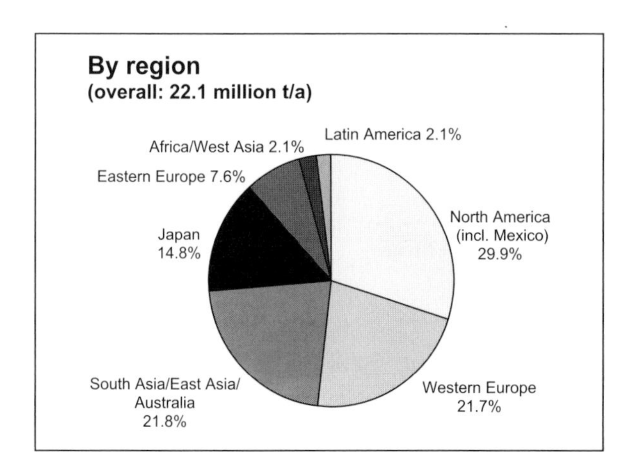

Fig. 17: Capacities for styrene, 1997[1]

The list of **top ten producer corporations** was led in 1997 by Dow, with 8.4% of world capacity, followed by a trio of companies with comparable capacities. These in turn were widely separated from the others in the list. Together the top ten firms accounted in 1997 for almost half the world production capacity (Fig. 18).

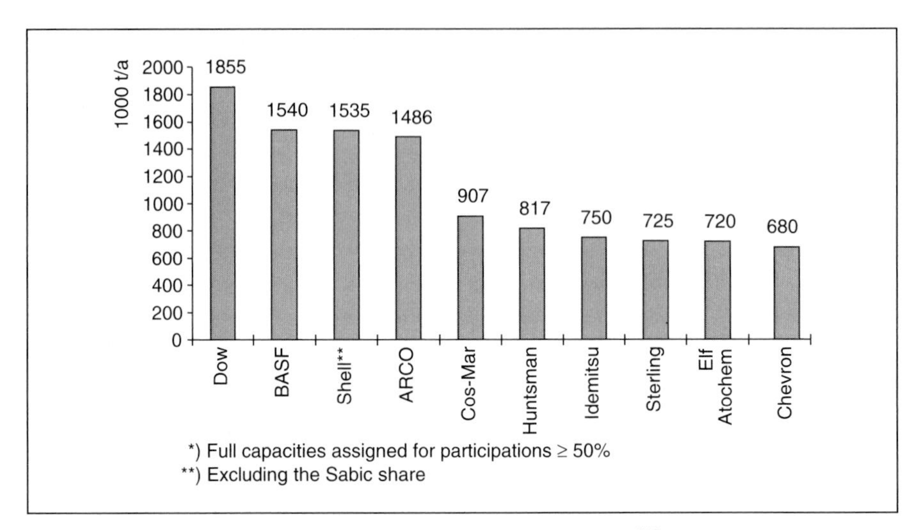

Fig. 18: Capacities* of the top ten producers of styrene, 1997[1]

Fibers

About 48.4 million metric tons of fibers were produced worldwide in 1997. Artificial fibers held a substantial lead over cotton, with an output of 27.3 million tons. Wool came in a distant third (Fig. 19).

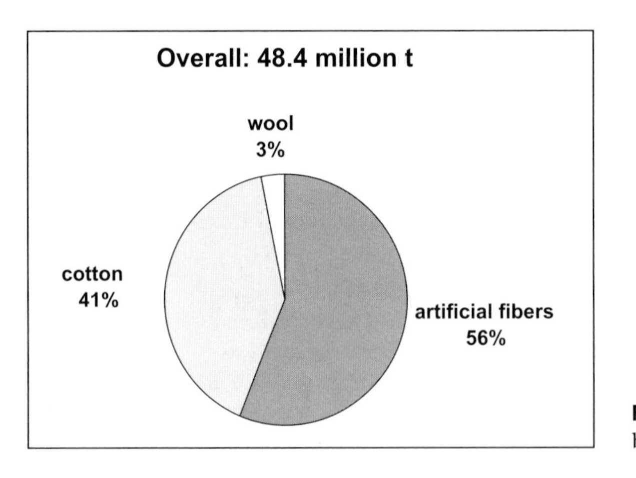

Fig. 19: Fiber production by type, 1997[2]

[2] Data from: Chemical Fibers Industrial Association (Germany); *Die Chemiefaserindustrie in Deutschland*, 1997.

Due to increases in population, cropland dedication to cotton is expected to face increased competition from demand for food crops.

Artificial Fiber Production at Record Levels

World production of artificial fibers, which has been increasing steadily for many years, achieved a new record of 27.3 million metric tons in 1997. Of this total, 24.4 million tons (also a record) was accounted for by synthetic fibers, with 2.9 million tons being material based on cellulose (preliminary data; Fig. 20a).

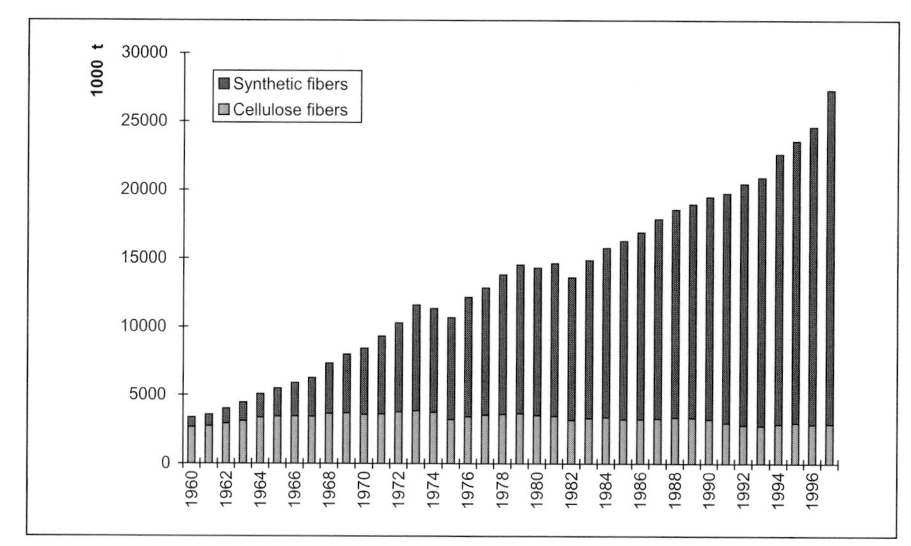

Fig. 20: a) Trends in artificial fiber production by type, 1960–1997

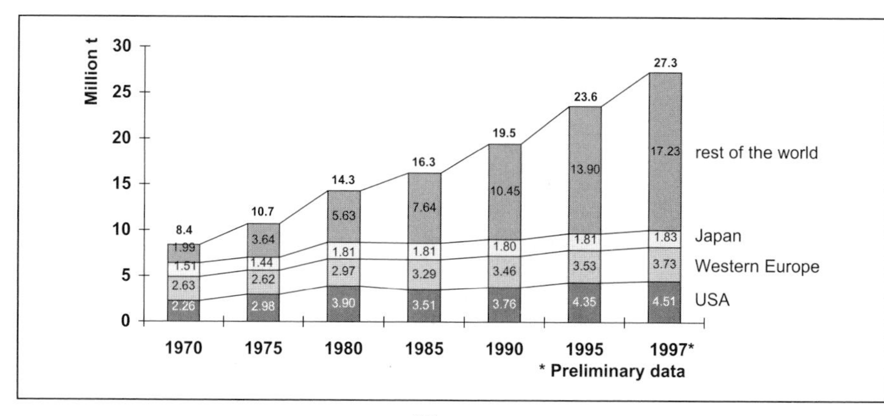

Fig. 20: b) Production trends by region[3]

---

[3] data from: Akzo Nobel: *Chemiefaserjahr 1997*, p 16f.

The principal driving force behind this growth was countries other than the United States, Japan, and the Western Europeans. Their share of world artificial fiber production climbed from 23.7% (2.0 million tons) in 1970 to 63.1% (17.2 million tons) in 1997 (Fig. 20b). Asian nations played a particularly important role in this development. As a consequence of persistent increases in China, Taiwan, and Korea, the **emphasis in artificial fiber production** shifted to this part of the world. Asia thus became responsible in 1997 for somewhat more than half the world production, with Western Europe and the United States claiming between them less than one-third (Fig. 21).

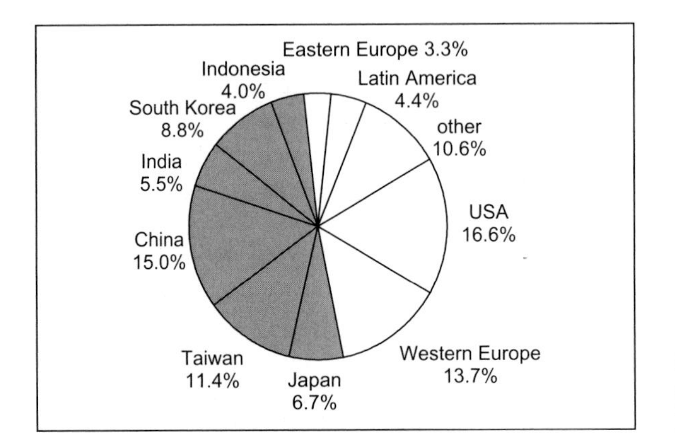

Fig. 21: Artificial fiber production by country, 1997[3]

Germany was the **leading production site** for artificial fibers **in Europe** during 1997, registering 1.07 million tons, of which about 18.3% was cellulose fiber. The lion's share of synthetic fiber production belonged to polyester (38.3%). Polyamides took second place (29.3%), followed by polyacrylic fibers (23.6%), as shown in Fig. 22.

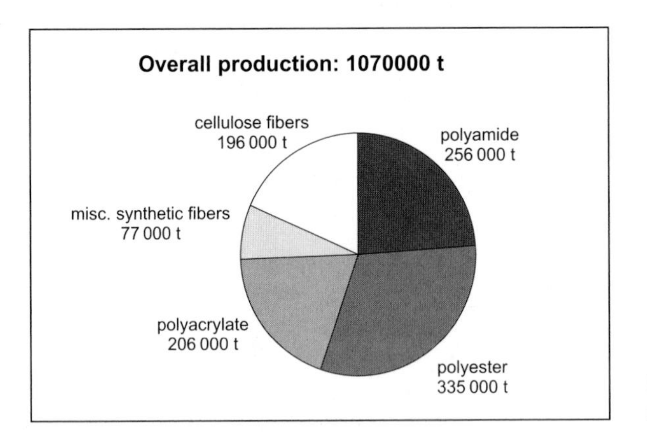

Fig. 22: German artificial fiber production, 1997[2]

Total sales of artificial fibers in Germany in 1997 amounted to $3.52 billion as the number of workers in this sector continued to decline (23 200 in 1994, 21 400 in 1995, 20 300 in 1996, and 19 800 in 1997).

Synthetic Fiber Production at Record Level

World production of synthetic fibers set a new record of 24.4 million metric tons per year in 1997, with an average annual growth rate since 1970 of 6.2% (Fig. 23).

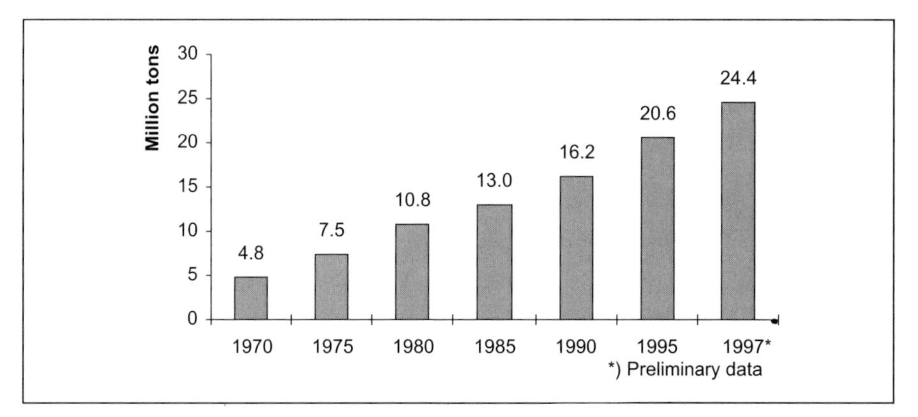

Fig. 23: Growth in world production of synthetic fibers, 1970–1997[3]

However, these data mask quite different growth rates among the three major classes of synthetics, ranging from 8.7% per year for polyester through 3.7% for polyacrylates to 2.7% for polyamides. As a result, polyester has increased its share of the market from 34% in 1970 (second behind polyamides) to 63% in 1997 (Fig. 24).[4]

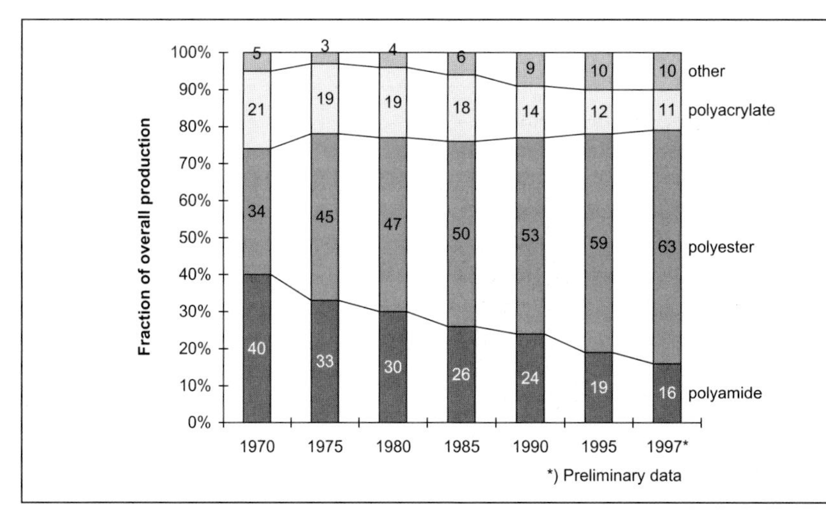

Fig. 24: Trends in synthetic fiber type, 1970–1997

Plastics

See the corresponding separate chapter.

Vitamins

Vitamins are low-molecular-weight substances that play a critical role in the life functions of both animals and humans, but substances that these species nevertheless cannot themselves synthesize in sufficient quantities. Their needs must therefore be met by dietary sources or through symbiotic relationships with microorganisms (intestinal bacteria).

The list of known vitamins now includes 13 entries (Fig. 25), with vitamin B_{12} (→ Synthesis, Masterful Achievements: Enzymes) being the one most recently isolated (1948). This is also the only vitamin which, because of its complex structure, is not prepared commercially by chemical means, but is rather produced by fermentation (Table 1).

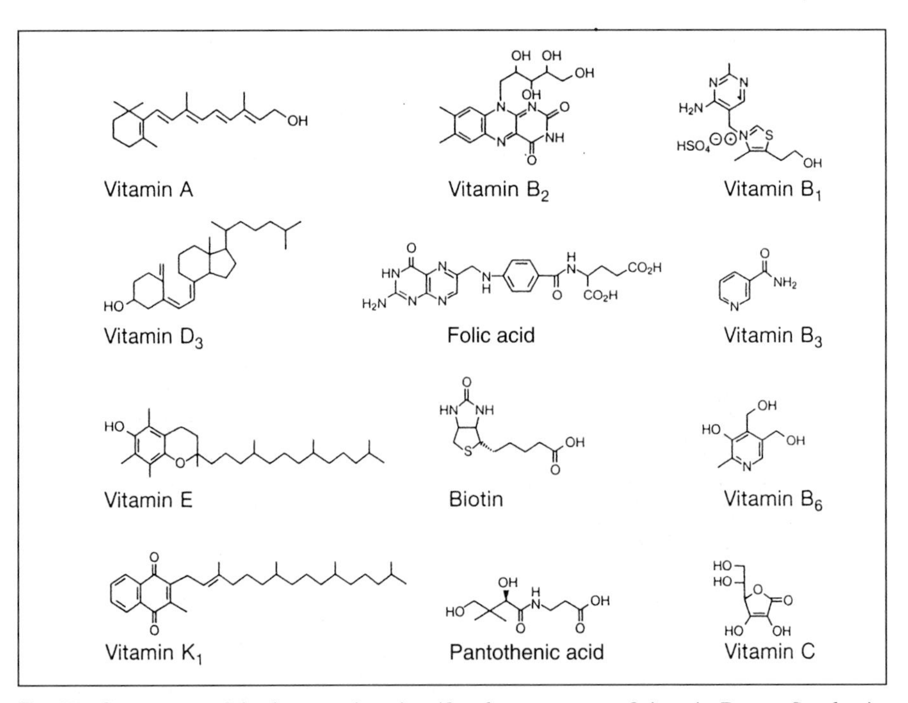

Fig. 25: Structures of the known vitamins (for the structure of vitamin B_{12}, → Synthesis, Masterful Achievements, Fig. 4)

The **world's foremost producers of vitamins** are Hoffmann–La Roche and BASF.

Table 1: Industrial processes for the preparation of vitamins

Vitamin	Synthesis	Fermentation	Isolation
vitamin A	✗		●
vitamin B_1	✗	●	
vitamin B_2	✗	✗	
vitamin B_6	✗	●	
vitamin B_{12}		✗	
vitamin C	✗	✗	
vitamin D_3	✗		●
vitamin E	✗		✗
vitamin K	✗		●
biotin	✗	●	
folic acid	✗	●	
niacin	✗		
pantothenic acid	✗	●	

✗ = utilized; ● = possible

The **world market** for vitamins adds (1994) up to about $3.1 billion, $1.4 billion of which is directed toward human nutrition.

From a quantitative standpoint, the market was dominated in 1994 by vitamin C (60 000 metric tons), vitamin E (22 500 tons), and niacin (21 600 tons), as shown in Fig. 26. The widely recognized vitamin C is primarily targeted at people (foods and vitamin supplements), whereas three-fourths of the annual production of both vitamin E and niacin is destined for animal nutrition.

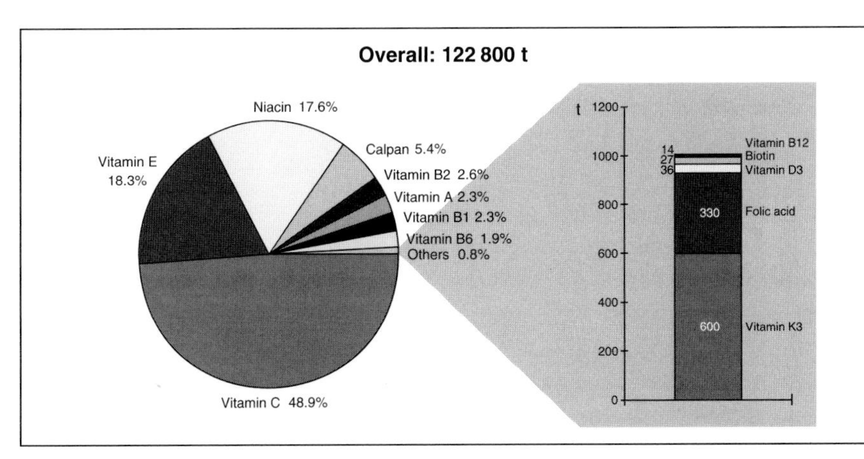

Fig. 26: The world market for vitamins, 1994[1]

The vitamin produced in the **smallest quantity** is B_{12}. Only 14 tons of this costly substance was consumed in 1994: 55% as animal nutrition and the rest in the human sector.

[1] BASF

Development Costs

From the first successful synthesis of a new active crop-protection ingredient, at least ten years will elapse before a commercially mature product has been developed. **Investment costs** associated with this process have increased five-fold in the past 20 years, and now average about 250 million German marks. In a relative sense the costs due to "research on side effects" have increased most rapidly – i.e., toxicology and ecotoxicology. This in turn sheds light on the importance now placed on questions of user and consumer safety with respect to crop-protection agents (Fig. 1).

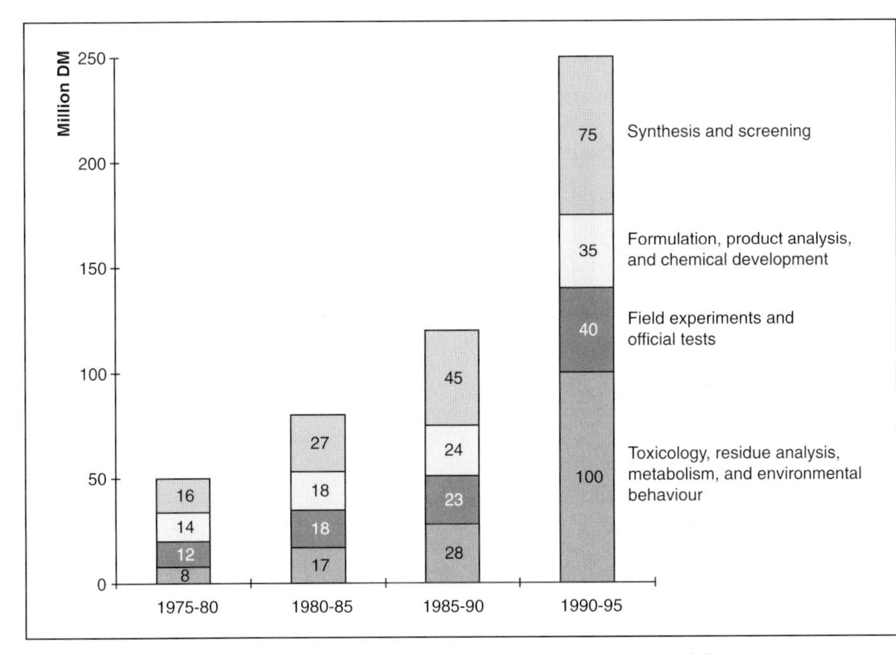

Fig. 1: Development costs for a crop-protection agent, 1975–1995[1]

Apart from pursuing analogies and investigating structure–activity relationships, much of the effort expended in this field is directed toward systematic screening of chemicals in the search for new active agents. This means submitting the largest possible number of newly synthesized compounds to multistep testing in the hope of discovering new activity, where unsuitable materials must be detected and eliminated at each step. From a statistical standpoint, 40 000 compounds must be examined in this way to identify a single new usefully active component.

[1] data from: *Folien des Fonds der chemischen Industrie*, text vol. 10.

The Most Important Crops

Fruits and vegetables were the crops in which the greatest investment was made for crop-protection agents in 1997. One-fourth of the world pesticide market was dedicated to this inhomogeneous sector that covers a wide range of products, with emphases that vary considerably by region (Fig. 2).

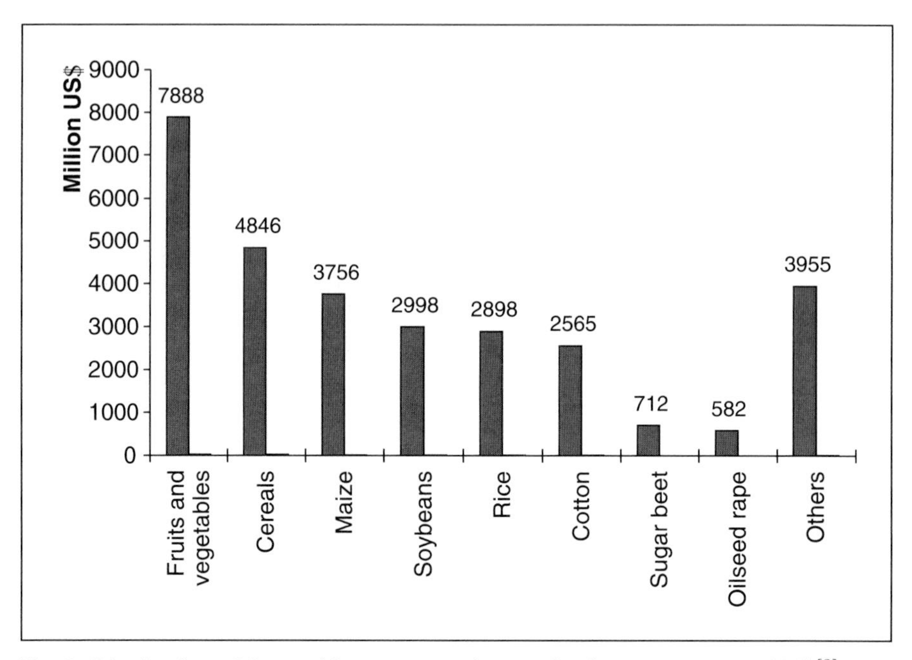

Fig. 2: Distribution of the world crop-protection market by crop category, 1997[2]

The second largest market segment was grain crops, lagging more than 10% behind fruits and vegetables, followed by rice and corn. Other crops accounted for less than 10% of the market for crop-protection agents.

The Most Important Classes of Compounds

Each of the three product categories in 1997 was dominated by a single class of compounds. In the case of insecticides it was the organophosphates (→ Poisons, Hit List: The Most Toxic Synthetic Poisons), with a market volume somewhat greater than the next two compound classes combined.

The situation was similar with respect to fungicides, where triazoles stood out as the leading compounds. Nevertheless, triazoles achieved a sales level only

[2] data from: *Wood Mackenzie Consultants*, Oct. **1998**, private communication.

half that of the most important class of herbicides, the amino acid derivatives (Fig. 3).[3]

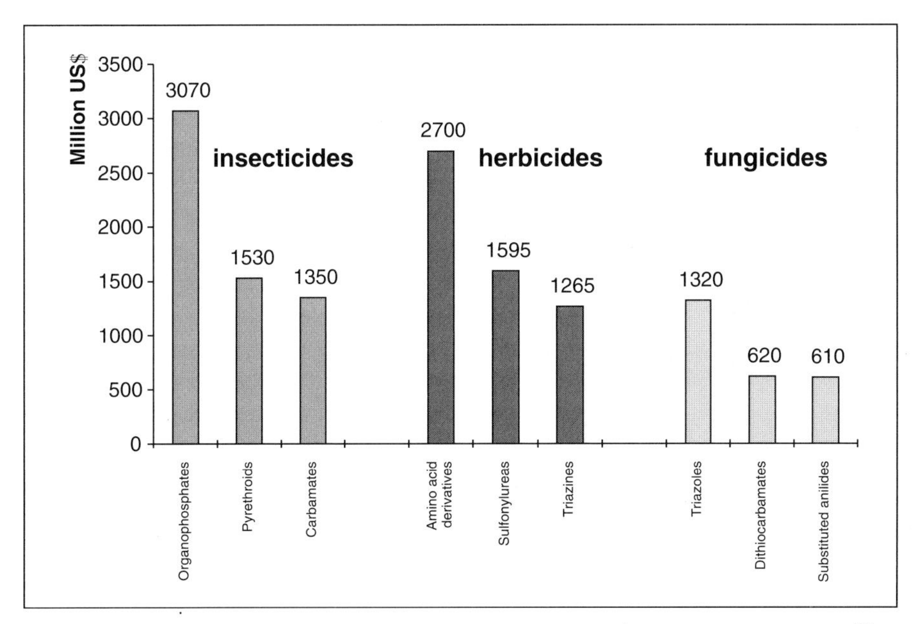

Fig. 3: Market values of selected classes of compounds, 1997 (based on end-users)[3]

Table 1 provides an overview of the ten leading crop-protection agents from the standpoint of total sales; structures of the relevant compounds are shown in Figure 4.

Table 1: The top ten active ingredients in crop-protection agents, 1997

Company	Substance	Trade Name	Sales ($ million)	Application
Monsanto	glyphosate	Roundup	2180	herbicide
Cyanamid	imazethapyr	Pursuit	540	herbicide
Zeneca	paraquat	Gramoxone	520	herbicide
Dow	chlorpyrifos	Dursban	475	insecticide
Bayer	imidacloprid	Admire	435	insecticide
Novartis	metolachlor	Dual	320	herbicide
Cyanamid	pendimethalin	Prowl	315	herbicide
Novartis	atrazin	Gesaprim	270	herbicide
Zeneca	fluazifop	Fusilade	265	herbicide
Rohm & Haas	mancozeb	Dithiane	265	fungicide

Source: Wood Mackenzie Consultants, 1998

[3] data from: Wood Mackenzie Agrochemical Service, Product Section, May **1998**, p. 7.

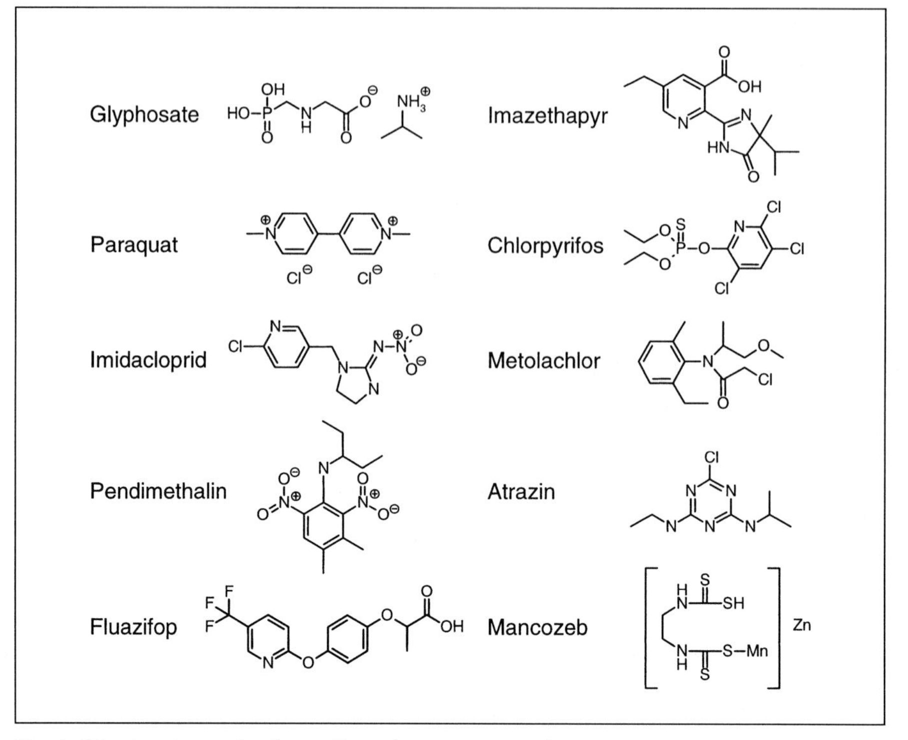

Fig. 4: The top ten active ingredients in crop-protection agents.

The Ten Leading Companies in the Production of Crop-Protection Agents

For many years, the list of the **largest companies with respect to agrochemicals turnover** was led by Ciba, with 1995 sales of $3.374 billion. The merger in 1995 of Ciba with eleventh-place Sandoz to form Novartis allowed the Swiss to increase their lead in 1996, with crop-protection sales in that year of $4.175 billion. Novartis' lead over second-place Monsanto amounted to a healthy $1.3 billion.

Monsanto was able to narrow this gap in 1997 to roughly $100 million, but Novartis continued to lead the top ten with sales of $4.173 billion.

Taken together, the ten top manufacturers of crop-protection agents increased their 1996 sales of $24.81 billion only slightly in 1997 to $25.05 billion. This list of ten included only one company devoted exclusively to the crop-protection market: AgrEvo (Fig. 5).

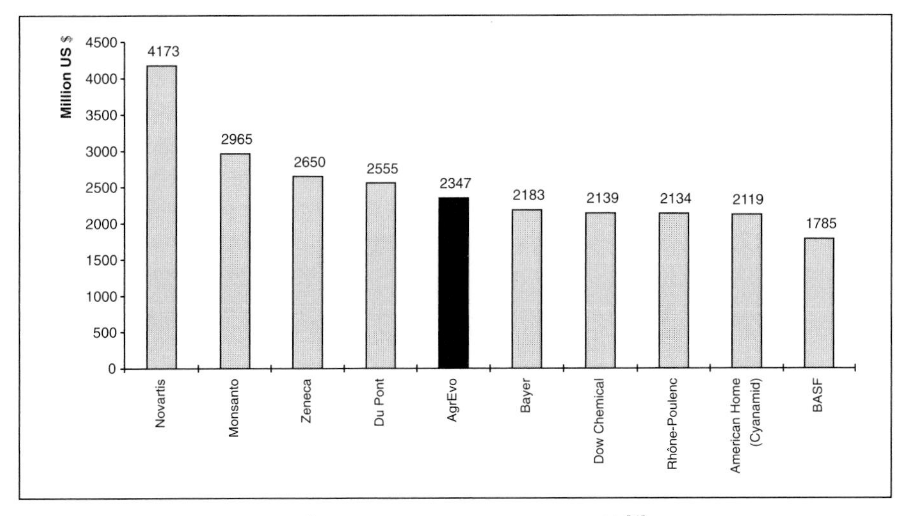

Fig. 5: Major companies 1997 agrochemicals turnover, 1997[4]

The Most Innovative Companies

The list of most innovative agrochemical companies from the standpoint of **newly introduced products** (new active ingredients) over the period 1987–1997 was led by the company recording the highest sales figures: Novartis. Nevertheless, smaller companies were also important sources of innovation. For example, the Japanese firms, which together in 1997 accounted for only slightly more than $100 million in sales, introduced during this period 36 new active ingredients. Sumitomo, was especially active, occupying second place ahead of its large American competitors (Fig. 6).

Over the period cited, 126 new products were introduced, 40% of them herbicides, 29% insecticides, and 25% fungicides.

[4] data from: Wood Mackenzie Agrochemical Service, Companies Section, August **1998**, p. 3.

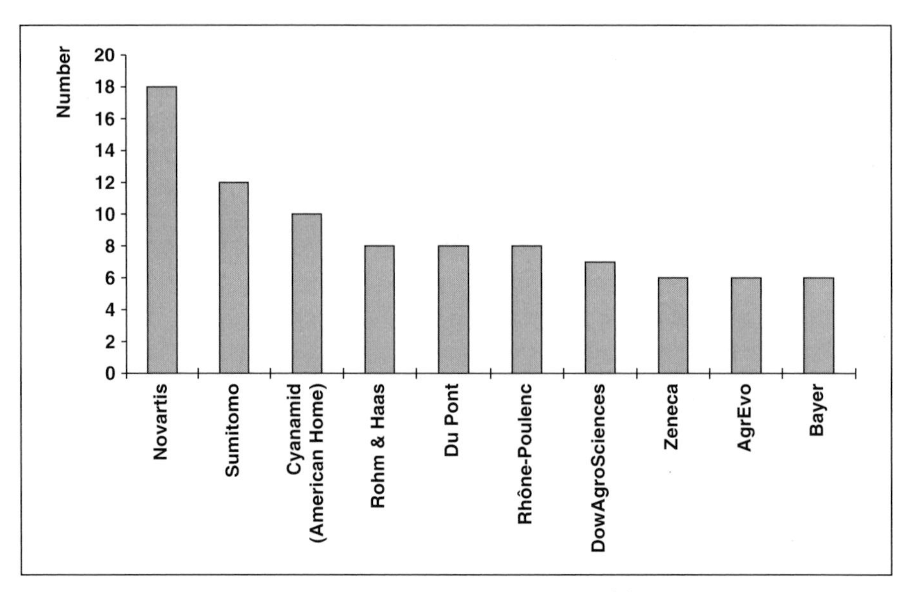

Fig. 6: New product introductions by company, 1987–1997[5]

The Most Important Consumer Regions

The **largest sales market** for pesticides in 1997 was North America (Fig. 7). This region was responsible for 34.5% of world consumption or $30.2 billion (end-user level).

It is generally assumed that Asia's importance in the world market will increase. Sales in the European Union depend in part upon the fate of the cropland set-aside program.

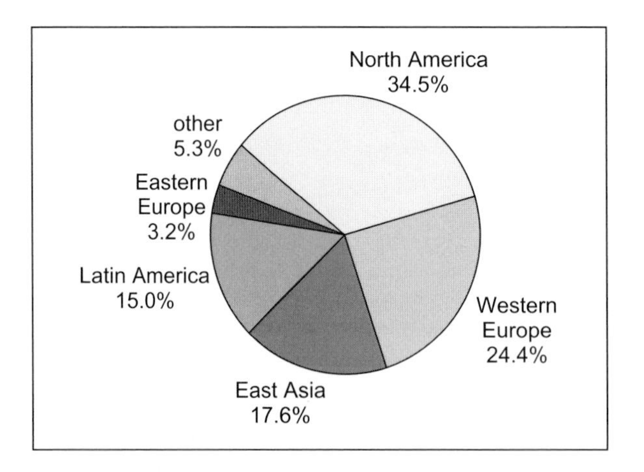

Fig. 7: World agrochemical market by region, 1997[2]

[5] Wood Mackenzie Agrochemical Service, *Development Products*, **1998**, p. 99.

The Chief Segments of the World Market

According to data from Wood Mackenzie Consultants, the **world market for crop-protection agents** in 1997 added up to $30.20 billion (end-user level). The most important segment of this market was that for herbicides, which was somewhat larger in 1996 than the markets for insecticides and fungicides combined (Fig. 8).

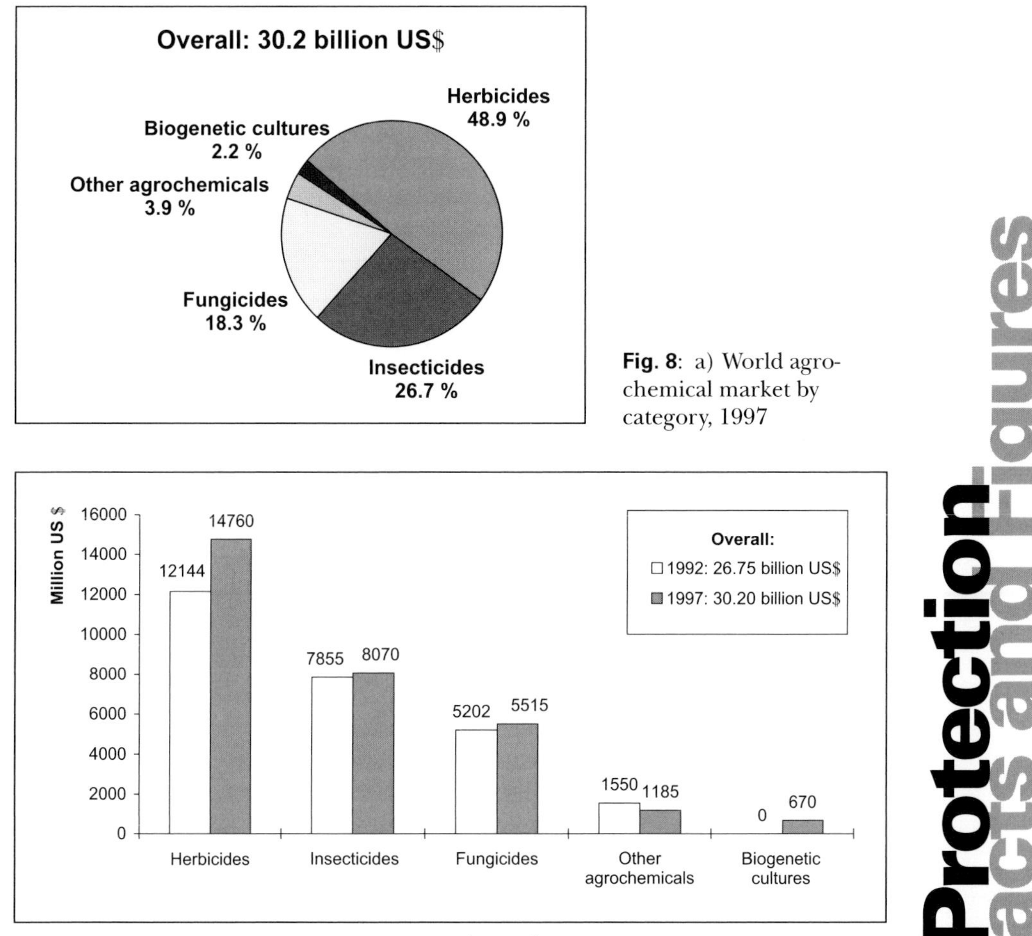

Fig. 8: a) World agrochemical market by category, 1997

Fig. 8: b) Market volume by category (in 1997, US $)

Over the period 1992–1997, reported herbicide sales grew by 4% annually to $14.76 billion. The leading performers in this market segment were amino acid derivatives (17.2% annually), sulfonylureas (11.4% annually) and aryloxyphenoxypropionates (10.7% annually).

The Highest and Lowest Melting Points

The physical properties of the known chemical elements are exceedingly diverse, and often quite astounding. At atmospheric pressure, hydrogen is the element with the **lowest melting point** (–259.34 °C; Fig. 1).

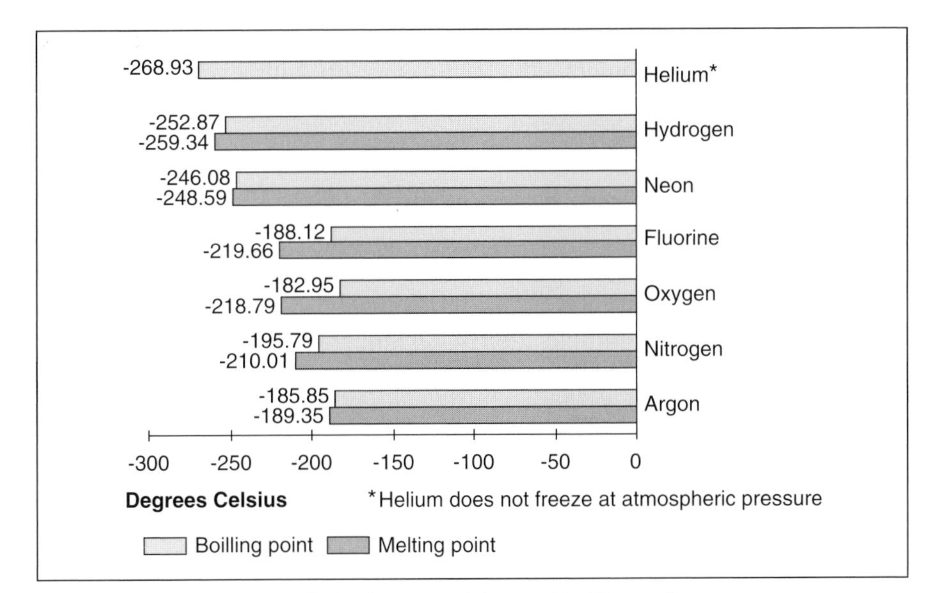

Fig. 1: a) The elements with the lowest melting and boiling points

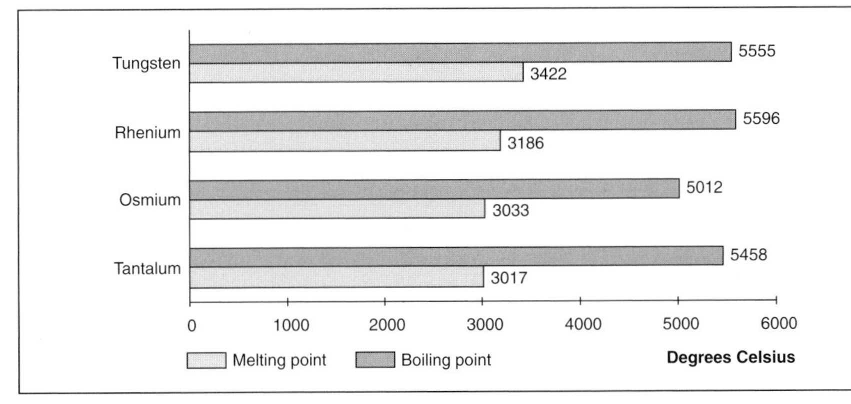

Fig. 1: b) The elements with the highest melting and boiling points

In second place is the noble gas neon (–248.59 °C), followed by fluorine (–219.66 °C), oxygen (–218.79 °C), nitrogen (–210.01 °C), and argon (–189.35 °C). Tungsten has the **highest melting point** (3422 °C),[a] followed by rhenium (3186 °C), osmium (3033 °C), and tantalum (3017 °C). Helium cannot be solidified at all under atmospheric pressure. Instead, at –271 °C – very near absolute zero – it becomes "superfluid," which means it is "more liquid than liquid," but certainly not superfluous. Its viscosity actually decreases to zero. A **superfluid** flows with absolutely no friction, and even creeps freely up the walls of any vessel that tries to contain it. In other words, an extremely cold beaker filled with superfluid helium would essentially empty itself.

The **lowest melting point of a solid element** under standard conditions is that of cesium (28.4 °C), with the second lowest being that of gallium (29.8 °C). The metal with the lowest melting point is mercury (→ Atoms and Molecules, $E = mc^2$), with a melting point of –38.8 °C. Bromine, the only element other than mercury that is a liquid under normal conditions, melts at –7.2 °C.

The Highest and Lowest Boiling Points

Helium has the **lowest boiling point** of all the elements (–268.93 °C). In second place is hydrogen (–252.87 °C), followed by neon (–246.08 °C), nitrogen (–195.79 °C), fluorine (–188.12 °C), argon (–185.85 °C), and oxygen (–182.95 °C). The **highest boiling point** for an element is that of rhenium (5596 °C), followed by tungsten (5555 °C), tantalum (5458 °C), and osmium (5012 °C). These four elements are also the record holders with respect to melting point.[b] The **smallest difference between melting point and boiling point for any element** is displayed by neon: 2.5 °C.

Density Records

The element with the **lowest density** is hydrogen (0.088 g/L under standard conditions; Fig. 2), followed by the noble gases helium (0.176) and neon (0.885).

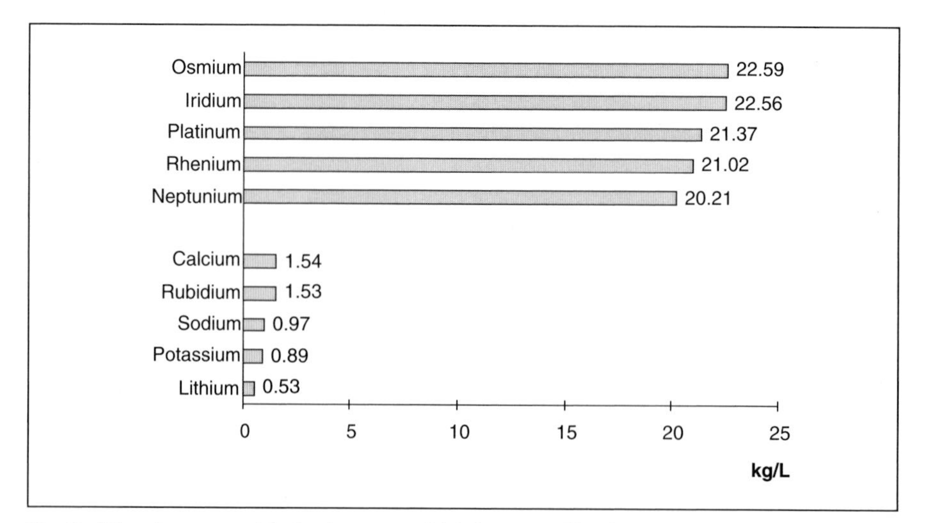

Fig. 2: The elements with the lowest and highest specific densities

The **most dense element** is osmium (22.587 kg/L).[1] Iridium is a very close second (22.562),[1] with platinum in third place (21.37), rhenium fourth (21.02), and neptunium[c] fifth (20.2). Among the solid elements, lithium is clearly the lightest; its density amounts to only 0.53 kg/L. Lithium is followed by potassium (0.89), sodium (0.97), rubidium (1.53), and calcium (1.54).

The Hardest and the Softest Element

The **solid element that classes as both softest and hardest** is carbon: in the two forms graphite and diamond. A single crystal of diamond scores the absolute maximum value on the Knoop hardness scale: 90 GPa.[2] Based on the somewhat less informative abrasive hardness scale of Mohs, diamond has a **hardness** of 10. Boron with a Mohs hardness of 9.5 is the second hardest element. The carbon allotrope graphite is an extremely soft material with a Mohs hardness of only 0.5 and a Knoop hardness of 0.12 GPa (→ Atoms and Molecules, Hardness Records).

Elements with the Highest and Lowest Thermal Conductivities

Carbon in the form of diamond is also a record-holder in the field of **thermal conductivity**. Values greatly exceeding 2000 W/m K have been reported at room temperature. It is generally the metals, however, that under normal conditions are the **best elemental conductors of heat** (Fig. 3).

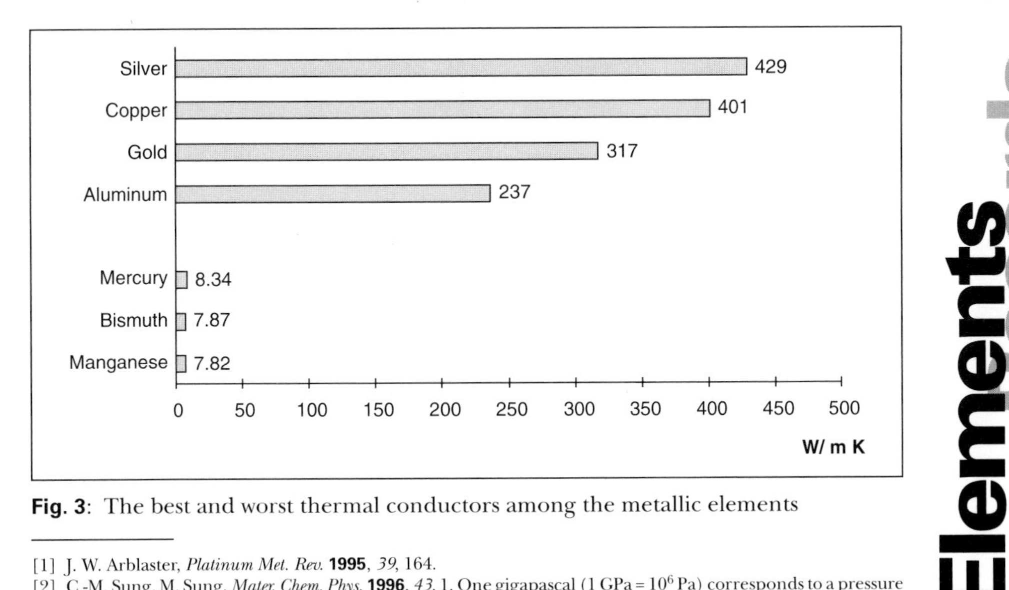

Fig. 3: The best and worst thermal conductors among the metallic elements

[1] J. W. Arblaster, *Platinum Met. Rev.* **1995**, *39*, 164.
[2] C.-M. Sung, M. Sung, *Mater. Chem. Phys.* **1996**, *43*, 1. One gigapascal (1 GPa = 10^6 Pa) corresponds to a pressure of 10 bar.

Elements Records

The leader here is silver (429 W/m K), ahead of copper (401), gold (317), and aluminum (237). The poorest thermal conductivities among the metals are displayed by manganese (7.82 W/m K), bismuth (7.87), and mercury (8.34). Much lower still, of course, are the thermal conductivities of the gaseous elements. Xenon is the **best of the known elemental thermal insulators**, with a value of 0.0055 W/m K, whereas hydrogen (0.187) and helium (0.157) are the gases with the highest thermal conductivities. The thermal conductivity of superfluid helium exceeds that of the "normal" liquid form by a factor of 10^6.

The Best and the Worst Conductors of Electricity

The element with the **lowest electrical resistance** (and thus the highest electrical conductivity) under standard conditions (Fig. 4) is silver, with a value of 1.617×10^{-8} Ω m, followed by copper (1.712×10^{-8}), gold (2.255×10^{-8}), aluminum (2.709×10^{-8}), and calcium (3.42×10^{-8}). The poorest conductors of electricity among the metals are manganese (144×10^{-8} Ω m), gadolinium (131×10^{-8}), and terbium (115×10^{-8}).

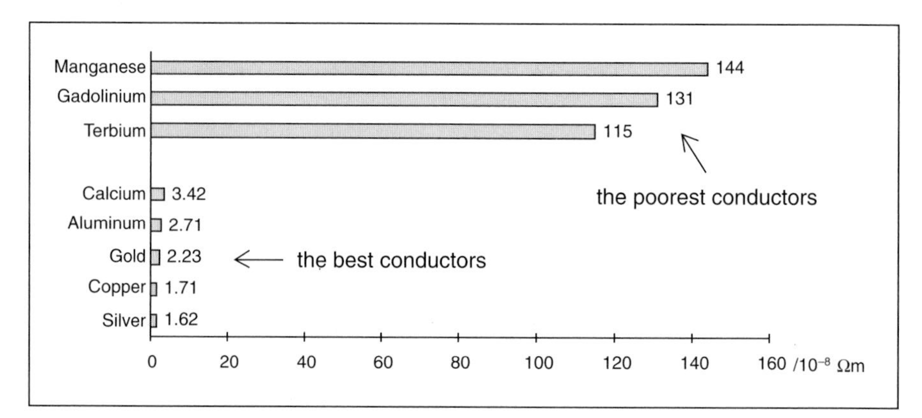

Fig. 4: The metallic elements displaying the highest and lowest values for electrical resistance

[a] D. R. Lide, *CRC Handbook of Chemistry and Physics*, 79th ed., CRC Press, **1998**, p. 4–92. It is also reported that at standard pressure, where graphite is the stable form of carbon, the metastable form diamond has a melting point of ca. 3550 °C (i.e., higher than that of tungsten; L. F. Trüb, *Die chemischen Elemente*, Hirzel, Stuttgart, **1996**, p. 259).

[b] All data from the *CRC Handbook of Chemistry and Physics*, 79th ed. The literature gives values that are in some cases much higher, including for rhenium (5870 °C), tungsten (5700), tantalum (5534), and osmium (5020); see A. F. Holleman, E. Wiberg, N. Wiberg, *Lehrbuch der anorganischen Chemie*, 33rd ed., DeGruyter, Berlin, New York, **1985**.

[c] D. R. Lide, *CRC Handbook of Chemistry and Physics*, 79th ed., CRC Press, **1998**, p. 4–72. The literature also contains the value 19.5 kg/L (L. F. Trüb, *Die chemischen Elemente*, Hirzel, Stuttgart, **1996**, p. 380).

The First Elements to Become Known

The **first element with which mankind became acquainted** and which was then utilized in a conscious way was carbon. Testimony to this includes cave drawings from prehistoric times that contain carbon in the form of soot (→ Pigments, Records: The Oldest Dye). Nevertheless, stone-age people were unaware of the elemental nature of carbon. Other **elements known at least since antiquity** are bismuth, iron, gold, copper, platinum, mercury, sulfur, silver, and tin. With the exception of the latter, all these elements are found in nature in elemental form. Lead and zinc, which are very easy to isolate from their ores, have been known at least since the Middle Ages. Working independently, Mendeleev and Meyer in 1869 both brought a certain amount of order into the confusion surrounding the then known elements, basing their schemes on physical and chemical data. Each thus developed a periodic system of the chemical elements, albeit with a number of gaps, which permitted prediction for the first time of the existence of still undiscovered elements along with their most important properties.[1]

Among the **predicted elements** were gallium (discovered in 1875 by de Boisboudran) and germanium (discovered in 1886 by Winkler). The last of the nonradioactive elements to be discovered were hafnium, identified in 1923 by Coster and v. Hevesy, and rhenium, the presence of which was established in mineral samples in 1925 by Tacke and Noddack; rhenium was then isolated in pure form in 1926. An important reason for the late discovery of these metals is the fact that no ores exist for either of them. The first mineral containing rhenium as a principal component was found in 1994 – on the edge of a volcanic crater. Small amounts of hafnium regularly accompany zirconium in ores of the latter, but separation of the two metals is extremely difficult because of their nearly identical properties.

Uranium was the **first radioactive element** to be discovered (Klaproth, 1789; first described as an element by Péligot in 1840), and the second was thorium (Berzelius, 1828). Their radioactivity of course remained unknown until the revolutionary (and in fact accidental) discovery of the phenomenon itself (Becquerel, 1896). Polonium and radium were the third and fourth unstable elements to be discovered: in this case actually *because* of their high radioactivity (Curie, 1898). The isolation by Marie Curie of **ca. 100 mg of radium chloride**[2] from two tons of Joachimsthal pitchblende (uraninite) under the most primitive of conditions was truly a masterful achievement and quite properly legendary. The husband-and-wife team of Marie and Pierre Curie received the 1903 Nobel Prize in physics along with Becquerel for their work on the investigation of radioactivity. In 1911 Marie Curie was also rewarded with the

[1] Regarding the development of the modern periodic table and predicting the properties of artificial elements see G. T. Seaborg, *J. Chem. Soc., Dalton Trans.* **1996**, 3899.

[2] N. N. Greenwood, A. Earnshaw, *Chemistry of the Elements*, Pergamon Press, Oxford, **1984**, p. 118. It has also been reported that the compound was actually radium bromide (L. F. Trüb, *Die chemischen Elemente*, Hirzel, Stuttgart, **1996**, p. 81).

Nobel Prize in chemistry for her discovery of polonium and radium and their characterization, isolation, and closer investigation (→ Appendix, Nobel Prizes: in Chemistry). The Curies thus joined the ranks of the shining heroes of science. They also ruined their health through their work, however, because so soon after the discovery of radioactivity nothing was yet known about its dangerous effects. In this "stone age" of male-dominated natural science Marie Curie was not until 1904 appointed for the first time to a paid position – as laboratory assistant to her husband, a year after receiving a Nobel Prize in physics.[a]

Nevertheless, despite two entries under her name, Marie Curie remains in the middle of the pack in the sport of "element discovery." The champion here is Klaproth, recognized as the (co)discoverer of eight elements: beryllium, cerium, chromium, strontium, tellurium, titanium, uranium, and zirconium. In second place on this hit list is Ramsay, who discovered the five nonradioactive noble gases (helium, neon, argon, krypton, and xenon). However if the notion of "discovery" is defined somewhat more broadly,[b] then Seaborg ranks above both of these, since his research team artificially synthesized nine elements between 1940 and 1958: plutonium, americium, curium, berkelium, californium, einsteinium, fermium, mendelevium, and lawrencium. It was later established that traces of plutonium are also present in nature. Hot on the heels of Seaborg is the research group of Armbruster, Hofmann, and Münzenberg, which between 1981 and 1996 enriched the periodic table with the **six most recently discovered elements** (107–112). This became possible with the aid of the linear accelerator UNILAC operated by the Society for Heavy Ion Research in Darmstadt, Germany, a device capable of electrostatically catapulting ions of all the chemical elements to 20% of the speed of light thanks to an accelerating potential of 150 million volts.

Table 1: The discoverers of the largest number of elements

Discoverer	No. of elements	Elements
Seaborg	9	Pu, Am, Cu, Bk, Cf, Es, Fm, Md, Lr
Klaproth	8	Be, Ce, Cr, Sr, Te, Ti, U, Zr
Armbruster, et al.	6	106–112
Ramsay	5	He, Ne, Ar, Kr, Xe

Records in the Recovery of Gold

Very few elements occur naturally in large quantity in elemental form. The main exceptions are sulfur, carbon, and the noble metals. The latter are isolated – sometimes with great difficulty – from rock containing them by tedious physical and chemical enrichment methods. A record with respect to **economical recoverability** is the lower limit for gold content, which for sources adaptable to open-pit mining can be as low as one gram per ton of rock. The **largest open-pit mine in the world for gold recovery** is the so-called "superpit" at Kalgoorlie in Western Australia, which in its final stages will be 5 km long, 2 km wide, and 500 m deep. The **largest gold deposit in the world** is found in South Africa near Johannesburg in a geological formation known as the Witwatersrand, once a lake. Ore is mined here at a depth as great as 4 km.

The Largest Open-Pit Mines

Those elements not found naturally in significant quantities in elemental form are recovered from suitable compounds. With some elements this is relatively unproblematic, the reason why copper was the **first metal to be isolated from ore**. By coincidence, the **largest open-pit mine for copper ore** is also the **largest open-pit mine for ore of any kind**. It is located in Chile, and currently has a surface area of 3.8 × 1.8 km with a mining depth of ca. 600 m. The **deepest open-pit mine in the world**, with a depth of 1200 m, is also the **deepest hole ever created by man**, and it lies in the middle of the South African metropolis Kimberley.[3] This pit was dug over the course of 43 years, and serves today as a monument in memory of the great diamond fever that broke out there on July 16, 1871. Before the mine's abandonment in 1914 it yielded 14 million carats of diamonds, and 25 million metric tons of soil and rock accumulated as overburden in piles extending in every direction around the hole. The **wettest and deepest open-pit mining regions in the world** are located in the Pacific South Seas and in the Atlantic, where manganese nodules are found on the ocean bottom. These probably developed through the action of microorganisms in the vicinity of undersea volcanoes. They contain mainly oxides of iron and manganese, but also certain amounts of nonferrous ores. At the present time their recovery is not economically practical. The **largest mining town** is Mount Isa in Northern Queensland, Australia. Its 40 467 square kilometers also make it the largest city in the world in terms of surface area. This is the site of the world's **largest lead and silver mine**.

The Most Difficult Elements to Isolate

The fact that the recovery of some elements from their compounds can be extremely complicated is demonstrated most clearly by what are known as the "rare earth metals" [scandium, yttrium, lanthanum, and the lanthanoids (cerium to lutetium)]. Chemically they differ very slightly from one another, and are found associated together in their minerals. (The separation of zirconium traces from hafnium is also extraordinarily difficult, and for similar reasons.) **Separation of the rare earth metals** requires a prodigious effort. It

[3] R. Gööck, *Alle Wunder dieser Welt*, Bertelsmann, Gütersloh, **1968**, pp. 178–179.

can be accomplished more readily today in part thanks to state-of-the-art automated ion extraction procedures. At the beginning of the 20th century such techniques did not yet exist, and time-consuming manual labor was the only option. Thus, an American by the name of James once recrystallized a thulium bromate sample 15 000 times in his effort to isolate pure thulium – probably a record for recrystallization! The rare earth metals proved so difficult to separate that over the course of the period from 1787, when the first rare earth mineral was discovered, until 1907, when the last stable rare earth element (lutetium) was identified, on roughly 100 occasions mixtures of the metals or even mere impurities were declared to be new elements.

Names: The Interesting and the Peculiar

The names of many elements are of geographical origin. It is not particularly surprising that elements have been named for heavenly bodies (e.g., uranium, neptunium), continents (e.g., europium, americium), countries (e.g., germanium, polonium), or at least capital cities (hafnium, lutetium, and holmium, derived from the Latin names for Copenhagen, Paris, and Stockholm). But four separate elements were named for the little town of **Ytterby**, located 30 km north of Stockholm: namely the rare earth metals yttrium, ytterbium, terbium, and erbium. In 1787 a new mineral, ytterbite, was discovered in the legendary feldspar mine at Ytterby. The Finnish chemist Gadolin in 1794 discovered in this mineral a new element, which was named yttrium from its place of origin. From then on, the mineral itself was called gadolinite in honor of Gadolin. Ten additional rare earth metals were later isolated from gadolinite, although it was really not possible to name *all* of them after the little village. **France** is the only country after which more than one element is named (gallium and francium). The **only country that was named after an element** is **Argentina** (from the Latin argentum = silver). The **most far-distant place to have an element named for it** is **Pluto**, the outermost planet in our solar system. Many elements are named for scientists. The first example was element 64, discovered in 1880 and designated gadolinium in honor of **Gadolin**. Many artificially prepared elements acquired their names in this way, including curium, einsteinium, fermium, mendelevium, nobelium, and lawrencium. The name suggested for element 106 by its discoverers, "seaborgium," found favor in 1995 with the appropriate IUPAC commission, so that **Seaborg**, the scientific father of nine elements, became the first person to have an element named for him during his lifetime.[c]

The Most Expensive Elements

The most expensive elements are not necessarily the ones that are most rare. In the case of certain especially costly elements the disproportionately high demand relative to supply is a function not only of practical utility, but also of a symbolic character nurtured by esthetic perception and mythical values. This is clearly apparent with the most expensive of all the commercially accessible natural elements: carbon. In one utterly useless form (at least from a practical standpoint), namely as a gemstone, the carbon allotrope diamond commands carat prices (1 carat = 0.2 g) of tens of thousands of dollars. The reasons why one would spend nearly \$100 000 for a gram of carbon of this type[d] are as old as the human psyche: diamonds, especially perfectly clear ones weighing many carats, are exceedingly rare. At the same time, despite the fact that they are thermodynamically unstable relative to graphite, they endure forever, at least on a human time scale ("diamonds are forever"; Mohs hardness of 10), and that for kinetic reasons. Their "cold fire" (high index of refraction, highest known thermal conductivity) fascinates the senses. For many they are simply sexy, and in the words of Marilyn Monroe "a girl's best friend." **Noble metals come next in the price list**. Currently the prices for gold and platinum are running neck-and-neck (Wall Street purchase prices as of April 3, 1997: \$363.00 per troy ounce for platinum, \$349.10 per troy ounce for gold; one troy ounce = 31.1 g). Significantly behind these two in third place is rhodium (\$225.00). Gold is another mythical element whose symbolic force continues to make itself felt in currency matters and in jewelry manufacture. Platinum and rhodium are of considerable industrial interest as important catalytic metals. It is astonishing that with appropriate authorization it is also possible to purchase such artificial elements as americium and berkelium. The isotopes ^{243}Am and ^{249}Bk cost roughly \$100 per milligram. The latter would be difficult to horde, however, since it has a half-life of only 314 days. By contrast, the half-life of ^{243}Am is 8.8×10^3 years.

The Purest Element and the Most Perfect Crystal

Generally speaking, removal of the last traces of impurities is very difficult, so the massive effort involved is justified only in special cases. One such case is the element silicon, which in ultrapure form is required in large quantities for semiconductor technology. Silicon can be singled out as the **purest commercially available element**.[e] Single-crystal, highest purity silicon, prepared by zone-melting techniques, contains less than 10^{-9} atom-% of metallic impurities.[4] If a sugar cube (2.6 g) were to be dissolved in 2.7 million liters of water – which roughly corresponds to the capacity of a small ocean-going tanker – then the sugar concentration would be in the same range as the silicon impurities above. Even a very sensitive palate would be stretched beyond its capabilities at this dilution (reminiscent of homeopathic remedies). The crystalline perfection of the material is almost unbelievable. One cubic centimeter of crystalline silicon contains 5×10^{22} Si atoms, but only ca. 10^4

[4] W. Zulehner, *Mater. Sci. Eng.* **1989**, *B4*, 1.

crystalline lattice errors;[5] only about every 1 000 000 000 000 000 000th (10^{18th}) atom is incorrectly placed. In 1996, 12 000 metric tons of single-crystal highest purity silicon was produced. Silicon single crystals can be prepared with a diameter of 30 cm, a length of 2.5 m, and a weight of over 100 kg. Thin wafers are sliced in salami-like fashion from the single crystals and then used for microelectronic purposes. The surface of such a wafer must be extraordinarily flat, so it is specially polished. If a typical silicon wafer (20 cm) were enlarged until it reached the diameter of the earth at the equator (12 900 km), it would be 47 km thick, but its surface would be so flat that in an area the size of France, Germany, and Italy combined the greatest difference in elevation would be only 20 m (by comparison, Florida would be regarded as mountainous). One atom on the surface of such an enlarged wafer would be roughly as large as a ping pong ball. The impurity level at the surface with respect to foreign metals is so low that it would correspond on the average to one single ping pong ball on a surface 70 m square. Obviously it is necessary to apply analytical methods that themselves challenge record sensitivity in order to detect near-record low levels of trace impurities.[f] The techniques that have stood the test in this context include ICP mass spectrometry,[g] total reflection X-ray fluorescence, and neutron-activation analysis, all with detection limits that suffice to identify one-hundredth or even one-thousandth of a nanogram of certain impurities (→ Atoms and Molecules, Record Atoms: Analytical Highlights).[6]

The Most Reactive Elements

The **most reactive element** is the nonmetal fluorine, both in the thermodynamic and in the kinetic sense. The **most reactive metal** is cesium. Whereas fluorine is the **most electronegative element**, cesium is the least electronegative ("most electropositive"). The **least reactive elements** are the noble gases (no "compounds" whatsoever in the usual sense of the word are known for helium, neon, or argon). The noble metals are also extremely reluctant to react. Thus, iridium, which classes as one of the heavy platinum metals, fails to dissolve in compact form even in aqua regia; at most it can be dissolved in hot euchlorine (the mixture known as "aqua regia" consists of three volumes of concentrated hydrochloric acid and one volume of concentrated nitric acid, and it is capable of dissolving even the "king of metals," gold; "euchlorine" is a mixture of concentrated chloric acid and fuming nitric acid).

---

[5] A. Ikari, K. Izunome, S. Kawanishi, S. Togawa, K. Terashima, S. Kimura, *J. Cryst. Growth* **1996**, *167*, 361.
[6] L. Fabry, S. Pahlke, L. Kotz, G. Tölg, *Fresenius J. Anal. Chem.* **1994**, *349*, 260.

The Elements with the Highest Oxidation States

The **highest formal oxidation state**, namely +VIII, is observed with ruthenium, osmium, and xenon, the **lowest formal oxidation state**, –IV, with carbon and silicon.[h] The most highly charged ion so far produced experimentally and observed (with the aid of an electron-beam ion trap) is U^{82+}.[7] Such highly charged ions can be manipulated only under special conditions in the gas phase. The most highly charged ions in a condensed phase are described as C^{4-} and Si^{4-}. These anions are to be found in salt-like carbides and silicides – although their characterizations should be taken with a grain of salt, because such high ionic charges cannot be accepted literally. Rather, they should be viewed as descriptions of extreme borderline cases. A completely ionic carbide or silicide would entail enormous coulombic forces brought about by the high ionic charges. Quadruply charged cations of certain metals (e.g., cerium and uranium) are known in strongly acidic aqueous solution, but these are of course not "naked" ions; instead they are solvated and present in the form of aquo complexes $[M(H_2O)_n]^{4+}$.

Miscellaneous Properties

Gold is the **most ductile and malleable metal**. Thinnest gold leaf is only 0.0001 mm thick.

Silver possess the **highest reflectivity of all metals**, hence its use in mirrors.

Palladium has the unique property for a metal that it is **extremely permeable to hydrogen**. It is therefore used as a purification filter for the production of highly purified hydrogen, since no other gas can diffuse through it.

[7] D. H. G. Schneider, M. A. Briere, *Phys. Scr.* **1996**, *53*, 228 and literature cited.

[a] Maria Goeppert-Mayer also received a Nobel Prize in physics (1963) for work she conducted without pay. Cf. S. Bertsch McGrayne, *Nobel Prize Women: Their Lives, Struggles, and Momentous Discoveries*, Carol Publishing, New York, **1993**.

[b] In the strictest sense, it is only possible to "discover" that which is hidden. Thus Columbus discovered America, but Edison made no discoveries whatsoever, having instead *invented* the light bulb, and in fact with the help of his own inventive skills. The artificial is thus invented, the natural is discovered. A similar linguistic distinction was once made in the German language between the accomplished (*Geschaffenes*) and that which had been accomplished for the first time (*Erschaffenes*), although the distinction is no longer rigorously observed (cf. J. Grimm, W. Grimm, *Deutsches Wörterbuch*, vol. 3, Hirzel, Leipzig **1862**, column 952).

[c] The names of the elements einsteinium and fermium, discovered in the course of nuclear weapons tests, were apparently decided upon while Einstein and Fermi were still alive, but since the nuclear research itself was "top secret" the official name designations were greatly delayed, and made known only after the deaths of the two physicists.

[d] Diamond prices are progressive as a function of size; that is, the price per carat increases with the number of carats.

[e] The information that follows was kindly provided by Dr. H. Fußstetter and Dr. L. Fabry (Wacker Siltronic, Burghausen, Germany).

[f] With respect to the general problems of ultratrace analysis, such as environmental concentrations of the elements and systematic error, see G. Tölg, *Naturwissenschaften* **1976**, *63*, 99.

[g] ICP is the abbreviation for "inductively coupled plasma," an ionization procedure based on a high-frequency field; see for example A. L. Gray in *Inorganic Mass Spectrometry* (F. Adams, R. Gijbels, R. Van Grieken, eds.), Wiley, New York, **1988**, pp. 257–300.

[h] It should be noted that a formal oxidation state provides no information whatsoever about the true electron density at an atom.

The First and Most Abundant Element in the Universe

The elements are the building blocks of chemistry. Of the 112 known elements, 94 occur in nature – albeit with some of the radioactive ones in the most miniscule amounts.[a] In the beginning[1] – immediately after the big bang[b] – was hydrogen, the **first element in the universe**. Starting with hydrogen, the isotope deuterium was formed through neutron capture. Nucleosynthetic "dimerization" of the deuterium led to helium, a process that consumed almost all the deuterium.[c] By a wide margin, hydrogen and helium are the **most abundant elements in the universe**. The relative atomic abundance of hydrogen amounts to 88.6%, that of helium 11.3%. The share left over for all the other elements in the universe (Fig. 1) thus adds up to one part per thousand (ca. 1% by weight). In other words, the nucleosynthesis of heavier elements, which takes place mainly in the interior of the stars and through stellar explosion processes, is not yet very far advanced.[2] In stars the size of our sun, only the lighter elements up through oxygen are able to form by fusion processes. Heavier elements up to iron can be synthesized in more massive stars.[3] Overall, subtle characteristics of nuclear physical processes are what determine the distribution frequency of the evolving elements. It is easy to understand that the iron isotope ^{56}Fe is particular prevalent among the heavy elements of the universe because it possesses **maximal nuclear binding energy**. This is the stable end product of the cosmic nuclear reactions. Irrespective of whether attempts are made to cleave it or build heavier elements from it, energy must in any case be expended; once a star creates a nucleus of iron it marks the end of the developmental phase in which the star produces energy from nuclear processes.[4] A very massive star, perhaps 20 times heavier than our sun, no longer has an opportunity to marshal the energy needed for stabilization once this fusion cascade terminating in iron is complete. It collapses within a few milliseconds. This event in turn leads to the most light-intense phenomenon in the universe, a supernova explosion in which as much radiant energy is released as from all the other stars in the universe together during the same period.[5] The shock wave from the supernova becomes the source of a host of new elements, since its enormous heat permits nuclear reactions that otherwise do not take place in stars. In this way, iron nuclei are transformed by neutron bombardment into gold, and these into lead (an alchemist's nightmare!). The lead is then converted into heavier, radioactive elements.[6]

---

[1] See for example S. Weinberg, *The First Three Minutes: A Modern View of the Origin of the Universe*, 2nd ed., Basic Books, New York, **1993**. A brief sketch of the development of the universe is given by P. J. E. Peebles, D. N. Schramm, E. L. Turner, R. G. Kron, *Sci. Am.* October **1994**, *271*, 28.

[2] Regarding the origin of the chemical elements see for example N. Langer, *Leben und Sterben der Sterne*, Beck, Munich, **1995**, pp. 105–119, as well as N. N. Greenwood, A. Earnshaw, *Chemistry of the Elements*, Pergamon Press, Oxford, **1984**, pp. 1–23. A brief survey is available in R. P. Kirshner, *Sci. Am.* October **1994**, *271*, 36.

[3] In order to accomplish all the possible phases of thermonuclear fusion, a star must have at least the so-called Chandrasekhar limiting mass, which corresponds to 1.4 times the mass of the sun (N. Langer, *Leben und Sterben der Sterne*, Beck, Munich, **1995**, p. 48).

[4] *Ibid.*, p. 75.

[5] *Ibid.*, p. 78.

[6] R. P. Kirshner, *Sci. Am.* October **1994**, *271*, 36.

All of these are accelerated by the force of the explosion out into the universe. The iron atoms in the hemoglobin of human blood are derived from just such supernova explosions, ones that occurred more than five billion years ago.[7] Many may take comfort in knowing that they are the descendants of such star dust. Apart from the "oldies" of the universe, hydrogen and helium, there are three additional elements formed almost entirely outside of stars: the light elements lithium, beryllium, and boron. These arise primarily in interstellar space by the action of cosmic rays from heavier atomic nuclei that have been literally demolished by the radiation.[8]

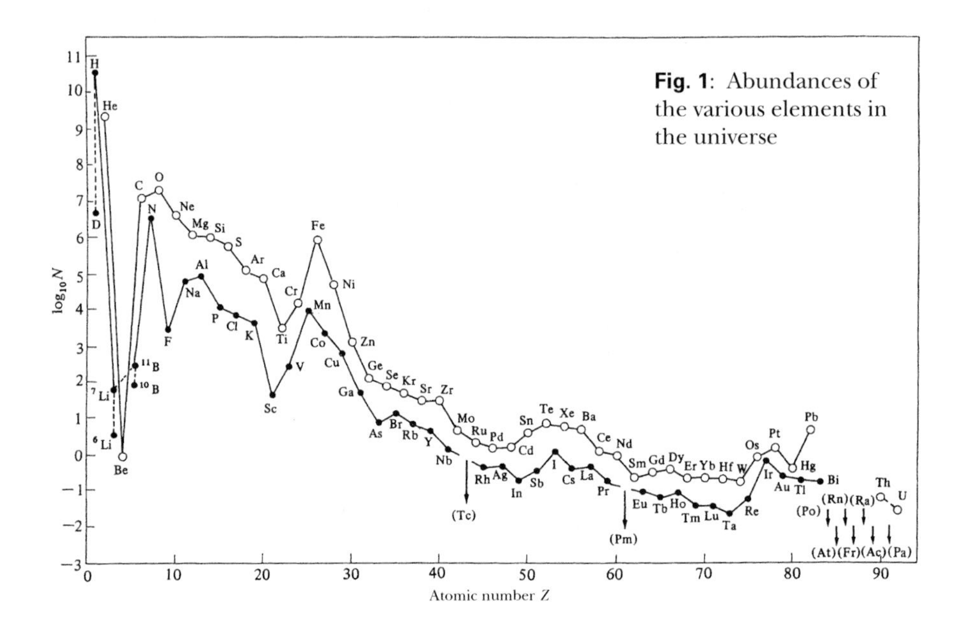

Fig. 1: Abundances of the various elements in the universe

The Most Abundant Elements on the Earth

Material is not distributed uniformly throughout the universe. On the earth, for example, hydrogen and helium are not an important factor;[d] instead it is the other elements that "matter," the ones that in the universe as a whole constitute only a "pitiful residue." The hit list for the **most abundant elements by weight** in all the earth's shells (Fig. 2) is led by oxygen (48.9 wt.-%), followed by silicon (26.3), aluminum (7.7), iron (4.7), and calcium (3.4). The three most abundant elements (O, Si, Al) are often present as a group within the earth's crust in the form of the widely distributed aluminum silicate minerals (e.g., feldspars, zeolites, and clays). Our own lives are inseparably bound up with the

[7] U. Borgeest, *Spektrum d. Wiss. Digest*, February **1996**, *4*, 6.
[8] N. Langer, *Leben und Sterben der Sterne*, Beck, Munich, **1995**, pp. 106–107.

most abundant element, oxygen. The earth is ca. 4.5 billion years old. There is evidence of very primitive forms of life on earth going back about 3.8 billion years. Enrichment of the earth's atmosphere with elemental oxygen O_2 was crucial to the development of higher forms of life.[9] O_2 was released very early in earth's history by primitive organisms carrying out photosynthesis in the oceans. Most of that oxygen reacted initially with dissolved metal ions present in low oxidation states (reduced). Once the reduced minerals in the sea had been oxidized, oxygen rapidly enriched the atmosphere. Recent investigations suggest that the increase in oxygen content of the atmosphere began abruptly about 2.1 billion years ago, marking the end of the anaerobic era. By 1.5 billion years ago the oxygen level of ambient air had nearly reached its present value.

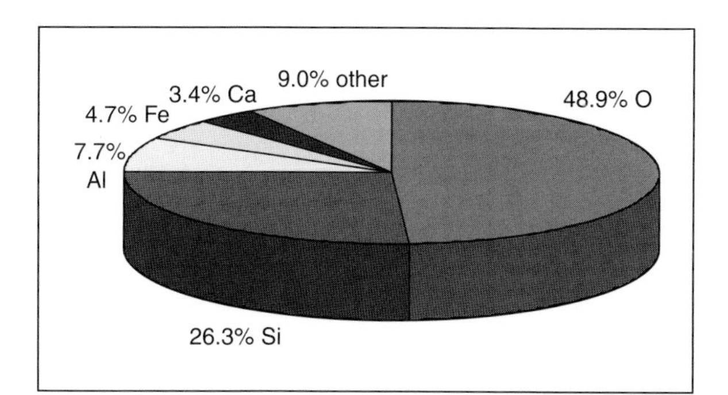

Fig. 2: The most abundant elements on the earth

The Least Abundant Elements on the Earth

Despite their collective designation, the least abundant of the natural elements are not the "rare earths" (scandium, yttrium, lanthanum, and the 14 lanthanoid metals), which actually appear near the middle of the abundance list between cerium (number 25, more abundant than copper) and thulium (number 63, more abundant than silver). The rarest of the nonradioactive elements in the earth's shells are krypton (1.9×10^{-8} wt.-%), iridium, and rhenium (both 10^{-7} wt.-%). Especially uncommon in nature are the radioactive elements with no long-lived isotopes. A few of these **least abundant natural elements** are present in such tiny amounts that it is almost impossible to express the traces in absolute terms. It has been estimated that the entire crust of the earth contains only ca. 1.2 metric tons of neptunium, 25 kg of plutonium, 12 kg of promethium, 100 g of francium, and a miniscule 45 mg of astatine.[e] Although the element americium has not yet been found in nature, many tons of it have already been prepared in nuclear reactors. Uranium, certainly the best-known of the radioactive elements, is very long-lived and roughly as abundant in the earth's crust as, for example, tin or tungsten. The half-life of

[9] See C. J. Allègre, S. H. Schneider, *Sci. Am.* October **1994**, *271*, 44. J. F. Kasting, *Science*, **1993**, *259*, 920.

the most abundant uranium isotope, ^{238}U, is 4.5 billion years, approximately the age of our solar system. **By far the least abundant of all elements** are those of which only a few atoms have ever been made – by the extravagant methods of high-energy physics – with all of these atoms decaying again almost immediately. Only two atoms have so far been "manufactured" of the most recently discovered element, number 112, and within a few milliseconds they took their leave of the researchers and decayed to the also highly unstable element 110.[10] An **antielement** has also been prepared already, when antiprotrons and positrons were combined to give at least 11 antihydrogen atoms – more than have ever been made of the heaviest "real" elements.[11] The antiatoms survived for only tiny fractions of a second under the experimental conditions. In other words, we are still light-years removed from a matter–antimatter engine like that powering the spaceship Enterprise,[f] or even from a systematic study of "antichemistry."

The Most and Least Abundant Elements in the Human Body

Just as on the earth, the **most abundant element** in the human body is oxygen, present to the extent of 65.4 wt.-% (Fig. 3). Second place goes to carbon (18.1), followed by hydrogen (10.1), nitrogen (3.0), calcium (1.5), phosphorus (1.0), and sulfur (0.25). The fact that the body has a high water content (on average, ca. 55–60%)[g] helps explain the relatively high values for oxygen and hydrogen. At the same time, however, these two elements along with carbon are also present in essentially all the organic biomolecules. Nitrogen, phosphorus, and sulfur are other elemental constituents of many biomolecules. A large amount of calcium and phosphorus is deposited in the bones, especially in the form of hydroxyapatite. A person weighing 70 kg typically contains ca. 45.5 kg of oxygen, 12.6 kg of carbon, 7.0 kg of hydrogen, 2.1 kg of nitrogen, 1.05 kg of calcium, 700 g of phosphorus, and 175 g of sulfur. The most abundant heavy metal is iron (4.2 g).[12] The **least abundant of the essential elements** are present in the human body to the extent of only a few milligrams (3–5 mg each of chromium, cobalt, and molybdenum).

The "humanest" of the elements is tin: because it can become "sick" (tin plague),[h] and also because when it is bent it "cries out" (tin cry). Whether or not tin is an essential element for humans is a subject of dispute.[12]

[10] S. Hofman, V. Ninov, F. P. Hessberger, P. Armbruster, H. Folger, G. Münzenberg, H. J. Schött, A. G. Popeko, A. V. Yeremin, S. Saro, R. Janik, M. Leino, *Z. Phys. A* **1996**, *354*, 229.

[11] G. Baur, G. Boero, S. Brauksiepe, A. Buzzo, W. Eyrich, R. Geyer, D. Grzonka, J. Hauffe, K. Kilian, M. LoVetere, M. Macri, M. Moosburger, R. Nellen, W. Oelert, S. Passaggio, A. Pozzo, K. Röhricht, K. Sachs, G. Schepers, T. Sefzick, R. S. Simon, R Stratmann, F. Stinzing, M. Wolke, *Phys. Lett. B* **1996**, *368*, 251.

[12] W. Kaim, B. Schwederski, *Bioanorganische Chemie*, 2nd ed., Teubner, Stuttgart, **1995**, p. 7.

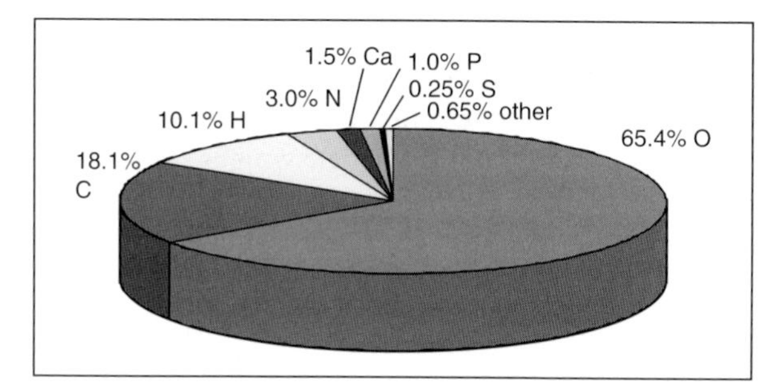

Fig. 3: The most abundant elements in the human body

---

[a] L. F. Trüb, *Die chemischen Elemente*, Hirzel, Stuttgart, **1996**, *passim.* Ninety elements are designated here as "stable" (p. 3), although this number is set too high on the basis of the half-lives of the longest-lived isotopes of the radioactive elements. The number 90 also appears in somewhat earlier works – incorrectly, based on the current state of knowledge; thus, the statement can be found that of the 92 elements from hydrogen to uranium, only 90 occur on earth (cf. for example N. N. Greenwood, A. Earnshaw, *Chemistry of the Elements*, Pergamon Press, Oxford, **1984**, p. 1).

[b] The term "big bang" probably sounds more "American" than it really is: it was first coined in 1950 by the British astronomer Fred Hoyle, who used it in a series of BBC interviews in order to poke fun at the concept itself and the associated theory of the expanding universe. It is documented that prior to Hoyle's remarks the Belgian canon Abbé Lemaître used the term "big noise," specifically in a lecture he delivered on the origin of the universe in the presence of Einstein at the Mt. Wilson Observatory in Pasadena (S. Ortoli, N. Witkowski, *Die Badewanne des Archimedes: Berühmte Legenden aus der Wissenschaft*, Piper, Munich, **1997**, pp. 145–151).

[c] The elements originated (or originate) first in the form of the corresponding atomic nuclei (protons, deuterons, etc.). A few minutes after the big bang the temperature of the universe was ca. one billion °C, and at such high temperatures atoms are not stable. With respect to the early cosmos see C. J. Hogan, *Sci. Am.* December **1996**, *275*, 36. The temperatures at the center of nucleosynthetically active stars are also extremely high; the interior of the sun, for example, is 15 million °C. Much higher temperatures then that are generated in supernova explosions.

[d] Elemental hydrogen (H_2) and helium are both so light that the earth's gravitational attraction cannot prevent them from escaping into space (R. P. Kirshner, *Sci. Am.* October **1994**, *271*, 36).

[e] For these reasons the literature also yields very different data regarding the least abundant elements – including, for example, the assertion that radon and plutonium occupy the leading positions here.

[f] Further details regarding the "warp drive" of the Enterprise, invented in 2061 by Zefram Cochrane, are available for example from M. Okuda, D. Okuda, D. Mirek, *The Star Trek Encyclopedia*, Pocket Books, New York, **1994**, pp. 371–373. The corresponding matter–antimatter reaction is controlled by a dilithium crystal (*ibid.*, p. 196).

[g] The water content decreases somewhat with increasing age. In the case of a slender individual it can be as high as 70%, for someone who is obese as low as 45% [E. Betz, K. Reutter, D. Mecke, H. Ritter, *Biologie des Menschen (Mörike/Betz/Mergenthaler)*, 13th rev. ed., Quelle & Meyer, Heidelberg, **1991**, chap. 11.1]. The water content of human blood is 80.5% (F. E. Davis, K. Kenyon, J. Kirk, *Science* **1953**, *118*, 276), which makes it the bodily fluid with the lowest water content – even lower than sperm (88.7% water; see C. Huggins, W. W. Scott, J. H. Heinen, *Am. J. Physiol.* **1942**, *136*, 467). The brain of an adult has a water content of ca. 77.4%, that of a newborn ca. 89.7% (E. M. Widdowson, J. W. T. Dickerson, *Biochem. J.* **1960**, *77*, 30). Many other organs also have a water content in the vicinity of 80%.

[h] Below 13.2 °C metallic β-tin transforms itself into the semimetallic, crumbly α-tin (gray tin). Once a tiny portion of metallic tin becomes so infected, at temperatures below the transformation temperature this so-called tin plague rapidly attacks surrounding material, with a maximum rate of transformation observed at –48 °C. For this reason, Napoleon's army stood literally in its shirtsleeves in Moscow in 1812 after a few days, because hundreds of thousands of tin uniform buttons disintegrated into gray powder.

Environmental Loads Imposed by the Manufacture of Chemicals

The large sums of money that the German chemical industry has invested in recent decades in the interest of environmental protection are clearly showing their effect. This can be illustrated by the mean **environmental load per ton of commercial products** generated by BASF (Fig. 1).

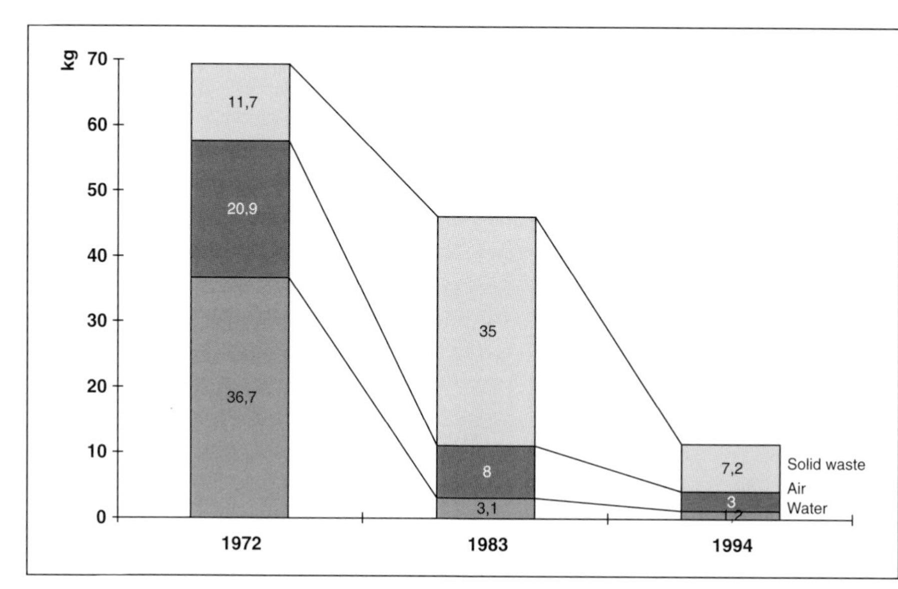

Fig. 1: Environmental load per metric ton for commercial products from BASF[1]

In 1972, the waste level per metric ton of gross product was still 6.9% as a result of the corresponding 69.3 kg of material not subject to reuse. That figure has in the meantime been reduced to 1.1%.

Emissions into streams accounted for the largest fraction of overall emissions in 1972. This was cut dramatically by the construction of a new wastewater treatment facility. The large amounts of resulting sludge, which were originally consigned to landfills, are now subject to complete combustion.

Air emissions were significantly reduced by the installation of filtering equipment. The result is that by 1994 the emissions per ton of product had been lowered to only 14.4% of the 1972 levels. Progress was even greater in the case of water emissions: reduction to only 3.3% of 1972 levels.

[1] BASF.

North America as the Largest Carbon Dioxide Emitter in the World

Carbon dioxide is together with methane and nitrous oxide among the most important greenhouse gases, which represent a threat to the world's climate. The World Climate Conference in Kyoto (Japan) in 1997 took first steps toward reducing the emission of these gases. The United States plays a significant part in this issue, since the US alone is responsible for one-fourth of the world emission of greenhouse gases. In the case of carbon dioxide, the United States produces the most emissions not only in an absolute sense, but also on a per capita basis, with a 1995 value of 15.4 metric tons – double that of Western Europe and four times the world average (Fig. 2).

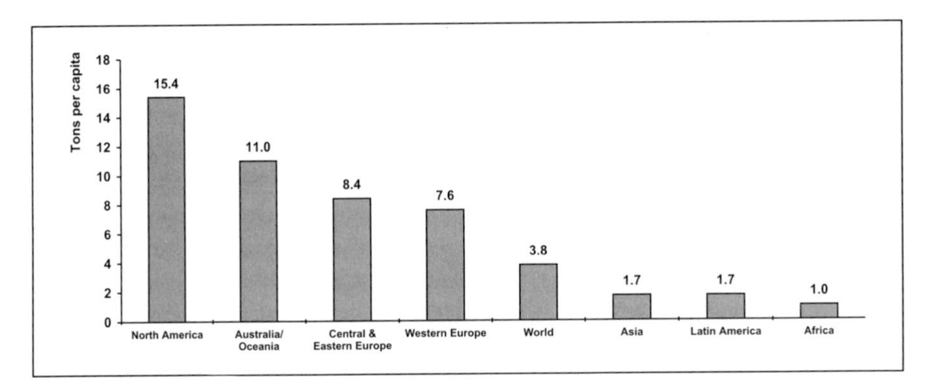

Fig. 2: a) Carbon dioxide emissions, 1995[2] a) metric tons per capita

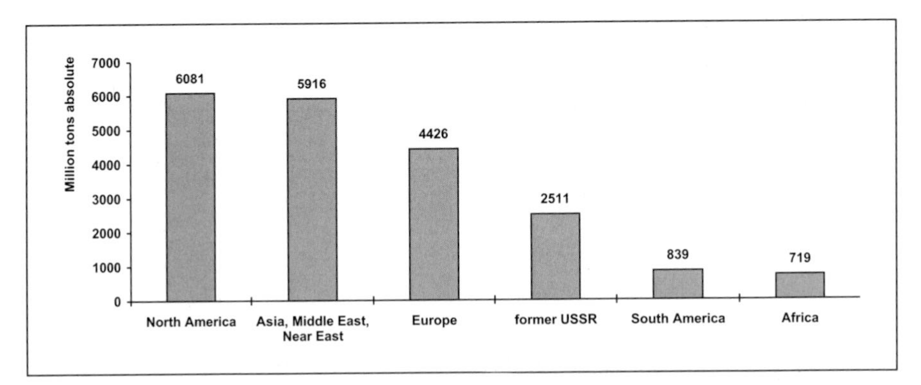

Fig. 2: b) Carbon dioxide emissions, 1995[2] b) absolute values in millions of tons (1995 or latest available data)

[2] data from: *Handelsblatt*, October 23 and December 12, **1997** (source: a) IWD, b) OECD).

Environmental Law in Germany

Environmental law is an instrument for the regulation of environmentally relevant societal behavior. Mandates and prohibitions are intended to preserve or improve the quality of the environment, thereby protecting both humans and animals from unnecessary risks.

Statutes to this end existed as early as Roman times (sewage and hygiene regulations). But it was only with modern industrialization in conjunction with unprecedented population growth that the need arose for further negotiation, which led ultimately to increased activity in the area of environmental law.

The result has been a veritable explosion of environmental laws and regulations in Germany starting in the early 1970s (Fig. 3), which of course also exerted a significant influence on economic activity in general. On one hand it meant an inevitable increase in production costs, but at the same time gave rise to new products and solutions and thus new marketing opportunities for a highly developed society.

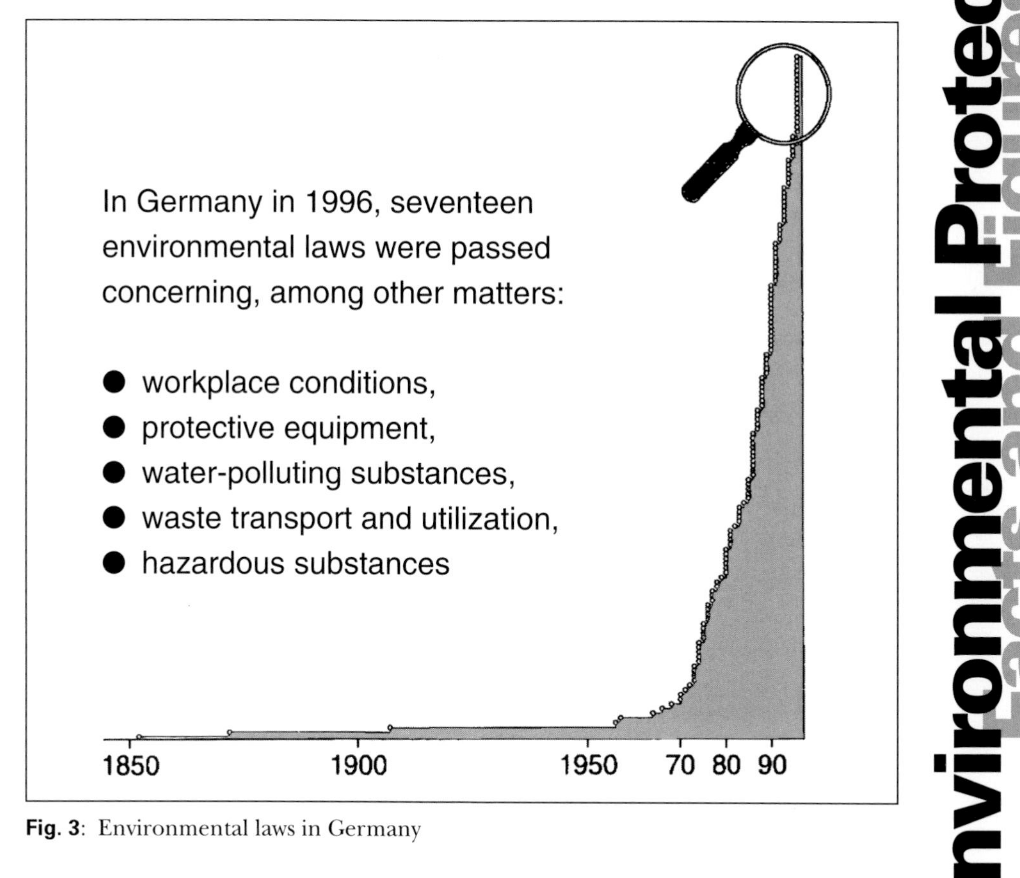

Fig. 3: Environmental laws in Germany

German Investment in Environmental Protection

The German chemical industry (i.e., that in the original 11 states) has exerted an enormous amount of effort in recent years to operate in a more environmentally friendly manner. Over the course of a single decade (1985–1994) over 13 billion German marks was invested in supplemental environmental protection, with emphasis on the protection of waterways and air quality maintenance.

The annual investment in environmental protection (Fig. 4) over and above plant investment in general reached a peak of roughly 2 billion German marks (16.2% of total investments in chemistry) by 1991 following a period of rapid growth. It then fell to 882 million German marks by 1994, although this still amounted to 10% of total investment. A significant share of this development can be attributed to the increasing importance of integrated environmental protection.

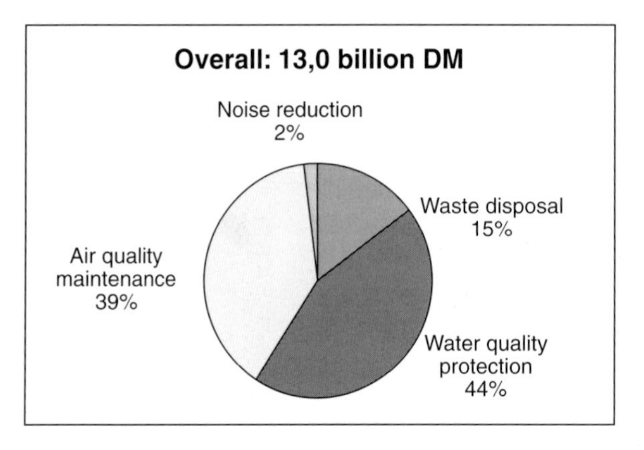

Fig. 4: a) Investment in environmental protection by the German chemical industry, 1985–1994[3]

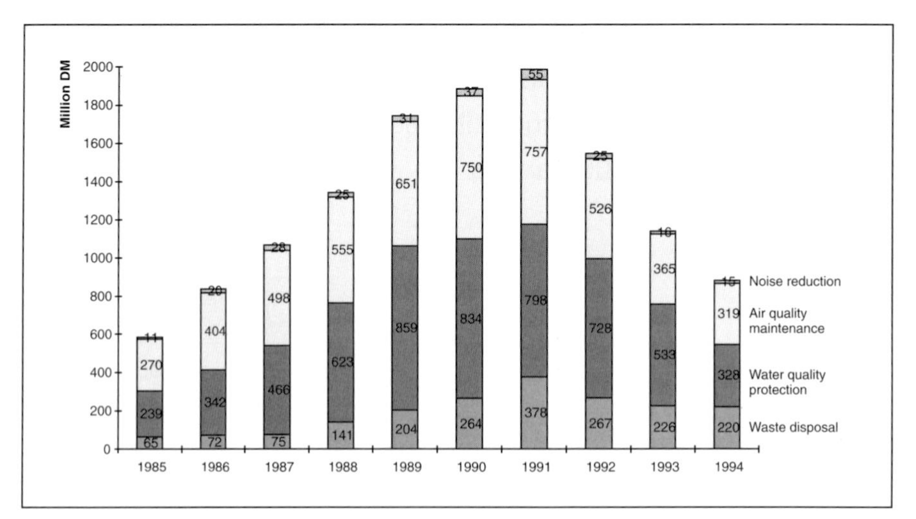

Fig. 4: (b) Annual development, 1985–1994[3]

As in previous years, most of the expenditure in 1994 was directed toward waterways protection, closely followed by air quality maintenance. Just as in 1985, the recipient of the third largest share of environmental investment (25%) was waste disposal, although the role played by waste disposal actually increased significantly over that time period.

In recent years environmental operating costs related to waste disposal have moved ahead of those for air quality maintenance. In 1994, 33% of such costs were associated with waste disposal, 42% with water management, and 24% with air quality, all of which together represented 6.22 billion German marks including write-offs.

In contrast to investment, operating costs have shown no definite declining trend since 1991, but appear instead to be holding steady at a rather high level.

Overall **environmental protection plant costs** for the German chemical industry over the time period 1985–1994 totaled an impressive 54 billion German marks (Fig. 5).

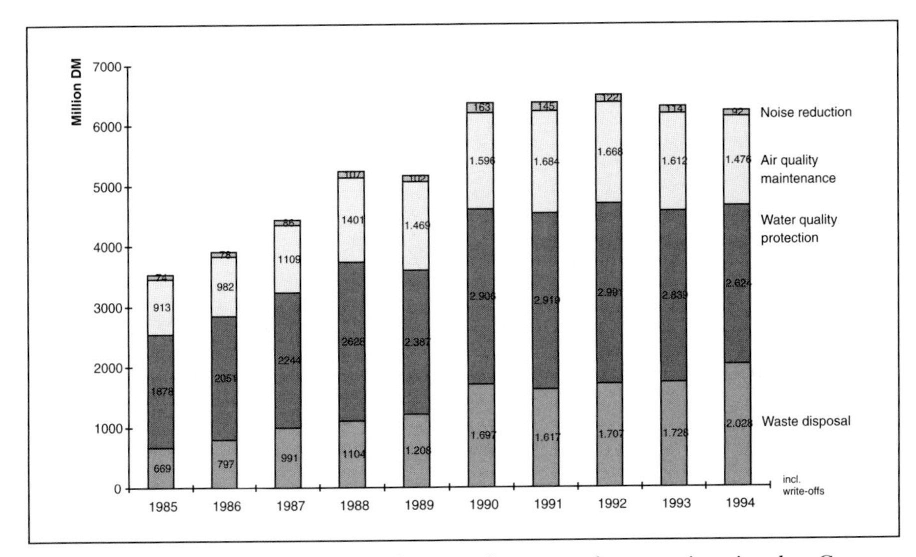

Fig. 5: Operating expenses related to environmental protection in the German chemical industry, 1985–1994[3]

[3] data from: Verband der Chemischen Industrie VCI, *Chemiewirtschaft in Zahlen.*

The Increase in World Population

September 1998: 5.946 billion people

2020: 8.05 billion people (of these, ca. 1.5 billion will be in China and 1.38 billion in India)

Every minute the world population increases by 170 people, so that every day adds the equivalent of one major city (250 000). The growth continues unabated at an average of 2.3% per year in the developing nations, which corresponds to a doubling in only 30 years.

The consequences: Urbanization (1950: only two cities with more than 10 million residents; today: 14; in 2020: probably 25), social conflicts, unemployment, migration, nutritional problems.

Increasing Demands on Land Utilization

The number of people on earth continues to rise without pause. Thus, within a quarter of a century the population grew from 2.5 billion (1950) to 4.3 billion (1975). By the year 2000 it will be 6.2 billion, and 25 years later it is estimated that 8.3 billion people will inhabit the blue planet. In other words, within three-quarters of a century the population of the earth will have more than tripled (Fig. 1).

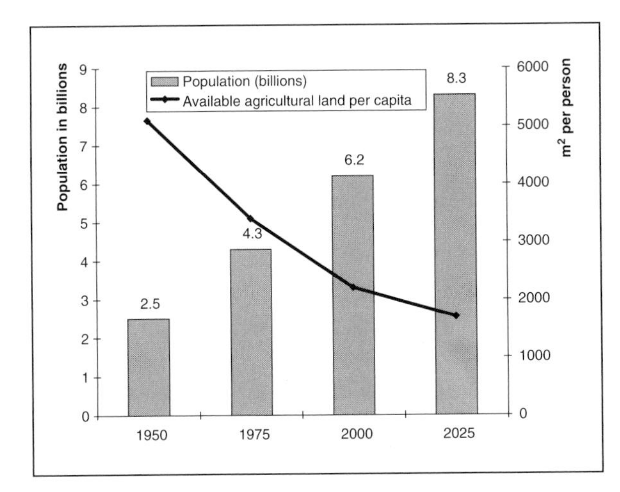

Fig. 1: a) Population growth and land availability for agricultural use[1]

[1] data from: Industrieverband Agrar e.V. and BASF.

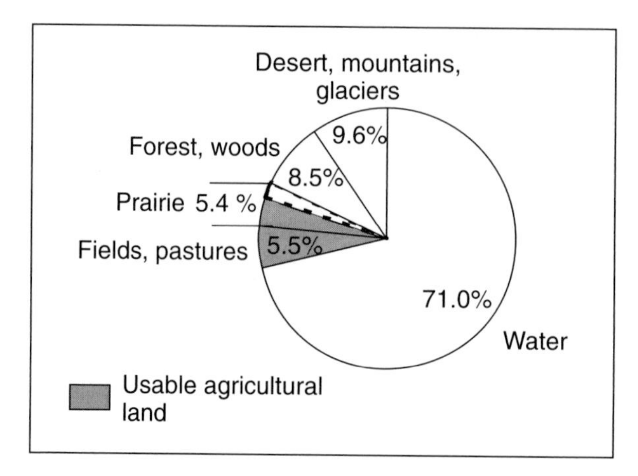

Fig. 1: b) The surface of the earth[1]

One of the most urgent problems this population explosion raises is that of **assuring mankind an adequate supply of nourishment**. Since arable land represents a very limited fraction of the earth's surface, and it cannot be significantly increased without cutting down all the remaining forests, it follows that the number of square meters of agricultural land available per capita is steadily decreasing. The trend will continue to the point that by 2025 three people will need to survive on the fruits of soil that was available to a single person in 1950. This presents an immense challenge in terms of land utilization, one that will need to be met primarily by application of the newest information available to biologists and chemists (in the areas of fertilizers, crop protection, plant genetics, etc.) with careful consideration of ecological demands (→ Biotechnology).

Surface Scarcity

The amount of surface area available for meeting nutritional needs is unlikely to increase to a meaningful extent. Only limited regions in South America and Africa are available for the purpose, and restrictions apply there as well.

1990: Agricultural land per capita	in Germany, 0.37 acres (at a high level of yield) in developing countries, 0.67 acres

By the year 2020, the amount of agricultural land available per capita in the developing countries will have fallen to only 0.37 acres as a result of population growth, a decrease of 45%. Currently there is a loss worldwide of ca. 12 million acres per year due to housing construction, industrial development, and road construction, but also as a consequence of erosion.

World Consumption of Fertilizers

In order to feed the steadily increasing world population it has been necessary in recent decades constantly to increase the yield per square meter of agricultural land. This has been achieved in part by **increased utilization of mineral fertilizers**. Within the past three decades their annual consumption (based on nutrient content) has risen at an average rate of 3.5% per year from 46.8 million metric tons in 1965/66 to 134.4 million tons in the agricultural year 1996/97 (Fig. 2).

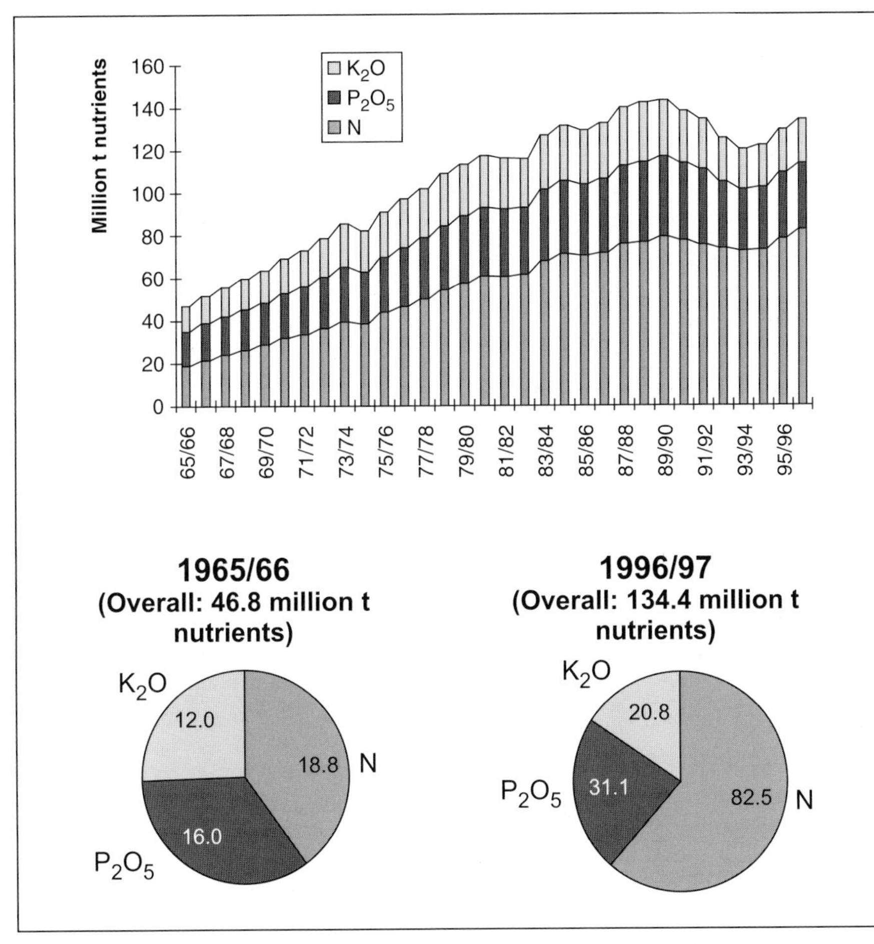

Fig. 2: Trends in the world consumption of fertilizers[2]

[2] data from: International Fertilizer Association (IFA), **1998**.

The growth was nearly continuous until 1989/90, when it peaked at 143.4 million tons. This long period of relatively steady **growth at an average annual rate** of 4.8% was followed by a distinct decline, retreating to a minimum of 120.4 million tons nutrients in 1993/94. The decrease was caused by a cut-back in fertilizer use in the developed countries, especially in Western Europe (environmental debates, land set-aside in the context of EU agricultural politics) and more dramatically in Eastern Europe and the former Soviet Union (collapse of agricultural markets). Total consumption of fertilizers in the crop year 1996/97 was 134.4 million tons.[1]

It is estimated that worldwide in 1996, 55% of the fertilizer was used for grain culture, mainly wheat. Rice, which serves as the basic nutrient for ca. one-third of the world's population, contributed only 13% to the fertilizer consumption (Fig. 3).

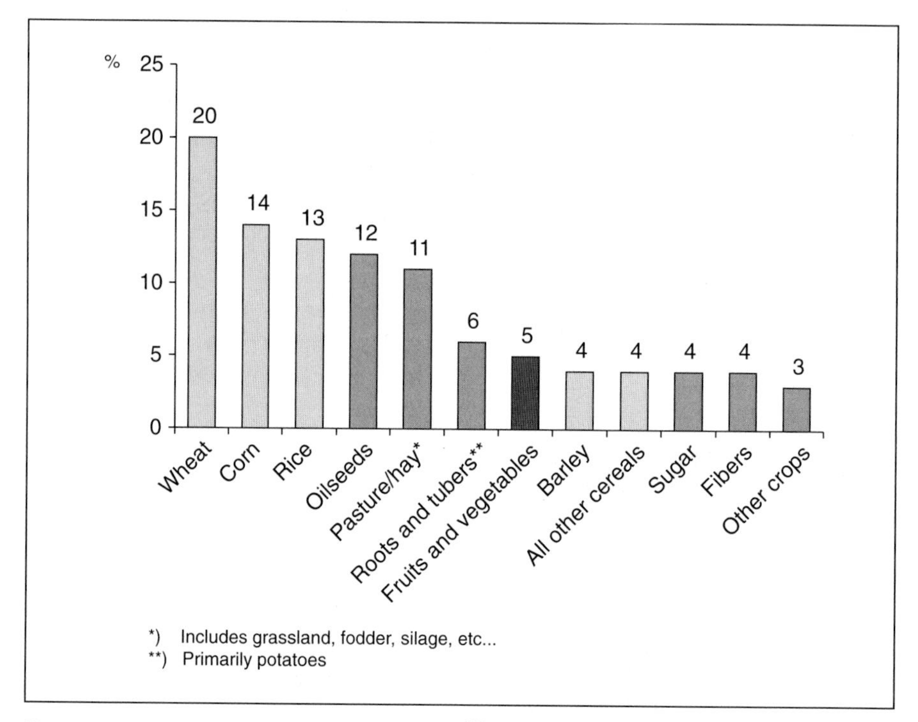

Fig. 3: Fertilizer consumption by crop, 1996[2]

The **most important fertilizer nutrient ingredient** is nitrogen (→ Chemical Products, Hit List: Ammonia). Its relative role has increased markedly in recent decades. Nitrogen (82.5 million metric tons) was responsible for ca. 61% of the world demand for fertilizers in the agricultural year 1996/97. This generated food for 2.6 billion people, or roughly half the world's population (980 million tons of grain units).

!

Providing for Nutritional Needs

Although progress has been made in meeting the nutritional needs of the world's peoples in the past 20 years, it is estimated that ca. 840 million people are still underfed worldwide. Decreasing world grain production in the last years together with depleted grain stores are a cause for serious concern. **Based on estimates by the FAO, the minimum necessary reserve levels – 18% of annual world consumption – are no longer available (in 1997 15%), which means that worldwide grain stocks are now sufficient to cover only 50 days of demand**. In the long term, measures must be taken to double food production by the year 2020. Such an increase in production will be severely restricted, however, by shortages of suitable land and water.

Important forms of fertilizer include urea, ammonium sulfate, potassium ammonium nitrate, superphosphates and ammonium phosphate, potassium chloride, potassium sulfate, and multicomponent fertilizers.

Principal Regions of Consumption

By far the **most important consuming region** for mineral fertilizers is Asia (Fig. 4).

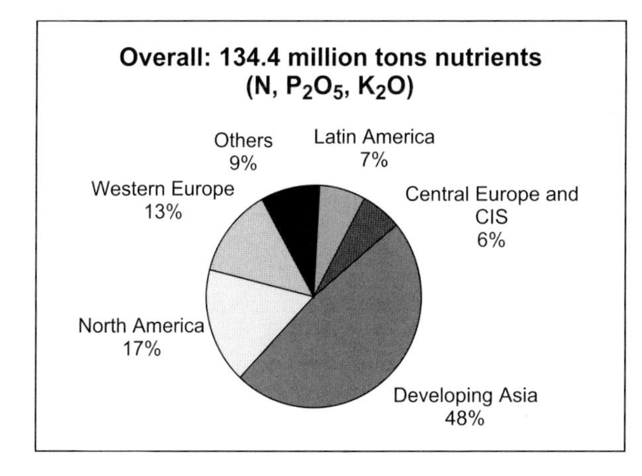

Fig. 4: Consumption of fertilizers by region, 1996/97[2]

A large share of this dominance is attributable to two developing nations: China and India. Both governments are attempting to increase agricultural yields considerably through the use of subsidies in the attempt to meet their own domestic needs. China has only 7% of the world's arable land, but must feed 22% of the world's population. Use of fertilizers in China has increased an average of 5.1% per year since 1980, although here too there have been dramatic fluctuations in growth. China and India together accounted for slightly more than one-third of the world's consumption of fertilizers in 1995/96, and the trend continues upward (Fig. 5).

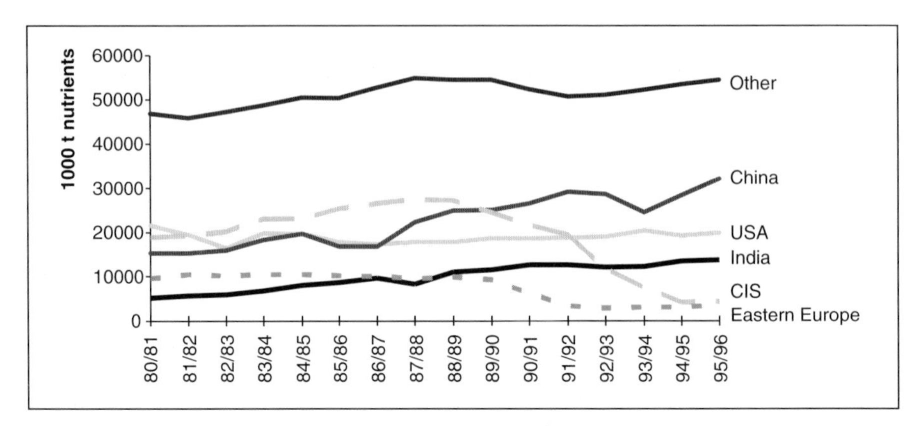

Fig. 5: Trends in fertilizer consumption by region[3]

The situation in the former Soviet Union and Eastern Europe generally presents a dramatic contrast to that in China and India. Fertilizer consumption in these countries collapsed during the 1990s as a result of political and economic changes. Agricultural organizations simply lacked capital to purchase the necessary fertilizing agents.

The Primary Raw Material: Ammonia

Ammonia, the raw material required to manufacture nearly all nitrogenous fertilizers, is now prepared almost exclusively (> 95%) by the Haber–Bosch process and its variants. The starting materials for this high-pressure, high-temperature reaction are nitrogen recovered from air and hydrogen. The amount of nitrogenous fertilizer available from 1 kg of ammonia suffices for the production of 12 kg of grain. Preparing 1 kg of ammonia in turn requires ca. 1 L of crude oil, enough to travel about 7.5 miles in an average automobile (→ Chemical Products, Hitlist: Ammonia).

[3] data from: *Current World Fertilizer Situation*, FAO, **1994**; International Fertilizer Industry Association, IFA, **1998**.

Developments in Nutrient Demand

Based on the increase in economic growth in developing countries (an average of 3% per year), it is estimated that in 2020 worldwide there will be three times the current purchasing power available, which will have a decisive effect on the demand for food. **Compared to the 1997 world trade volume of 170 million metric tons of grain, a volume of about 800 million tons is anticipated**. In particular, import demands by China (1995: ca. 17 million tons of grain) will have a powerful impact on the future development of the world grain market. In the next few decades it is estimated that China's import demand will rise to more than 100 million tons of grain. These assumptions are quite plausible when one realizes that just a single bottle of beer per year for each Chinese citizen would translate into a demand for 270 000 tons of grain.

In conjunction with the expected urbanization of the world population (60% in 2020 compared with 45% today), patterns of consumption will also change. Thus, there will be a greater need for refined products and more ready-to-use items prepared in advance by the food industry.

Records and Curiosities Associated with Scientific Publications

Chemists spend a substantial amount of their time putting their results on paper, either to protect them in the form of patents (→ Patents) or to publish them in one of the wide assortment of technical journals. Among the many functions served by publication, documentation of a set of achievements in a particular research area is certainly the oldest, and even in today's scientific community, dominated by a "publish or perish" mentality, it is still the most important. Nevertheless, what are blithely termed "papers" do have other roles, especially in that they indelibly associate the authors' names with the research field in question.

The Most Pithy Concluding Sentence

Moses Gomberg in 1900 underscored his inherent right to conduct further research in the area of the triphenylmethyl radical with an openness rarely equaled in all the years since it appeared (Fig. 1). Today, nearly a hundred years later, such an **emphatic closing comment** would surely fall victim to the censorial pen of the editor.

> "This work will be continued and I wish to reserve this field to myself."
>
> M. Gomberg, "The Instance of Trivalent Carbon: Triphenylmethyl", *J. Am. Chem. Soc.* **1900**, *22*, 757 – 771.

Fig. 1: The most unusual concluding sentence

A Political Statement

The fact that publications can also be used for **political statements** is illustrated by a contribution from László von Szentpály, who expressed his displeasure over the dismissal of 430 of his Hessian colleagues in the form of a most unusual dedication (Fig. 2) of a paper he published in the *Journal of the American Chemical Society* (JACS).

"This article is dedicated to the 430 former assistant professors dismissed in Hessen State (FRG) between 1978 and 1980 and cum sano gralis to Hans Krollmann (Wiesbaden) who signed responsible for this deed."

L. v. Szentpály, "Carcinogenesis by Polycyclic Aromatic Hydrocarbons", *J. Am. Chem. Soc.* **1984**, *106*, 6021–6028.

Fig. 2: The most unusual dedication

Are There More Compounds with Even Numbers of Carbon Atoms than with Odd?

A team of Indian and Swiss chemists[1] recently noted that the database of the Beilstein Information System, containing ca. 7 million organic compounds, includes significantly more substances with an even number of carbon atoms than with an odd number. Statistical analyses of smaller sets of organic compounds (e.g., from the *Cambridge Crystallographic Database*, in the *CRC Handbook of Chemistry and Physics*, and even from the catalogues of commercial distributors of chemicals) led to the same results. A possible explanation[2] for the observed asymmetry might be that organic compounds are ultimately derived from biological sources, and nature frequently utilizes acetate, a C_2 building block, in its syntheses of organic compounds. It may therefore be that the manufacturers' and synthetic chemists' preferential use of relatively economical starting materials derived from natural sources has left permanent traces even in chemical catalogues and databases.

[1] J. A. R. P. Sarma, A. Nangia, G. R. Desiraju, Jack D. Dunitz, *Nature* **1996**, *384*, 320.
[2] B. J. Gaede, *Chem. Eng. News*, January 20, **1997**, p. 4.

The Longest Footnote

There is one absolutely prime candidate for the **longest footnote ever published**: J. R. Murdoch, in one of his papers (again in JACS) entitled "Theory of Nuclear Substitution and the Hemistructural Relationship" required for a "brief" discussion of perturbation theory a single footnote covering more than *two* complete pages and a total of 134 lines![1]

[1] J. R. Murdoch, "Theory of Nuclear Substitution and the Hemistructural Relationship," *J. Am. Chem. Soc.* **1982**, *104*, 588–600.

The Publication with the Most Authors

A similarly remarkable demand for space was exacted in two physics journals in order to list all the scientists who wanted to present the results of a set of experiments conducted at two particle accelerators. Whereas the list of 406 (!) authors of a contribution to *Physical Review Letters* with the title "First Measurement of the Left–Right Cross Section Asymmetry in Z-Boson Production by e^+ e^- Collisions" consumed two pages, the editors of *Physical Reviews* succeeded in restricting the list of 271 authors of the paper "Limit on the Top Quark Mass from Proton–Antiproton Collisions at $\sqrt{s}$ = 1.8 TeV," complete with institutional affiliations, to a single page.[2] In the face of this **enormous number of authors** it is a welcome development that the labor-saving abbreviation "et al." has now taken root in the citation of scientific work. Since chemical investigations usually involve a much more modest investment in personnel, the number of coauthors acknowledged in chemical journals tends to be smaller by a factor of ten. A good candidate for the **chemical publication with the most coauthors** is the posthumous account of the total synthesis of erythromycin achieved in the laboratory of R. B. Woodward (→ Appendix, Nobel Prize Winners: in Chemistry),[3] which involved 49 people. It will be intriguing to see, however, how many authors will be acknowledged in the summary publication emerging from the ongoing Human Genome Project.

The Most Frequently Cited Publication

One measure of the importance of a piece of scientific work, at least with respect to the interest it arouses in the scientific community, is the frequency with which the effort is cited by other scientists. The **most frequently cited article** of all time – by a wide margin – is a paper that appeared in 1951 by O. H. Lowry and coworkers on the photometric determination of proteins ("Protein Measurement with the Folin Phenol Reagent," *J. Biol. Chem.* **1951**, *193*, 265–275).[4] Based on statistics developed by Eugene Garfield,[5] this classic has been cited more than 245 000 times since it first appeared.

---

[2] "First Measurement of the Left–Right Cross Section Asymmetry in Z-Boson Production by e^+ e^- Collisions," *Phys. Rev. Lett.* **1993**, *70*, 2515–2120. The vice-champion is a publication with 271 authors: "Limit on the Top Quark Mass from Proton–Antiproton Collisions at $\sqrt{s}$ = 1.8 TeV," *Phys. Rev. D: Part. Fields* **1992**, *45*, 3921–3948.

[3] R. B. Woodward et al., *J. Am. Chem. Soc.* **1981**, *103*, 3210–3213.

[4] O. H. Lowry, N. J. Rosebrough, A. L. Farr, R. J. Randall, "Protein Measurement with the Folin Phenol Reagent," *J. Biol. Chem.* **1951**, *193*, 265–275.

[5] E. Garfield, *The Scientist* **1996**, *10* (17), 13–16.

Records Related to Scientific Journals

It is not only individual publications that have potentially record peculiarities to their credit: The same applies to the journals in which papers appear. The Institute for Scientific Information in Philadelphia annually evaluates about 3400 scientific journals on a statistical basis and then publishes the results in part in the form of *Science Citation Index*. From an analysis of the corresponding data for 1994 it turns out that the *Journal of Biological Chemistry* (JBC) was the **journal in which the most articles appeared** (Table 1), closely followed, however, by the American *Proceedings of the National Academy of Sciences*. With roughly 4900 articles in 1994 alone it is no wonder that JBC and the *Proceedings* also lead the list of **most frequently cited journals** for 1994 (Table 2). The fact that these two journals are at the top of both lists is surely an indication that the many articles published there are also reaching the target readership. Apart from the most active journals (as identified in Table 1), Table 2 also includes the weekly publications *Nature* and *Science*, whose solid reputations confer a seal of approval on papers they accept.

Table 1: The most productive journals (number of articles)[1]

Journal	Number of articles
J. Biol. Chem.	4915
Proc. Natl. Acad. Sci.	4894
Tetrahedron Lett.	2448
J. Am. Chem. Soc.	2134
J. Chem. Phys.	2107

[1] E. Garfield, *The Scientist* **1996**, *10* (17), 13–16.

Table 2: The most frequently cited journals[1]

Journal	Number of citations
J. Biol. Chem.	265 300
Proc. Natl. Acad. Sci.	259 300
Nature	246 500
Science	190 900
J. Am. Chem. Soc.	153 000

[1] E. Garfield, *The Scientist* **1996**, *10* (17), 13–16.

If one takes into consideration data like those in Tables 1 and 2 along with a few other criteria, what results is a so-called **impact factor** (Table 3) intended to describe the influence a particular journal exerts within the scientific community – but which can also serve its publisher as a powerful sales argument. The two top spots with respect to this statistic have been assumed by journals that publish review articles. In particular, the impact factor for *Chemical*

Reviews stands well apart from those of all the other chemical print media. Among journals publishing mainly descriptions of original chemical research, the international edition of the Wiley–VCH flagship *Angewandte Chemie* leads the way – even ahead of the (British) Royal Chemical Society's review journal and JACS, the pride of the American Chemical Society.

Table 3: The most influential journals (impact factors)[1]

Journal	Impact factor
Chem. Rev.	14.5
Acc. Chem. Res.	8.8
Angew. Chem. Int. Ed. Engl.	7.0
Chem. Soc. Rev.	5.6
J. Am. Chem. Soc.	5.3

[1] *Journal Citation Reports 1995*, Science Citation Index.

The Scientific Literature of Various Countries

Under the title "The Scientific Wealth of the Nations," Robert M. May recently published in the journal *Science*[6] a statistical analysis of the scientific publications for various countries, based on data drawn from *Science Citation Index*. Some of the results of this analysis are presented in Table 4.

Table 4: Countries responsible for the most publications[1]

Country	Fraction of scientific publications [%]	Fraction of citations [%]	Citations per publication (rank)
United States	34.6	49.0	1.42 (1)
Great Britain	8.0	9.1	1.14 (5)
Japan	7.3	5.7	0.78 (18)
Germany	7.0	6.0	0.86 (15)
France	5.2	4.5	0.87 (14)

[1] R. M. May, *Science* **1997**, *275*, 793–796.

The United States was the country that according to this survey generated the **most scientific publications**, with 34.6% of the total, significantly ahead of Great Britain (8%), Japan (7.3%), Germany (7.0%), and France (5.2%). The countries of the European Union taken together threatened the dominance of the USA, however, serving as the source of 32% of all publications. The same

[6] R. M. May, *Science* **1997**, *275*, 793–796.

national rank order applies as well to the **frequency with which papers from the various countries were cited**. The rankings change drastically, however, when *both* number of citations and number of papers are taken into account. The United States still leads the list after consideration of this "quality factor" – albeit one with limitations – but places two through five are now occupied by Switzerland, Sweden, Denmark, and Great Britain. Germany is relegated to an unspectacular 15th place. When the statistical evaluation is restricted to chemical papers, the list by number of citations is led by a top five consisting of the United States, Japan, Germany, Great Britain, and France. It is also interesting to look at the number of **publications per resident**. From this perspective Switzerland (167) has a clear lead, followed by Israel (152) and Sweden (147). Once again, Germany managed to reach only 17th place, at the lower end of midrange.

The Longest Printed Index

The tome known as *Chemical Abstracts Twelfth Collective Index, Volumes 109–115, 1987–1991* is presumably the longest index ever printed. It consists of 115 bound volumes with more than 200 000 pages containing in excess of 35 million entries. This gigantic opus requires roughly 6.7 meters of shelf space sturdy enough to handle a total weight of 246.7 kg. A less "weighty" alternative – only 45 g – is provided by an equivalent set of four CD-ROM disks.[1] The *Thirteenth Collective Index*, scheduled to appear at the end of 1997, will certainly surpass these records.

[1] *Chem. Int.* **1993**, *15*, 103.

The Chemical Abstracts Service of the American Chemical Society

In an attempt to master the flood of publications released every year, so-called abstract services have proved invaluable, providing brief summaries of original works. The idea of making such summaries available to chemists reaches back into the early 19th century, when the **Pharmaceutische Centralblatt** was introduced by the Leipzig publisher Leopold Voss on January 14, 1830. The first editorial entry published by Voss was a biographical sketch of Carl Wilhelm Scheele (1742–1786) delivered before the "Assembly of Associations in the North of Germany" by E. F. Aschoff on September 8, 1829. From 1897 this publication was continued as **Chemisches Zentralblatt** under the editorial management of the German Chemical Society, later also by the Chemical Society of the German Democratic Republic (East Germany) and the

Academies of Science of Göttingen and Berlin. It ceased publication in 1969, in its 140th year.

Certainly by this point the scene was very much dominated by the American equivalent of the *Zentralblatt*: *Chemical Abstracts Service* (CAS), sponsored by the American Chemical Society, with an electronically searchable database (CAS-Online) playing a key role. The first summary of a scientific article prepared by CAS appeared on January 1, 1907, and described a piece of work by A. Kleine from the *Zeitschrift für Angewandte Chemie* about an apparatus for the determination of sulfur and carbon. The first chemical compound registered by CAS was "carbonyl-hydroferrocyanic acid" $H_3FeCO(CN)_5 \cdot H_2O$, the sodium salt of which is familiar as sodium carbonylprussiate and whose enthalpy of formation was determined by J. A. Muller.[7] From these humble beginnings, CAS has been busily setting its own records, especially in the 1950s (cf. Fig. 3). Whereas in the first year only about 12 000 abstracts were published, and it took precisely 30 years to reach a total of a million abstracts, by 1995 the abstracting rate surpassed 687 789 per year, and the total number of abstracts published since 1907 had reached 16.2 million, with another million added roughly every 18 months.[8]

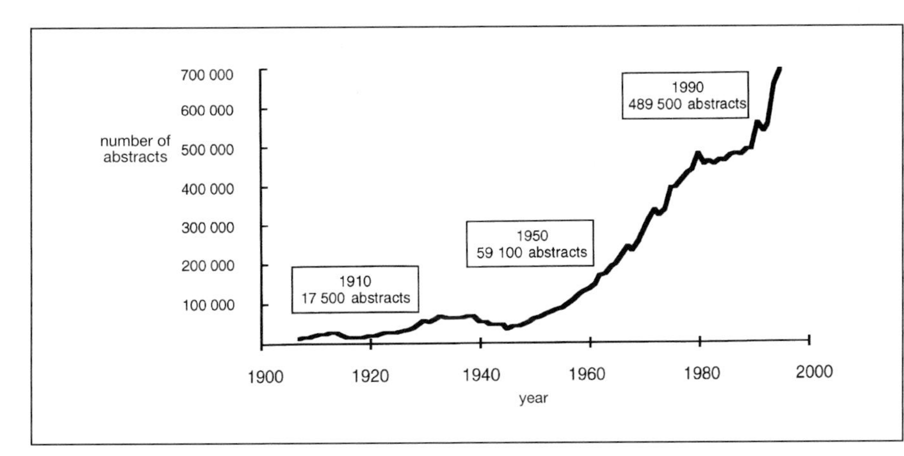

Fig. 3: Rampant rise in the number of publications processed annually by Chemical Abstracts Service

Not only is the **number of published abstracts** steadily increasing, so too is the number of compounds, with more than a million added to the registry every year (1995: 1 186 334 compounds) (→ Chemical Industry, Company Hits: Scientists: Busy Bees). Because chemical substances have been systematically entered into a registry file only since 1965, the current recorded inventory still does not take into account all the compounds actually described in Chemical

[7] J. A. Muller, *Ann. chim. phys.* **1906**, *9*, 263–271.
[8] *CAS Statistical Summary 1907–1995*, Chemical Abstracts Service, Columbus, Ohio (USA), **1996**.

Abstracts. The **total number of registered compounds** is estimated to be 16.8–17.3 million,[9] of which 94.5% contain carbon (including such compound classes as carbides and organometallic CO complexes).

Chemists as Artists and Literary Figures

Chemists don't necessarily limit their creations to patent documents and papers in scientific journals. Some, for example, also write textbooks. Probably the most revered and certainly the **most comprehensive of the current German-language chemistry textbooks** is the famous "Holleman/Wiberg," the first edition of which appeared in 1900. The 101st edition was published in 1995 and contains more than 2000 pages. The subject index alone for this book covers 163 pages with small type. Unlike earlier editions the latest index no longer includes the keyword "God," but it still provides a great many refreshing and detailed entries, such as "Wieliczka's crackling salt" or the "Bologna bottle."

A small subset of chemists is engaged even more actively in literary pursuits than the textbook authors. First mention should surely be reserved here for **Elias Canetti**. Canetti studied chemistry in Vienna, where he acquired his doctorate in 1929. He became world-famous as a novelist, playwright, and master of aphorisms, and was awarded the Nobel Prize for literature in 1981. There appear to be no scientific publications whatsoever attributable to Canetti with the exception of his doctoral dissertation. Canetti presents a vivid contrast to **Yuri T. Struchkov**, the **author with the most chemical publications in a single decade**. Between 1981 and 1990 Struchkov generated 948 scientific papers, corresponding to a publication rate of 0.26 per day. As a result he was awarded in 1992 the Anti-Nobel Prize (IgNobel Prize[a]) for literature. Canetti is not the only chemist to establish a reputation as an artist. A few other recent examples are **Primo Levi**, **Johannes Mario Simmel**, and **Rafik Schami**. These three did labor for several years as chemists, but then abandoned the profession for various reasons, certainly to the benefit of the arts. Other literary figures with a chemical past include the great parodist **Robert Neumann**, the novelist and film writer **Alain Robbe-Grillet**, the famous Hollywood director **Frank Capra**, the British mystery writer **Agatha Christie**,[10] and even **Friedrich von Schiller**,[11] **August Strindberg**,[12] and **Theodor Fontane**, who after an apprenticeship in a drugstore studied pharmacy, but rather than taking over his father's pharmaceutical shop prescribed for himself a career in writing. Another of the chemically trained authors of his day was **Georg Christoph Lichtenberg**, who was among other things an experimental physicist, a philosopher, a satirist, an aphorist, and a scholarly journalist. Lichtenberg is credited with the saying: "He who understands only chemistry does not

---

[9] Dr. W. Val Metanomski, Senior Editor, Chemical Abstracts Service, personal communication.
[10] O. Krätz, *Das Rätselkabinett des Doktor Krätz*, VCH, Weinheim, **1996**, pp. 157–160.
[11] Ibid., pp. 130–133.
[12] Ibid., pp. 144–148 and 193.

understand that properly either." His famous philosophical contemporaries **Jean-Jacques Rousseau**[13] and **François Marie Arouet** (**Voltaire**)[14] also engaged in chemical activities. The chemist and physician **Alexander Borodin** became famous as a musician. There is actually a chemical reaction named for him: the "Borodin silver salt decarboxylation," developed in 1861. Nevertheless, Borodin is probably better known even among chemists for his Polovtsian Dances.

Only a very few chemists today manage to straddle both art and science; the days of great generalists and universal geniuses are long past. Among those rare, gifted individuals who have published successfully in both areas are the chemistry Nobel Prize winner (→ Appendix, Nobel Prize Winners: in Chemistry) **Roald Hoffmann** and **Carl Djerassi**. Two biochemistry professors, **Erwin Chargaff**[b] and **Isaac Asimov** (who died in 1992), were also very active authors. The literary productivity of the latter in particular was almost unbelievable: Asimov was the author of over 400 books! His literary output was also extraordinarily diverse, extending from textbooks and popularizations of scientific material to science-fiction novels and even limericks for children.

The American chemist **Ebenezer Emmet Reid** is a candidate for two records in this context – "oldest author" and "most original title." At the age of 100 Reid published an autobiography with the truly fitting title "My First Hundred Years."[16]

[13] *Ibid.*, pp. 126–129.
[14] *Ibid.*, pp. 13–15 and 178.
[15] *Ibid.*, p. 188.
[16] E. E. Reid, *My First Hundred Years*, Chemical Publishing Co., New York, **1972**.
[a] IgNobel Prizes are awarded for research that cannot (or should not) be reproduced. Regarding the origin of these anti-Nobel Prizes see S. Mirsky, *Sci. Am.* December **1994**, *271*, 17. Additional information about the IgNobel Prizes can be obtained from the Internet (http://walk.pci.on.ca/dgilbert/ignobel).
[b] Just like Canetti, Chargaff was also strongly influenced by Karl Kraus during his studies in Vienna.

Spectacular Blunders

To err is human. Scientists are human. Ergo, is making a mistake scientific? A miscast syllogism, perhaps, but not entirely false. All the sciences have been subject to mistakes, so they occur in chemistry too. Forthright scientists assume mistakes will be made. After all, "trial and error" constitutes the core of the scientific method, a central tool in empirical science. Many errors in science are a result of insufficient data, which can lead to premature and perhaps false conclusions. It becomes more critical when mechanisms of the human psyche get involved, mechanisms that interfere with the detection of errors and sometimes take on pathological characteristics. The classic symptom is a loss of objectivity. There are other characteristics of "pathological science"[1] as well.[2] Phenomena under investigation are sometimes somewhat sensational, and run counter to current teachings, so they cause an uproar and the scientists involved end up in the public spotlight. The experimental effects under investigation with such phenomena are usually so minute that they lie at the observational limits, which makes it easy to interpret background noise subjectively as a legitimate effect. "Infected" scientists display a strong tendency simply to dismiss established theories that run counter to their observed effects – either by ignoring them or by substituting new, often revolutionary theories in their place. Debates with advocates of the established wisdom then turn into veritable holy wars fought under the motto: "many enemies, much honor," virtually excluding the possibility of objective and (self-) critical confrontation with the underlying issues. Note, however: what is at issue here is not intentional scientific fraud,[a] but rather self-delusion. The infected parties no longer believe what they see, but see what they believe. For this reason it is often the most elementary experiments, ones that could provide a clear solution to the dispute, that they themselves fail to carry out, and would ignore or dispute in principle if conducted by doubters or opponents.

Polywater and **cold fusion** are two especially prominent chemical examples from the recent past.[2] During the 1960s, Soviet scientists (including the respected Derjaguin) reported a new form of water that was allegedly formed from normal water inside glass capillaries. Unfortunately, it was possible to prepare only miniscule quantities of this so-called polywater experimentally. Compared with normal water, polywater was supposed to display most unusual properties, including a boiling point of ca. 300 °C and an incredibly high viscosity. This sensational new form of water was investigated intensively worldwide for several years before it finally became clear that "polywater" was nothing but a conglomeration of concentrated impurities in normal water. Even famous research institutes allowed themselves to be deceived royally by

[1] This term was coined by Langmuir (→ Appendix, Nobel Prize Winners: in Chemistry). See I. Langmuir (R. N. Hall, Ed.), *Physics Today* **1989**, October, 36.
[2] D. L. Rousseau, *Am. Sci.* **1992**, *80* (January–February), 54.

stopcock grease and the sweat of the experimenters. Polywater was not only a sensational research target, but also an economical one.[3]

A similar set of circumstances applied to **cold fusion**. The requisite electrochemical facilities for its investigation could be assembled in any reasonably well-equipped high school laboratory. In the course of electrolyzing a solution of lithium deuteroxide (LiOD) in heavy water (D_2O) with a platinum anode and a palladium cathode, the electrochemists Pons and Fleischmann detected an evolution of heat that in one case was reportedly so intense it caused an electrode to melt. The researchers explained their experimental observation by assuming that deuterium in the platinum electrode, formed during the electrolysis, underwent nuclear fusion at room temperature to produce helium, tritium and neutrons – a very exothermic process. But given the incredibly powerful electrostatic repulsion that must be overcome during fusion of two deuterium nuclei, an enormous amount of kinetic energy would be required at the time of collision – in excess of 10 000 electron volts – a condition compatible only with an extremely high temperature (ca. 100 million °C, as in the interior of a star, or upon the explosion of a thermonuclear bomb).[4] The theory advanced by Pons and Fleischmann was that cold fusion was induced by electrochemical compression exerted on absorbed deuterons by the palladium crystal lattice, corresponding to a calculated pressure of 10^{24} bar.[5] Another physicist by the name of Jones was working on similar experiments at the same time as Pons and Fleischman,[b] and he measured not the heat evolution during electrolysis but rather the accompanying neutron flux. Indeed, he arrived at conclusions very similar to those of his competitors. There then developed a mighty publication contest between the two research groups, not always conducted in the most candid ways, with both groups vying for priority in announcing sensational results. One victim in this race, which was narrowly won by Pons and Fleischmann,[6] was of course the scientific facts. Two pages of errata were the result – a rather sad record. The findings reported by Pons and Fleischmann, but also those from Jones, failed to stand up under critical scrutiny. In order to salvage them, various bold theories were advanced. For example, Pons and Fleischmann suggested as an explanation for the pitifully low level of neutron emission in their experiments that it was a consequence of a hitherto unknown nuclear reaction. The conflict between skeptics and "believers" is still not entirely resolved,[c] but one can say with certainty that so far cold fusion has produced considerably more agitation than energy.

The famous/infamous **phlogiston theory**, developed in 1697 by Stahl, belongs to a totally different category of mistakes. According to this theory, every flammable material contains a gaseous substance, called phlogiston, that

---

[3] See for example F. Franks, *Polywater*, MIT Press, Cambridge, Mass., **1981**.
[4] See J. W. Schultze, U. König, A. Hochfeld, *Nachr. Chem. Tech. Lab.* **1989**, *37*, 707.
[5] See the editorial comment in *Phys. Unserer Zeit* **1989**, *20*, 93.
[6] M. Fleischmann, S. Pons, *J. Electroanal. Chem.* **1989**, *261*, 301.

escapes during combustion. The theory provided a direct explanation for why only a small residue of ash remains after the combustion of many materials – they simply happen to be especially rich in phlogiston. It also shed light on why "burned" metals can often be restored to their original state after combustion by treatment with such "phlogiston-rich" substances as charcoal. Stahl was the first to provide a solid basis for the process of combustion and its reverse, one that went beyond the affected materials themselves, identifying at the same time what we would today describe as oxidation and reduction.[7] His interpretation of the phenomena is almost the complete reverse of what we now understand, and that is in fact one of the reasons why his ideas proved so enduring: Many scientific theories retain their internal consistency despite a change in sign. We can thank Lavoisier for the downfall of the phlogiston theory. It was he who introduced the analytical balance into chemical experimentation, thereby transforming chemistry into an exact science. Lavoisier showed in 1777 that oxygen, recently discovered by both Scheele and Priestley, is a necessary constituent in all combustions, and that the combustion products taken together weigh more than their precursors. In other words, there is no loss of weight during combustion, although such a loss would be anticipated from the phlogiston theory.[d] At this point the phlogiston theory was essentially dead. Attempts were made to save it, however, by assuming that oxygen is "dephlogisticated air," and that it has a powerful tendency to extract the phlogiston from other substances. Thus, phlogiston itself must have a negative mass. In order to signal the end of the phlogiston theory in the most ostentatious way he could devise, the otherwise modest and reserved Lavoisier arranged a public auto-da-fé at which his wife, as a white-clad embodiment of oxygen, burned Stahl's books.[8]

The phlogiston theory certainly does not fit in the category of "pathological science." Rather, its fate demonstrates in a very pointed way how "healthy" science normally functions. At the time of its formulation this theory corresponded with the state of current knowledge, and it possessed a considerable amount of explanatory power with respect to many experimental observations. But it was also testable, and it was finally refuted by more precise experimental data.[e] Additional hypotheses proposed by its adherents in an attempt to save the original theory proved impotent. A paradigm shift finally occurred – the old theory was replaced by a newer, better theory.[9] The phlogiston theory thus went the way of many other theories. Even the tenacious sojourn it enjoyed in the heads of many 18th century chemists is not particularly unusual. Through its considerable success over the course of decades it had already acquired the status of dogma. Established theories are slow to die – but perhaps this is itself nothing more than one of those established theories?

[7] G. Prause, T. v. Randow, *Der Teufel in der Wissenschaft*, Rasch und Röhring, Hamburg, **1985**, pp. 189–192.
[8] M. Speter in *Das Buch der grossen Chemiker* (G. Bugge, Ed.), Vol. 1, 6th reprint of the 1st ed., VCH, Weinheim, **1984**, p. 331. J. Dettmann, *Fullerene: die Bucky-Balls erobern die Chemie*, Birkhäuser, Basel, **1994**, pp. 20–22.
[9] See T. S. Kuhn, *The Structure of Scientific Revolutions*, 3rd ed., University of Chicago Press, Chicago, **1996**.

One interesting mistake in analytical chemistry had especially far-reaching consequences: **overstatement of the iron content of spinach by a factor of ten.**[10] Millions and millions of babies and small children suffered from the results directly – in some cases to the serious detriment of the clothing or the household furnishings of the feeders as well. In this case the error was not one of analysis, however, but rather a piece of careless typing. The decimal point was mistakenly shifted one place to the right in the transcription of an analytical result. So it was not in fact a *chemical* error!

---

[10] W. Krämer, G. Trenkler, *Lexikon der populären Irrtümer*, 10th ed., Eichborn, Frankfurt am Main, **1996**, pp. 294–295 and references cited.

[a] Intentional, full-scale scientific frauds today are rather rare, but they do occur. Black sheep exist everywhere. See W. Broad, N. Wade, *Betrayers of the Truth*, Simon and Schuster, New York, **1982**.

[b] Russian research groups also busied themselves with cold fusion, which they referred to as "fractofusion." One of the very active investigators was Derjaguin, already familiar from polywater. An especially complete bibliography on the subject is maintained at the Chemical Institute of the University of Aarhus (Denmark), and it can be accessed via the Internet (http://www.kemi.aau.dk/~britz/fusion).

[c] With respect to the continuing controversy as well as background information see for example F. D. Peat, *Cold Fusion: The Making of a Scientific Discovery*, Contemporary Books, Chicago, **1990**. E. F. Mallove, *Fire from Ice: Searching for the Truth Behind the Cold Fusion Furor*, Wiley, New York, **1991**. F. E. Close, *Too Hot to Handle: The Race for Cold Fusion*, Princeton University Press, Princeton, **1991**. J. R. Huizenga, *Cold Fusion: The Scientific Fiasco of the Century*, Oxford University Press, Oxford, **1993**. G. Taubes, *Bad Science: The Short Life and Weird Times of Cold Fusion*, Random House, New York, **1993**.

[d] Lavoisier even recognized that the overall mass remains constant during a combustion reaction. His postulate of the conservation of mass was the first conservation law in chemistry. The same conservation law was proposed independently at about the same time by Lomonossov.

[e] According to Popper, an acceptable theory is characterized by the fact that it can in principle be falsified. See K. R. Popper, *The Logic of Scientific Discovery*, 14th ed., Routledge, New York, **1996**, *passim*.

The Most Strained Molecules

If the ideal bond angle about a carbon atom in a particular hybridization state is deformed, a *strain* is produced that significantly raises the energy content of the molecule. This situation is often encountered in the construction of small-ring systems. The relationship between ring closure and ring strain was first recognized by Adolf von Baeyer, who formulated his famous strain theory[1] in 1885. This concept, which has been amplified several times in the meantime, proved to be extraordinarily fruitful in organic chemistry, because among other things it led to a better understanding of chemical bonding (→ Molecular Form, Bonding Records), facilitated the interpretation of intramolecular interactions, assisted considerably in the clarification of reaction mechanisms, and served thereby as an important link between experimental and theoretical chemistry. Moreover, highly strained compounds are themselves extremely interesting from a chemical standpoint.

The strain energy of a molecule is defined as the difference between its measured enthalpy of formation and the enthalpy of formation of a "strain-free" model compound with the same number of atoms arranged in the same way. The dilemma associated with this definition is of course the fact that different strain-energy values are obtained depending upon what one selects as the model system, which in contrast to reality must itself be completely free of strain. In the case of carbocyclic molecules it has become customary to designate cyclohexane as a strain-free system and then to determine energy increments for those atomic groupings most commonly found in hydrocarbons.[2] Given the enthalpy of formation of some particular molecule of interest, this method makes it possible quickly to calculate the energy content of an appropriate strain-free model compound and thus determine the extent of the strain energy in the target.

What, then, are the **possible structures associated with the highest values for strain energy**? It would be relatively easy to list some hit parade of substances in response to this question, but comparisons of this sort are not in fact terribly useful since the overall strain energy of a molecule increases, for example, with the number of atoms involved. Thus, buckminsterfullerene C_{60} (**1**), with a strain energy of ca. 480 kcal mol^{-1}, certainly has one of the highest strain energies ever computed.[3] Since there are 60 atoms, however, this averages to only ca. 8 kcal mol^{-1} of strain per carbon atom, which is not particularly unusual. Devising a uniform normalization process applicable to all compounds is not very practical, so strain energies for various molecules end up serving only as clues to relative energy content. For this reason we limit ourselves in what follows to introducing a few especially strained molecules with no attempt at providing a rank order. (Fig. 1).

---

[1] A. von Baeyer, *Ber. Dt. Chem. Ges.* **1885**, *18*, 2278–2280.
[2] J. L. Franklin, *Ind. Eng. Chem.* **1949**, *41*, 1070.
[3] A. Hirsch, *The Chemistry of the Fullerenes*, Thieme, Stuttgart, **1994**, p. 186.

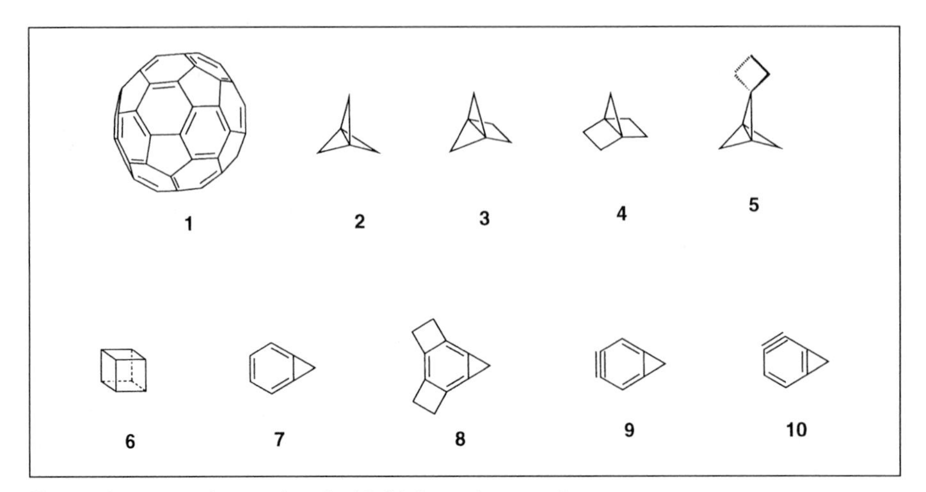

Fig. 1: Compounds associated with high strain energies

Two compounds that unquestionably deserve inclusion in a list of the most strained systems are sensational as well because of their unusual geometries (→ Molecular Form, Symmetry Highlights). One of these, [1.1.1]propellane (**2**), synthesized first by Wiberg[4] and later by Szeimies,[5] has a calculated strain energy of 98 kcal mol^{-1}.[6] The enormity of this value is immediately apparent when one recalls that the bond energy of the C–C single bond in ethane represents only 88 kcal mol^{-1} (→ Molecular Form, Bonding Records). Nevertheless, **2** is stable at room temperature. It is interesting to note that the first synthesis of this unusual compound was undertaken after quantum chemical calculations suggested its possible stability. Contrary to chemical intuition, increasing the lengths of bridges in the system in a stepwise fashion is not paralleled by a corresponding increase in stability: the higher homologues **3** and **4** are not stable at room temperature,[7] and the recently synthesized tetracyclic spiro compound **5** polymerizes rapidly at room temperature.[a]

Another molecule whose existence is due more to kinetic than to thermodynamic stability is the cube-shaped hydrocarbon cubane (**6**), first synthesized by Eaton's research group.[8] As Eaton himself expressed it: "**6** is kinetically a rock, but thermodynamically a powerhouse!" A somewhat less colorful way of making the same point is to note that **6** has in the meantime been routinely synthesized in kilogram quantities despite the fact that the compound is assigned a strain energy of 155 kcal mol^{-1}.[6]

---

[4] K. B. Wiberg, F. H. Walker, *J. Am. Chem. Soc.* **1982**, *104*, 5239–5240.
[5] M. Werner, D. S. Stephenson, G. Szeimies, *Liebigs Ann.* **1996**, 1705–1715.
[6] K. B. Wiberg, *Angew. Chem.* **1986**, *98*, 312–322; *Angew. Chem. Int. Ed. Engl.* **1986**, *25*, 312.
[7] K. B. Wiberg, *J. Am. Chem. Soc.* **1983**, *105*, 1227–1233.
[8] P. E. Eaton, T. W. Cole, Jr., *J. Am. Chem. Soc.* **1964**, *86*, 3157–3158.

The equivalence of all the C–C bonds in benzene has repeatedly challenged chemists to try to influence delocalization of the benzene π-electron system by annelation with small rings and thus alter the reactivity of the aromatic system. Cyclopropabenzene **7**,[9] with a strain energy of ca. 68 kcal mol^{-1}, marked a first step in this endeavor. The question of whether or not relatively high strain energy has anything to do with the unusual olfactory qualities of the compound (→ Sensors, Odor Hit List) would surely warrant investigation. In fact, however, despite the demonstrated toxicological safety of **7**, all research projects concerned with its properties had to be canceled at the University of Heidelberg because the smell of even tiny traces of the substance proved intolerable. Anellation of a single three-membered ring onto the benzene core actually represents the end of the line in this particular "strain series." Although it has proven possible to add two four-membered rings to the compound, leading to dicyclobutacyclopropabenzene **8**,[10] annelation with a second three-membered ring has never been accomplished. Even so, one noteworthy advance beyond cyclopropabenzene has occurred: starting with halogenated derivatives of **7**, the team of Brian Halton[11] managed to trap the highly strained arynes **9** and **10** in the form of cycloadducts. It was not possible for them to establish experimentally the enthalpies of formation of **9** and **10**, but theoretical investigations suggest strain energies of 170 (**9**) and 173–177 kcal mol^{-1} (**10**).

[9] (a) E. Vogel, W. Grimme, S. Korte, *Tetrahedron Lett.* **1965**, 3625–3631. (b) W. E. Billups, A. J. Blakeney, W. Y. Chow, *J. Chem. Soc. Chem. Commun.* **1971**, 1461–1462.
[10] W. E. Billups, B. E. Arney, L.-J. Lin, *J. Org. Chem.* **1984**, *49*, 3436–3437.
[11] Y. Apeloig, D. Arad, B. Halton, C. J. Randall, *J. Am. Chem. Soc.* **1986**, *108*, 4932–4937.
[a] Despite the high strain energy of their three-membered rings, cyclopropane derivatives most definitely occur in nature. Esters of chrysanthemic and pyrethric acid, found in plants of the genus *Pyrethrum* and harmless with respect to humans, are used as natural insecticides (called "pyrethroids"; → Crop Protection).

Explosive Records

It is of course quite reasonable to put compounds with high strain energy (→ Molecular Energy, Strain) to practical use in the form of explosives or blasting agents. Indeed, plans even exist to try the octanitro derivative of cubane **6** – which hasn't even been synthesized yet – as a high-energy rocket fuel.[1] Nevertheless, explosives generally must display other application-oriented characteristics beyond simply a high energy content (including safe manipulation, high density, a high rate of detonation, and the release after detonation of a large amount of gas) before they become of serious interest to the construction and mining industries or to the military.

The Oldest Explosive

Black powder, an explosive mixture of potassium nitrate, charcoal, and sulfur, was in use in Europe as early as the 13th century, so this can be regarded as the **oldest explosive**. The use of black powder was superceded near the close of the 19th century after Alfred Nobel (→ Nobel Prize) succeeded in 1866 in "taming" nitroglycerin. The substance was invented in 1847, but it is highly sensitive to shock and its utility was not established until Nobel found a way to combine it with kieselgur in a handy form that he successfully marketed under the trade name "Dynamit."

The Cheapest Explosive

But despite this economically significant development, nitroglycerin eventually relinquished its lead position from a quantitative standpoint. The **cheapest explosive** today, with a market share of around 80%, is the so-called ANFO explosive, a mixture of ammonium nitrate and diesel oil.[2]

The Most Powerful Explosive

Alongside these mass-market explosive agents there is also a demand for particularly powerful explosives for special applications. Two parameters serve as criteria for ranking energy-rich compounds with respect to their explosive power: detonation rate, which describes the spread of the shock wave generated by an explosion, and physical density of the material as a measure of "energy concentration" in the substance. Both parameters assume large values for a

[1] P. Eaton, lecture in Heidelberg, November 22, **1996**.
[2] A. Homburg, N. Fiederling, *Spektrum der Wissenschaft* **1996**, issue 8, 92–95.

compound with high explosive power. A summary of known explosives is provided in Table 1.[3]

Table 1: The most powerful explosive agents[a]

Year of introduction	Compound	Detonation rate [m s^{-1}]	Density [g/cm^3]
1870	nitroglycerin	7580	1.58
1910	TNT (**11**)	6930	1.63
1940	RDX (**12**)	8754	1.80
1955	HMX (**13**)	9110	1.89
1990	CL20 (**14**)	9380	1.98

[a] Compiled by H. Schubert, Fraunhofer Institut für Chemische Technologie, Pfinztal (Germany).

It is clear from the data that the most recent century has witnessed continuous efforts to maximize both parameters, and astounding progress has been made. The current "best in class" entry is hexanitroisowurtzitane (**14**; Fig. 1), developed a few years ago in the United States and known by the pseudonym CL20. However, its synthesis for large-scale applications is still rather complicated. With an astonishingly high detonation rate of 9380 m s^{-1} the shock wave from explosion of this compound expands at a rate in excess of 21 000 miles per hour! For comparison purposes, the speed of sound in air is only about 675 miles per hour, or a factor of 30 lower. The **most powerful explosive** that can currently be produced on a commercial scale is tetramethylenetetranitramine **13**, known as octogen or HMX. The smaller and also less expensive homologous compound hexogen (RDX, **12**) is an important additive in rocket propellants. Trinitrotoluene (**11**), introduced in the 19th century, is not limited to use in military applications, because despite its great explosive power it has a melting point above 80 °C that can still be determined with relative safety.

Not all the effects of explosives are destructive in nature: they can also act in life-saving ways. The most commonly encountered example of an **explosive device in everyday life**, albeit (hopefully!) only on rare occasions, is based on sodium azide, NaN_3.[4] The airbag restraint system incorporated into most modern automobiles inflates following a collision within the extremely short period of a fraction of a second thanks to release of a cloud of nitrogen gas, thereby protecting occupants of the front seats from serious injuries in the event of an accident. The requisite nitrogen comes from explosion of a powdered mixture of sodium nitrate and amorphous boron, which is initiated by electronic priming with sodium azide.

[3] H. Schubert, Fraunhofer Institut für Chemische Technologie, personal communication. See also H. Schubert, *Spektrum der Wissenschaft* **1996**, issue 8, 97–101.
[4] P. W. Atkins, J. A. Beran, *General Chemistry*, Freeman, New York, **1992**.
[5] H. Schildknecht, *Angew. Chem.* **1970**, *82*, 17–25; *Angew. Chem. Int. Ed. Engl.* **1970**, *9*, 1.

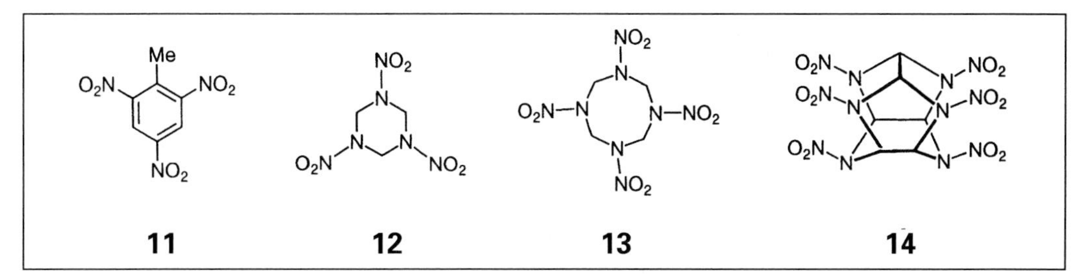

Fig. 1: Explosive substances for special applications

Explosive substances are not the exclusive prerogative of humans, by the way. Nature makes use of explosive mixtures as well. In what is surely a unique defense strategy, the bombardier beetle *Brachynus explodans*[5] frightens away potential aggressors by catalytically decomposing hydrogen peroxide inside a small bladder with the aid of the enzyme catalase, thereby generating water and oxygen. In a parallel reaction, facilitated by the enzyme peroxidase, hydroquinone is caused to react with H_2O_2 to give quinone (Fig. 2). Gas generation within the bladder becomes so violent that the corrosive mixture is explosively ejected with a loud bang and propelled in the direction of any pursuer.

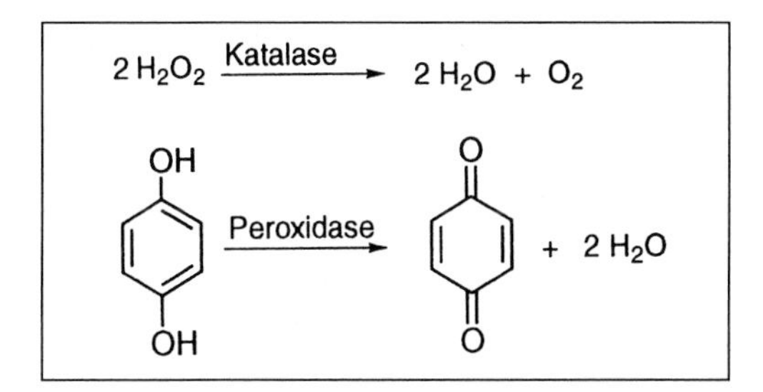

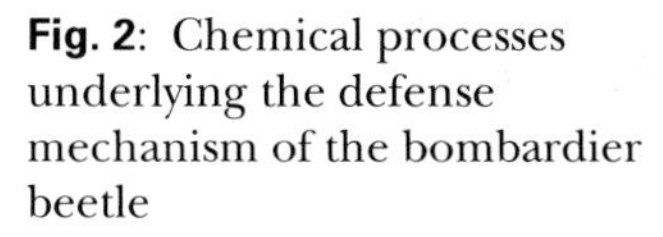

Fig. 2: Chemical processes underlying the defense mechanism of the bombardier beetle

Primary Applications for Industrial Explosives

Commercial explosives are substances capable of being detonated for blasting purposes, as propellants for firearms (gun powder), as igniters, or in pyrotechnic devices.

The most important area of direct application for these materials is in the mining industry, with coal mine operators consuming a 58% share of world explosives production. In addition to their use in quarries, the other primary role for industrial explosives, especially in the Western world, is in underground construction (Fig. 3; the extent of the latter application has not been indicated in the figure).

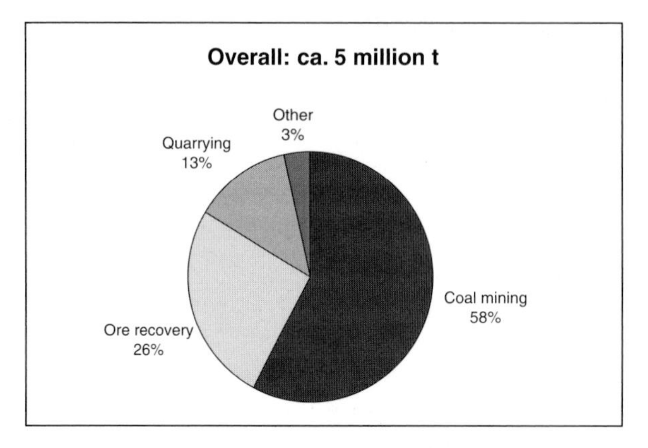

Fig. 3: Principal applications of industrial explosives, 1995[6]

Airbags

An interesting illustration of the fact that explosive materials can be used to save lives is the airbag. Airbags were first introduced about ten years ago as restraint systems in automobiles, and if all cars were so equipped on both the passenger side and the driver's side ca. 1000 lives a year could be saved.

The operation of an airbag is such that, if a crash occurs, a solid propellant is ignited within a fraction of a second by a sensitive control device. The resulting explosion-like burn releases nitrogen gas, which causes the airbag itself to inflate.

The chief producers of industrial explosive agents are ICI and Dyno Industries.

[6] data from: *Chemical Economic Handbook*, SRI, November **1995**.

Bonding Extremes in Carbon Compounds

The enormous number of carbon compounds newly synthesized each year (→ Literature) is impressive testimony not only to the importance of the element carbon in chemistry, but also to the versatility this element displays in bonding with other atoms and with itself (→ Molecular Form, Chains and Rings). The structural diversity exhibited by carbon compounds first became fully apparent when it became practical to apply to organic compounds the principles of X-ray diffraction, introduced originally by Max von Laue and the Braggs (→ Appendix, Nobel Prize Winners: in Physics). Today, thanks to refined techniques and more rapid data processing, the X-ray structural investigation of crystals has matured into a standard method of structure determination. This development has in turn provided us with extensive and detailed knowledge regarding the bonding parameters in carbon-containing molecules, including bond lengths and angles.

Probable Record C–C Bond Lengths

With the introduction of more precise techniques for structural investigations it rapidly became clear that the length of carbon–carbon bonds is not uniform in all compounds. Instead this key dimension depends upon the hybridization of the carbon atoms (sp^3, sp^2, and sp in singly, doubly, and triply bonded systems) as well as on such other structural factors as angle strain, electron delocalization, steric hindrance, etc. A few **extreme values for C–C bond lengths** are presented (with no claims of completeness) in Table 1. Examination of the structures of compounds **1–6** (Fig. 1), from which the data in Table 1 are derived, clearly reveals that in most cases the record-holders are subject to special structural relationships that dictate the observed extremes. Thus, angle strain present in cubylcubane (**1**)[1] causes rehybridization of the carbon atoms, and this leads to an extremely small value [1.458(8) Å] for the length of the exocyclic bond. (The average bond length for a C–C single bond is 1.530 Å.[2]) On the other hand, the **longest formal C–C single bond** so far

Table 1: Extreme values for C–C bond lengths

	Bond lengths [Å]		
	Minimal value (compound)	Standard value[a]	Maximal value (compound)
C_{sp^3}–C_{sp^3}	1.458(8) (**1**)	1.530 **C–C**	1.724(5) (**2**)
C_{sp^2}–C_{sp^2}	1.294(3) (**3**)	1.316 **C=C**	1.416(2) (**4**)
C_{sp}–C_{sp}	1.158(4) (**5**)	1.181 **C≡C**	1.248(1) (**6**)

[a] Crystallographic average values from F. H. Allen, O. Kennard, D. G. Watson, L. Brammer, A. G. Orpen, R. Taylor, *J. Chem. Soc. Perkin Trans. 2* **1987**, pp. 1–19.

Molecular Form
Bonding Records

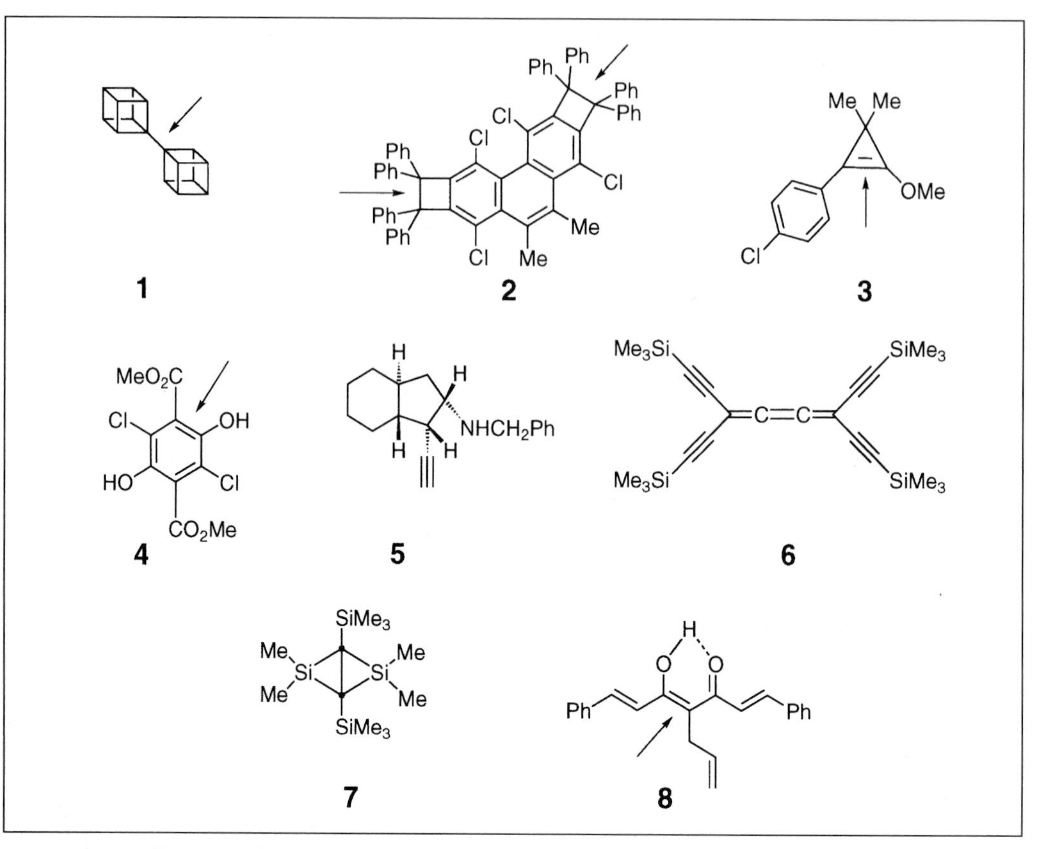

Fig. 1: Molecules characterized by bond-length extremes

detected was found in the substituted disilabicyclo[1.1.0]butane **7**,[3] with a length of 1.781(15) Å. A bit of caution is in order here, however; one should in this case probably speak instead of a C–C *distance*, because the bond between the carbon atoms in question has a bond order of only 0.5. Moreover, d-orbital participation from the silicon atoms causes the bonds between silicon and the bridgehead carbons to have at least partial double-bond character. True record-setting potential for a long C–C bond appeared recently as a result of the successful synthesis and structural characterization by Toda's group[4] of the di(cyclobuta)phenanthrene derivative **2**. The indicated C–C bonds (Fig. 1) in crystals of this unsymmetrical molecule have lengths in the vicinity of 1.710(5)

---

[1] R. Gilardi, M. Maggini, P. E. Eaton, *J. Am. Chem. Soc.* **1988**, *110*, 7232–7234.

[2] Average values from an evaluation of the Cambridge Crystallographic Database by F. H. Allen, O. Kennard, D. G. Watson, L. Brammer, A. G. Orpen, R. Taylor, *J. Chem. Soc. Perkin Trans. 2* **1987**, pp. 1–19.

[3] G. Fritz, S. Wartanessian, E. Matern, W. Hönle, H. G. v. Schnering, *Z. Anorg. Allg. Chem.* **1981**, *475*, 87–108.

[4] F. Toda, K. Tanaka, Z. Stein, I. Goldberg, *Acta Crystallogr.* **1996**, *C52*, 177–180.

[5] For a critical review of exceptionally long C–C bonds see G. Kaupp, J. Boy, *Angew. Chem.* **1997**, *109*, 48–50; *Angew. Chem. Int. Ed. Engl.* **1997**, *36*, 48–49.

and 1.724(5) Å – remarkably high indeed.[5] These large internuclear distances are a combined result of steric overloading due to the vicinal phenyl substituents, ring strain in the cyclobutabenzene subunits, and hyperconjugative effects.

The Longest and the Shortest C–C Double Bonds

The spread in known bond lengths for C–C single bonds is thus about 0.3 Å, but for C–C double bonds it is significantly less because these bonds are stronger and the corresponding molecules are less free to adapt. Just as with single bonds, the shortest known C–C double bond is associated with a small-ring compound, cyclopropene derivative **3**,[6] with a length of 1.294(3) Å. The **longest known C–C double bond** joins two phenyl carbon atoms in the terephthalic acid dimethyl ester **4**[7] across an internuclear distance of 1.416(2) Å. In this compound the unusual bond length is undoubtedly a consequence of delocalization of the π electrons of the aromatic ring system. Conjugation effects may also play a role in the acyclic derivative **8**,[8] although the main reason for a C–C bond 1.413(2) Å long in this enol is tautomerism with the neighboring ketonic oxygen atom.

The Longest and the Shortest C–C Triple Bonds

The variation in bond lengths for **C–C triple bonds** is even a bit smaller than that for C–C double bonds. Still surprising, however, is the perhydroindane derivative **5**,[9] with a C–C triple bond only 1.158(4) Å long. We call this result surprising in that neither the constitution of the molecule nor its X-ray investigation explains why the triple bond in this particular molecule should be so exceptionally short. Conjugative or hyperconjugative effects may explain the relatively long alkyne bond [1.248(1) Å] in the tetraalkynylated 1,2,3-butatriene **6**.[10] Overall, however, the bond lengths in alkyne derivatives are far less variable than the separations associated with C–C single bonds. The overall recorded span for triple bonds is only 0.09 Å.

[6] J. Søtofte, I. Crossland, *Acta Chem. Scand.* **1989**, *43*, 168–171.
[7] Q.-C. Yang, M. F. Richardson, J. D. Dunitz, *Acta Crystallogr.* **1989**, *B45*, 312–323.
[8] C. H. Görbitz, A. Mostad, *Acta. Chem. Scand.* **1993**, *47*, 509–513.
[9] R. M. Borzilleri, S. M. Weinreb, M. Parvez, *J. Am. Chem. Soc.* **1994**, *116*, 9789–9790.
[10] J.-D. van Loon, P. Seiler, F. Diederich, *Angew. Chem.* **1993**, *105*, 1235–1238; *Angew. Chem. Int. Ed. Engl.* **1993**, *32*, 1187.

The Largest and the Smallest C–C–C Bond Angles

C–C bonds exhibit extremes not only in bond lengths; crystallographically determined bond angles also sometimes vary considerably from idealized values for tetrahedral, trigonal-planar, and linear environments (Table 2). It is

Table 2: Extreme values for C–C–C bond angles

	Bond angles [°]		
	Minimal value (compound)	Standard value	Maximal value (compound)
C–C_{sp^3}–C tetrahedral	50.7(1) (**3**)	109.4	127.6(3) (**9**)
C–C_{sp^2}–C trigonal-planar	61.9(1) (**3**)	120	176.9(1) (**10**)
C–C_{sp}–C linear	145.8(7) (**11**)	180	–

not entirely surprising that the record-holders with respect to these bonding parameters are all carbocylic systems, in which strain-induced deformations lead to especially large deviations from the norm. The **minimal values for C–C–C bond angles** for both sp^3- and sp^2-hybridized carbon are again found in the cyclopropene derivative **3** (Fig. 1), already cited above for an extremely short C–C bond. The geometry of this three-membered ring dictates considerable compression of the internal C–C–C angle to a value slightly over half that of the idealized standard angle. The **largest angle about an sp^3 carbon atom** is found in the barium salt of the spiro[3.3]heptane dicarboxylic acid **9** (Fig. 2),[11] where the spiro connection of two four-membered rings leads to an angle expansion to 127.6(3)°. Even more dramatic is the expansion of the sp^2 bond angle in 1,2-dihydrocyclobuta[*a*]cyclopropa[*c*]benzene **10**.[12] The C–C–C angle about the carbon atom closest to the four-membered ring and shared by the six-membered and three-membered rings is nearly that of a straight line compared to the norm (176.9°)! A highly strained ring also forces the record deformation about an sp-hybridized carbon atom. Since a linear arrangement cannot be achieved by the triple bond in the thiacycloheptyne derivative **11**,[13] the C–C–C bond angle is reduced to a mere 145.8(7)°.

[11] L. A. Hülshoff, H. Wynberg, B. van Dijk, J. L. de Boer, *J. Am. Chem. Soc.* **1976**, *98*, 2733–2740.
[12] R. Boese, D. Bläser, W. E. Billups, M. M. Haley, A. H. Maulitz, D. L. Mohler, K. P. C. Vollhardt, *Angew. Chem.* **1994**, *106*, 321–324; *Angew. Chem. Int. Ed. Engl.* **1994**, *33*, 313.
[13] Determined by electron diffraction. J. Haase, A. Krebs, *Z. Naturforsch. Teil A*, **1972**, *27*, 624–627.

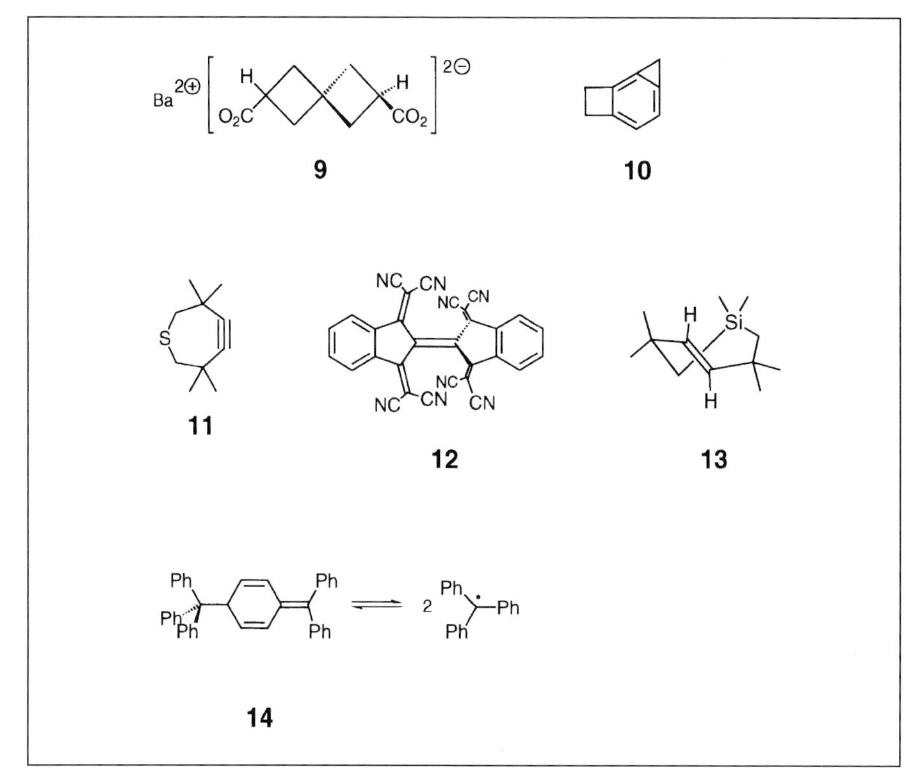

Fig. 2: Compounds with unusual bond angles

The Most "Twisted" Double Bonds

A trigonal-planar environment about sp^2-hybridized carbon ordinarily requires a coplanar arrangement of the four substituents on a C–C double bond. If this is sterically impossible, the geometry becomes distorted, a distortion that can be described in terms of a dihedral angle. Thus, a dihedral angle of 90° would mean that the plane of one end of the double bond was perpendicular to that of the other end. The **most severely twisted double bonds** so far known are found in compounds **12**[14] and **13**[15] (→ Molecular Form, Chains and Rings), with dihedral angles of 49.7° and 49.0°, respectively.

[14] A. Beck, R. Gompper, K. Polborn, *Angew. Chem.* **1993**, *105*, 1424–1427; *Angew. Chem. Int. Ed. Engl.* **1993**, *32*, 1352.

[15] A. Krebs, K.-I. Pforr, W. Raffay, B. Thölke, W. A. König, J. Hardt, R. Boese, *Angew. Chem.* **1997**, *109*, 159–161; *Angew. Chem. Int. Ed. Engl.* **1997**, *36*, 159–160.

A "Twisted" Molecule

A new record for the twisting of a polycyclic aromatic hydrocarbon (PAH) has been reported by R. A. Pascal, Jr., et al.[1] According to a structural analysis, the orange compound they synthesized (**1**) has an *end-to-end twist* of 105°. This angle is larger by a factor of 1.5 than any previously observed. Each of the benzene rings of the four naphthacene units plays an approximately equivalent role in the overall twisting of the system. Nevertheless, the molecule still maintains a conjugated π-electron system. Despite steric strain, **1** is very stable: The compound shows no signs of decomposition even at temperatures around 400 °C.

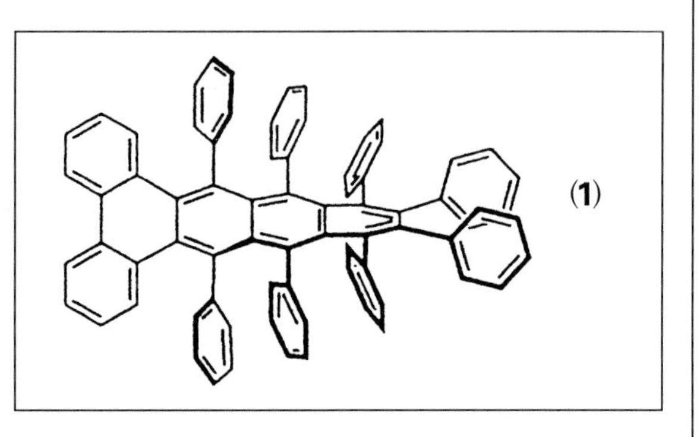

[1] X. Qiao, D. H. Ho, R. A. Pascal, *Angew. Chem.* **1997**, *109*, 1588–1589; *Angew. Chem. Int. Ed. Engl.* **1997**, *36*, 1531–1532.

The Strongest C–C Single Bond

The collection of bonding extremes presented above clearly underscores how flexible carbon can be in its compounds. Its unique position relative to the other elements in the periodic table is also reflected in the bond energies (more precisely: bond-dissociation enthalpies) of C–C single bonds (→ Molecular Energy, Strain).[16] Whereas the value 368.2 kJ mol^{-1} for ethane testifies to a relatively strong bond, the bond in **14**, the famous Gomberg triphenylmethyl radical dimer (→ Literature), earns a value of only 50.2 kJ mol^{-1}. The highest C–C bond enthalpy so far determined is that of dicyan, NC–CN:[17] 603 ± 21 kJ mol^{-1}.

[16] C. Rüchardt, H.-D. Beckhaus, *Angew. Chem.* **1980**, *92*, 417–429; *Angew. Chem. Int. Ed. Engl.* **1980**, *19*, 429.
[17] J. A. Dean (Ed.), *Lange's Handbook of Chemistry*, McGraw–Hill, New York, 14th ed. **1992**, p. 4.25.

The Largest Molecules

There is one question that of course should under no circumstances go unanswered in a book about records in chemistry: Which is the **largest molecule**? More precisely, which molecule has the highest molecular weight? In order not to get lost in the trackless wilds of polymer aggregates it is best to qualify the question a bit: The only permissible candidates should be single, well-defined molecules (i.e., no molecular distributions) that are entirely covalent in their construction. This avoids, for example, all those proteins in the realm of biopolymers in which various noncovalently-bound subunits are held together by hydrogen bonds. Also not open to consideration are synthetic macromolecular systems (→ Chemical Products; Chemical Industry; Plastics) that consist of mixtures of closely related but slightly differing molecules.

As a rule, molecules are extremely small, as illustrated by the following typical example.[a] If one were to pour a shot glass of brandy (32% alcohol by volume) into the ocean, then once the alcohol it contained (ca. 5 g) had been thoroughly distributed, a randomly drawn shot glass of ocean water would be found to include on the average one molecule of the original alcohol, regardless of which ocean it came from or at what depth it was collected.

The **largest molecules that have ever been found** are diamonds (→ Elements), which can be regarded as gigantic molecules made up of a three-dimensional network of carbon atoms.[b] The largest of all was the so-called Cullinan diamond, found on January 25, 1905, in South Africa and sporting a weight of 621.6 g (3106 carats).[1] Unfortunately, this diamond only existed for three years; in 1908 it was broken up into nine large diamonds and 95 smaller fragments.[c] The **largest man-made synthetic molecule** is an artificial diamond of 38.4 carats, the growth of which required 25 days. Artificial diamonds of this size are made under what are presumably record-setting extreme conditions, namely in high-pressure presses at temperatures between 1500 and 1800 °C and pressures between 70 000 and 100 000 atmospheres.[2] They can also be produced in the **smallest high-pressure chemical reactors** in the world: in the interior of tiny 50–100 nm "carbon onions" consisting of a number of fullerenes nested inside one another.[3]

[1] N. N. Greenwood, A. Earnshaw, *Chemistry of the Elements*, Pergamon, New York, **1989**, p. 300.
[2] L. F. Trueb, *Die Chemischen Elemente*, Hirzel, Stuttgart, **1996**, pp. 262–263.
[3] F. Banhart, P. M. Ajayan, *Nature (London)* **1996**, *382*, 433.

The Largest Biomolecules

If asked to provide a spontaneous answer to the question regarding the largest molecule, most would probably have chosen some record holder from the realm of biomolecules. This category does indeed supply a list of compounds of truly gigantic proportions. Among proteins only a very few have molecular weights exceeding 200 000 Daltons,[4] but one of these stands out by itself. So far, the **largest known protein** from the standpoint of molecular weight is titin,[5] also known as connectin, a muscular protein that together with actin and mysosin filaments contributes heavily to the elasticity of muscles. It consists of a linear sequence of 26 926 amino acids, has a molecular weight of nearly three million (more precisely: 2 993 000) Daltons, and in molecular terms the enormous length of over 1 µm.

Very few biomolecules are in a position to challenge the lead of the heavyweight titin. One example, and thus perhaps the bronze medallist, is the ribosome,[6] a ubiquitous, covalently constructed ribonucleoprotein that plays a key role in protein biosynthesis. Ribosomes of the bacterium *Escherichia coli*, for example, consist of three ribonucleic acid strands (with a total of 4560 nucleotides) and 55 proteins for a combined molecular weight of ca. 2.7 million Daltons.

[4] M. Barinaga, *Science* **1995**, *270*, 236.
[5] S. Labeit, B. Kolmerer, *Science* **1995**, *270*, 293–296.
[6] J. A. Lake, *Ann. Rev. Biochem.* **1985**, *54*, 507–530.
[a] Freely adapted from A. F. Holleman, *Lehrbuch der anorganischen Chemie (Holleman–Wiberg)*, 101st ed., W. de Gruyter, Berlin **1995**, p. 47. The daily alcohol influx into seawater simply as a consequence of seasick ship passengers considerably exceeds the quantity under consideration here (S. Hehn, private communication).
[b] The free valences of carbon atoms at the surface are satisfied mainly by hydrogen and oxygen.
[c] The Cullinan diamond was the source of the famous diamonds "Star of Africa" (530 carats) and "Cullinan II" (371 carats). (J. Dettmann, *Fullerene: die Bucky-Balls erobern die Chemie*, Birkhäuser, Basel, **1994**, p. 29). The large diamonds cut from the Cullinan diamond are now among the British crown jewels.

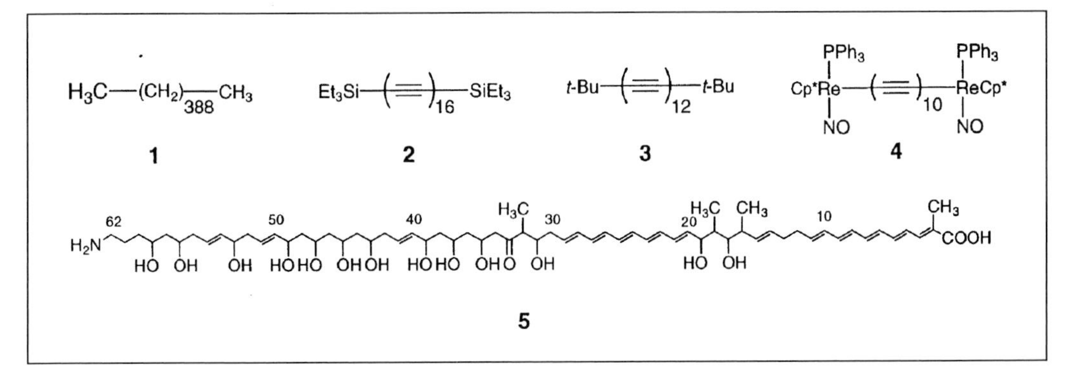

Fig. 1: Compounds containing exceptionally long chains of carbon atoms

The Longest Carbon Chains

For the existence of "organic" life based on carbon we can thank among other things the fact that carbon forms stable bonds with other carbon atoms (→ Molecular Form, Bonding Records). Together with a valency of four, this bonding principle opens the way to open-chain and ring-shaped molecules of a wide variety of lengths and sizes.

The **longest, linear hydrocarbon synthesized as a single molecule** was introduced in 1985 by Bidd and Whiting.[1] The compound is nonacontatrictane **1**, with a carbon-atom chain of 390 atoms and the molecular formula $C_{390}H_{782}$. The targeted synthesis of this paraffin proceeded from C_{12} precursors via six successive Wittig reactions (→ Synthesis; Appendix, Nobel Prize Winners: in Chemistry) and was undertaken to establish the crystallization behavior of long-chain hydrocarbons (are they linear or folded?).

In contrast to the conformationally rather flexible chain of an alkane, a polyalkyne, in which the carbon atoms are held together by alternating C–C triple and single bonds, is quite stiff and thus of interest with respect to the design of **molecular wires** for nanotechnology. One problem in the realization of this concept is the relatively high energy content of the triple bond (→ Molecular Energy, Strain), which means that most alkynes are rather reactive compounds. It is therefore perhaps no surprise that the **record for a continuous string of carbon atoms in a polyalkyne** has not changed since 1972 when it was set with a chain length of 32 carbon atoms.[2] Kinetic stabilization was achieved in compound **2** by terminal silyl groups, but even this trick was insufficient to permit isolation of the compound in analytically pure form. Compounds of this type become easier to work with when the chain length is decreased, as demonstrated by molecule **3**, also prepared in the research group of D. R. M.

[1] I. Bidd, M. C. Whiting, *J. Chem. Soc. Chem. Commun.* **1985**, 543–544.
[2] R. Eastmond, T. R. Johnson, D. R. M. Walton, *Tetrahedron* **1972**, *28*, 4601–4606.

Walton.[2] Here the twelve alkyne units are protected by terminal *tert*-butyl groups, and the compound could be characterized in a straightforward way both chemically and physically. The idea of a molecular wire has recently been resurrected by Gladysz and coworkers, who introduced organometallic fragments as terminal groups on a polyalkyne chain. The record with respect to their efforts is held so far by compound **4**,[3] in which two rhenium centers are separated by a chain of twenty carbon atoms. Yet the metal fragments are still able to communicate with each other electronically.

The Biomolecule with the Longest Chain

Potential record chain lengths are certainly not the exclusive property of synthetic compounds. Nature, too, gives birth to molecules with impressively long, linear carbon skeletons. The linearmycins A and B[4] are carboxylic acid derivatives with linear, nonconjugated polyene backbones respectively 60 and 62 carbon atoms in length. The longer of the two, linearmycin B (**5**), is illustrated in Fig. 1. These long-chain natural products were discovered only recently in micelles of streptomycetes, and they are the first linear polyene antibiotics to show not only antibacterial but also antifungal activity.

The Largest Natural Macrocycles

Apart from linear carbon skeletons, cyclic systems constitute another widely encountered structural motif in natural products. A possible record-holder in the category **largest natural macrocycle** is isoswinholid A (**6**, Fig. 2) with a 46-membered bislactone ring system. Kitagawa and coworkers[5] were able to isolate **6**, which shows promising cytotoxic characteristics *in vitro*, from the sea sponge *Theonella swinhoei* off the coast of Okinawa (Japan).

---

[3] T. Bartik, B. Bartik, M. Brady, R. Dembinski, J. A. Gladysz, *Angew. Chem.* **1996**, *108*, 467–469; *Angew. Chem. Int. Ed. Engl.* **1996**, *35*, 414–417.

[4] M. Sakuda, U. Guce-Bigol, M. Itoh, T. Nishimura, Y. Yamada, *J. Chem. Soc. Perkin Trans. I* **1996**, 2315–2319.

[5] M. Kobayashi, J. Tanaka, T. Katori, I. Kitagawa, *Chem. Pharm. Bull.* **1990**, *38*, 2960–2966.

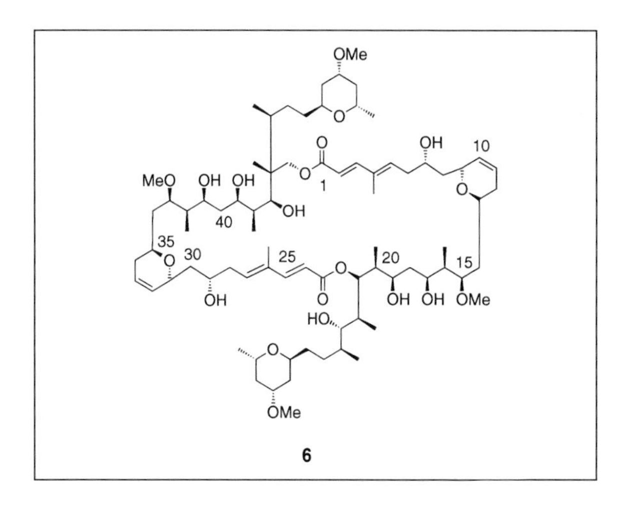

Fig. 2: Perhaps the largest natural macrocycle

The Largest Synthetic Rings

Among synthetic hydrocarbons, the largest ring systems represented in *Chemical Abstracts* incorporate 288 carbon atoms. This category includes two substances: the cyclic alkane cyclooctaoctacontadictane $C_{288}H_{576}$ (**7**, Fig. 3)[6] and a macrocyclic polyalkyne $C_{288}H_{480}$[7] with 24 triple bonds.

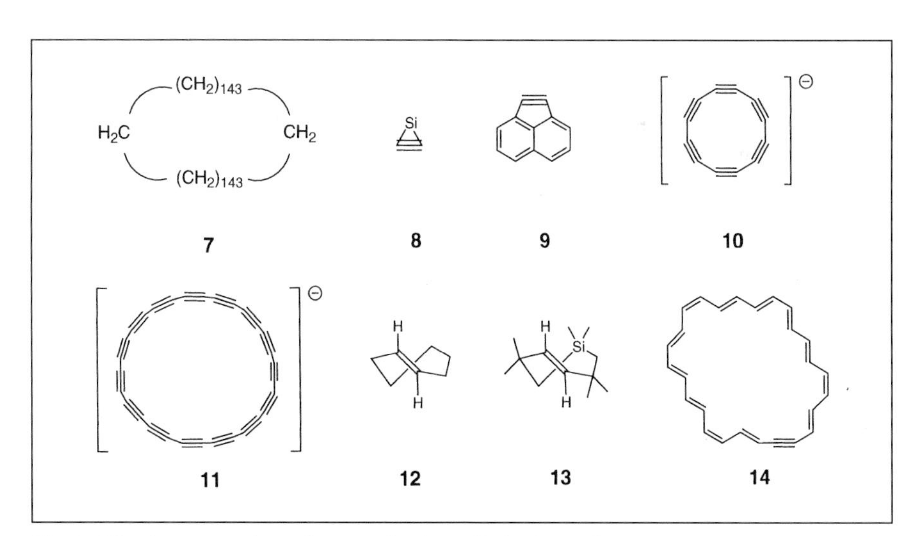

Fig. 3: The largest and smallest synthetic cyclic hydrocarbons

[6] K. S. Lee, G. Wegner, *Makromol. Chem. Rapid Commun.* **1985**, *6*, 203–208.
[7] G. Schill, C. Zürcher, H. Fritz, *Chem. Ber.* **1978**, *111*, 2901–2908.

The Smallest Cyclic Alkyne

The challenge of introducing a C–C triple bond into a ring has repeatedly spurred chemists on to greater heights of synthetic achievement. Maier et al.[8] succeeded only a few years ago in presenting matrix-spectroscopic proof of the existence of silacyclopropyne **8**, the **smallest cyclic alkyne**. The preparation of such a three-membered ring system depended on the incorporation of a silicon atom in the ring, because the presence of this heteroatom permits a significant increase in the bond lengths and bond angles (→ Molecular Form, Bonding Records) relative to a pure carbocycle. It is for this reason that the much more strained cyclopropyne itself remains unknown. The series of carbocyclic alkynes begins with five carbon atoms. Cyclopentyne, in the form of the acenaphthyne **9**, has been prepared photochemically, and its existence was demonstrated through a trapping reaction.[9] The smallest unsubstituted cyclic hydrocarbon of this type that is stable under normal laboratory conditions is the eight-membered cyclooctyne.

Cyclo[*n*]carbons

The "gold-rush fever" unleashed by the discovery of fullerenes (→ Molecular Form, Symmetry Highlights) encouraged researchers to try synthesizing other carbon allotropes, ones based on alkynes. Compounds of this type, the so-called cyclo[*n*]carbons[10] (where *n* is the number of carbon atoms), are nothing other than rings of carbon atoms bound to one another. The difficulty in this undertaking will be apparent from the fact that so far all the entries in the category are prepared starting with suitable precursors and detected only as ions in a mass spectrometer. The species with the smallest value for *n* is the C_{12} system **10**, prepared by Tobe and coworkers in the form of a singly charged anion.[11] The record for the largest cyclo[*n*]carbon goes to the C_{30} ion **11** from the research group of Diederich.[12] Interestingly, $\mathbf{11}^+$ coalesces in the mass spectrometer to a cation with double its mass, for which a fullerene structure (C_{60}) could be verified experimentally.

[8] G. Maier, H. P. Reisenauer, H. Pacl, *Angew. Chem.* **1994**, *106*, 1347–1349; *Angew. Chem. Int. Ed. Engl.* **1994**, *33*, 1248.

[9] O. L. Chapman, J. Gano, P. R. West, M. Regitz, G. Maas, *J. Am. Chem. Soc.* **1981**, *103*, 7033–7036.

[10] F. Diederich, *Nature (London)* **1994**, *369*, 199–207.

[11] Y. Tobe, H. Matsumoto, K. Naemura, Y. Achiba, T. Wakabayashi, *Angew. Chem.* **1996**, *108*, 1924–1926; *Angew. Chem. Int. Ed. Engl.* **1996**, *35*, 1800–1802.

[12] S. W. McElvany, M. M. Ross, N. S. Goroff, F. Diederich, *Science* **1993**, *259*, 1594–1596.

The Smallest Cyclic Alkenes

Introduction of C–C double bonds into cyclic hydrocarbons as a limiting structural element is much less problematic. Indeed, this time the smallest possible representative, **cyclopropene** (→ Molecular Form, Bonding Records), is a compound that can be comfortably manipulated at low temperature, although it does polymerize readily. Nevertheless, there are interesting records to be sought even with **cyclic alkenes**. The geometry about sp^2-hybridized carbon is such that in small cycloalkenes the *Z*-configuration of the double bond (as in a *cis*-alkene) is the most thermodynamically favored. Small rings with *trans*-alkene linkages can be prepared only with some difficulty. Thus, the **smallest *trans*-alkene** that can be characterized by UV and NMR spectroscopy is cycloheptene (**12**),[13] which is obtained by photochemical transformation of the corresponding *cis* isomer. Recently it proved possible by methylation and incorporation of a silicon atom to synthesize a derivative of this system stable at room temperature: *trans*-1,1,3,3,6,6-hexamethylsilacyclohept-4-ene, **13**.[14] The smallest unsubstituted *trans*-cycloalkene stable at room temperature is cyclooctene, first synthesized by Ziegler.[15]

The Largest Conjugated Cycloalkene

The question of how many double bonds can be incorporated in conjugated fashion into a carbocyclic compound has also been addressed preparatively. The **largest conjugated carbocyclic alkenes** have been synthesized in the research group of Sondheimer, who with his annulenes made groundbreaking contributions to experimental verification of the Hückel rule. Despite one small flaw – the presence of a single triple bond (which technically renders it a monodehydro[26]annulene) – the largest representative of this family is the [26]annulene **14**,[16] the aromatic character of which (consistent with its π-electron count) is evident, for example, through its diamagnetic ring current.

[13] M. Squillacote, A. Bergman, J. De Felippis, *Tetrahedron Lett.* **1989**, *30*, 6805–6808. See also E. J. Corey, F. A. Carey, R. A. E. Winter, *J. Am. Chem. Soc.* **1965**, *87*, 934–935.

[14] A. Krebs, K.-I, Pforr, W. Raffay, B. Thölke, W. A. König, I. Hardt, R. Boese, *Angew. Chem.* **1997**, *109*, 159–161; *Angew. Chem. Int. Ed. Engl.* **1997**, *36*, 159–160.

[15] K. Ziegler, H. Wilms, *Liebigs Ann.* **1950**, *567*, 1–43.

[16] B. W. Metcalf, F. Sondheimer, *J. Am. Chem. Soc.* **1971**, *93*, 5271–5272.

The Smallest Cyclic Cumulenes

The **best-known annulene,** benzene, which can also be termed [6]annulene, is the starting point for attempts to arrange the three double bonds of the six-membered ring system in a (formally) nonconjugated way: that is, in a cumulated fashion. The reactive benzene isomers 1,2,3-cyclohexatriene (**15**; Fig. 4)[17] and 1,2,4-cyclohexatriene (**16**)[18] were prepared by an ingeniously selected elimination reaction and verified by trapping reactions. But this still does not constitute the record-holder for the **smallest cyclic cumulene**: that honor belongs instead to the 3,4-didehydrothiophene **17**,[19] whose existence was recently verified through cycloadducts.

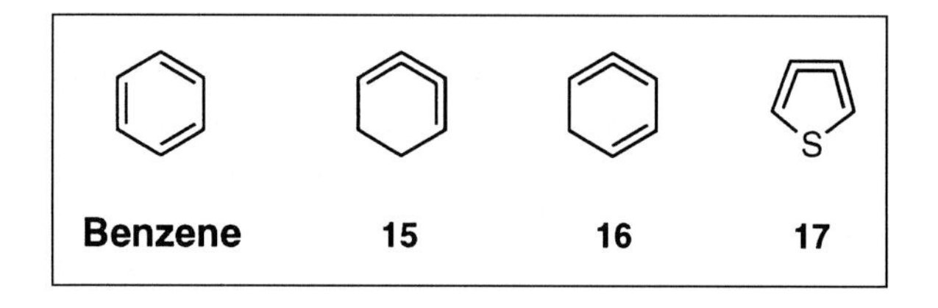

Fig. 4: Cyclic cumulenes

[17] W. C. Shakespeare, R. P. Johnson, *J. Am. Chem. Soc.* **1990**, *112*, 8578–8579.
[18] M. Christl, M. Braun, G. Müller, *Angew. Chem.* **1992**, *104*, 471–473; *Angew. Chem. Int. Ed. Engl.* **1992**, *31*, 473.
[19] X.-S. Ye, W.-K. Li, H. N. C. Wong, *J. Am. Chem. Soc.* **1996**, *118*, 2511–2512.

Symmetrical Molecular Structures

Aesthetics as it relates to the form of highly symmetrical molecules has been a fertile source of inspiration for organic chemists, repeatedly eliciting from them masterpieces of synthetic achievement in their quest to expand the limits of the theoretically and preparatively attainable. The reward for a chemist lies not only in the successful completion of a synthesis of an intriguing compounds; work directed toward the assembly of a compelling symmetrical structure frequently leads to new molecular architectures, some of them with potentially record-setting functions and properties.

Platonic Bodies and Fullerenes

The attempt to prepare in the form of hydrocarbons such **basic Platonic figures** as the tetrahedron, the cube, and the dodecahedron began quite early. Thus, Maier et al.[1] succeeded toward the end of the 1970s in synthesizing tetra-*tert*-butyltetrahedrane **1** (Fig. 1), a tetrahedral hydrocarbon composed of four equilateral triangles. The bulky alkyl groups around the periphery proved necessary for stabilization of this highly strained system, and to date no one has succeeded in isolating the parent hydrocarbon, **tetrahedrane** itself. Also highly strained, but nevertheless isolable as an unsubstituted hydrocarbon, is the symmetrical box-shaped **cubane** (**2**) (→ Molecular Energy, Strain), first prepared by Eaton and Cole.[2] The bonding relationships forced upon this compound by its unusual geometry lead in some of its derivatives to extremely short C–C bonds (→ Molecular Form, Bonding Extremes in Carbon Compounds).

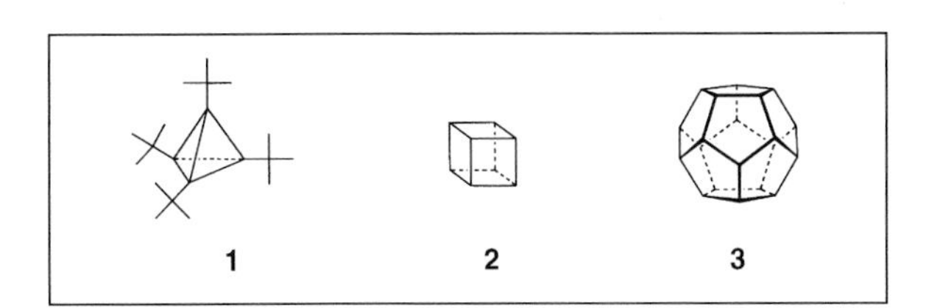

Fig. 1: Platonic bodies based on hydrocarbon skeletons

Another synthetic masterpiece was the construction of **dodecahedrane** (**3**), a hydrocarbon consisting exclusively of five-membered rings. The achievement in this case is credited to the research group of Paquette,[3] and it involved 23

[1] G. Maier, S. Pfriem, U. Schäfer, R. Matusch, *Angew. Chem.* **1978**, *90*, 552–553; *Angew. Chem. Int. Ed. Engl.* **1978**, *17*, 520. G. Maier, *Angew. Chem.* **1988**, *100*, 317–341; *Angew. Chem. Int. Ed. Engl.* **1988**, *27*, 309.

[2] P. E. Eaton, T. W. Cole, Jr., *J. Am. Chem. Soc.* **1964**, *86*, 3157–3158.

[3] R. J. Ternasky, D. W. Balogh, L. A. Paquette, *J. Am. Chem. Soc.* **1982**, *104*, 4503–4504. L. A. Paquette, *Chem. Rev.* **1989**, *89*, 1051–1065.

steps starting from cyclopentadiene. Prinzbach et al.[4] developed an alternative strategy for dodecahedranes known as the "pagodane route." This highly symmetrical compound (point group I_h), supports 120 different symmetry operations that regenerate the identical structure.

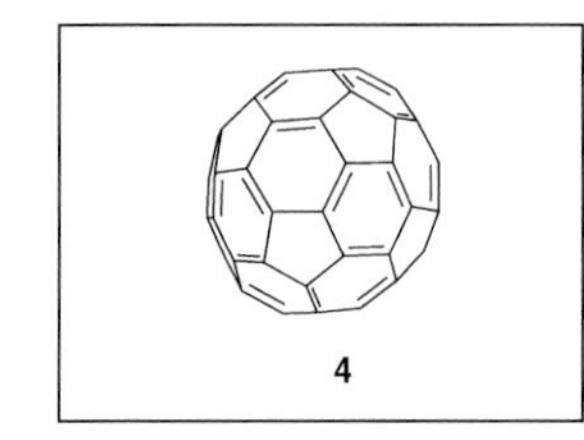

Fig. 2: Buckminsterfullerene, C_{60}

Another compound in the same point group as **3** is buckminsterfullerene C_{60} (**4**; Fig. 2), first isolated in 1990,[5] the discovery of which led to a Nobel Prize for chemistry (→ Appendix, Nobel Prize Winners: in Chemistry). The three-dimensional form of this hydrocarbon sphere results from linking 20 hexagons with 12 pentagons in an arrangement exactly like that of an ordinary soccer ball. Despite its structural complexity, the synthetic demands are minimal: Fullerene C_{60} can be isolated by rapidly cooling graphite vaporized in an electric arc, and it is even available commercially. This synthetic method also produces small amounts of other more or less spherical hydrocarbons (C_{70}, C_{76}, C_{78}, C_{80}, ...), which differ from one another only in the number of six-membered rings present. The cavity within the hydrocarbon sphere has of course tempted chemists to punch holes in the outer shell with the goal of stuffing other molecules into the interior, thereby producing endohedral complexes (keywords: drug carriers, molecular probes, etc.). The largest hole so far cut into a fullerene involved expansion of one of the six-membered rings of **4** into a ring with 15 sides.[7] It appears that it is only a matter of time until the first "filled fullerene" is created by the methods of chemical synthesis.

The Sharpest Needle

The reaction mixture from the preparation of **4** permits isolation not only of spherical molecules, but also of long, hollow carbon tubes, so-called *bucky tubes,* that amount to a miniaturized form of graphite capillaries. Since such tubes display electrical conductivity (→ Elements, Records) and have a tip diameter of only 5–20 nm, bucky tubes have been successfully introduced into scanning

[4] H. Prinzbach, K. Weber, *Angew. Chem.* **1994**, *106*, 2329–2348; *Angew. Chem. Int. Ed. Engl.* **1994**, *33*, 2239–2257.
[5] W. Krätschmer, L. D. Lamb, K. Fostiropoulos, D. R. Huffmann, *Nature* **1990**, *347*, 354–358.
[6] H. W. Kroto, J. R. Heath, S. C. O'Brian, R. F. Curl, R. E. Smalley, *Nature* **1985**, *318*, 162–163.
[7] M.-J. Arcre, A. L. Viado, Y.-Z. An, S. J. Khan, Y. Rubin, *J. Am. Chem. Soc.* **1996**, *118*, 3775–3776.

tunneling microscopy and scanning force microscopy as the **sharpest available needles for probing surfaces**.[8]

Probable Record Structures Based on Benzene Rings

The symmetrical, hexagonal form of the benzene nucleus is conducive to systematic construction of molecular architectures whose diversity in form is limited only by the chemist's own fantasy (Fig. 3). The molecule with the **longest linear array of anellated benzene rings** is the one known as heptacene (**5**),[9] with seven such rings. The phenanthrene motif has also been utilized in the design of benzenoid molecular structures; so far an array of 11 benzene nuclei is the most to be coupled in this fashion to give tetrapentyl-[11]phenacene (**6**).[10] In the absence of alkyl substituents, extremely low

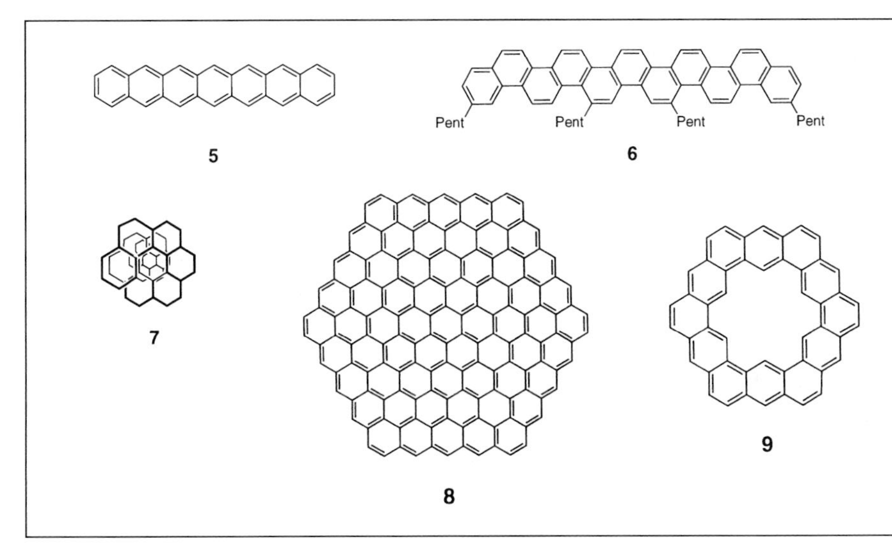

Fig. 3: Noteworthy molecules composed of anellated benzene rings

solubility obstructs the synthesis of phenacenes with more than seven rings. If one maintains a consistent rotational strategy (i.e., *ortho*-anellation) during the construction process, one arrives quickly at structures in which the benzene rings begin to overlap, as in the so-called **helicenes**. For these molecules the record currently belongs to a compound with 14 benzene nuclei, [14]helicene (**7**), the synthesis of which was carried out similarly to that of the phenacenes

[8] H. Dai, J. H. Hafner, A. G. Rinzler, D. T. Colbert, R. E. Smalley, *Nature* **1996**, *384*, 147–150.
[9] W. J. Bailey, C.-W. Liao, *J. Am. Chem. Soc.* **1955**, *77*, 992–993.
[10] F. B. Mallory, K. E. Butlar, A. C. Evans, E. J. Brondyke, C. W. Mallory, C. Yang, A. Ellenstein, *J. Am. Chem. Soc.* **1997**, *119*, 2119–2124.

(photocyclization reactions starting with stilbene derivatives).[11] Not only structural properties, but also certain physical properties confer record-setting status on helicenes. If benzene anellation were to be pursued *ad infinitum*, then the ultimate result would be the surface of graphite, which as a polymer is no longer classifiable on the basis of number of rings. The **largest carbon ring system** (apart from fullerene derivatives) recorded in the database of the Chemical Abstracts Service (CAS) (→ Literature) is hydrocarbon **8**, $C_{150}H_{30}$, the 61 rings of which Dias[12] investigated in a theoretical study. Along the periphery of this molecule there are 54 carbon atoms. In the same publication Dias also describes the larger $C_{170}H_{32}$ with 58 atoms along its outer edge, but CAS appears to have capitulated rather than register such a large polycyclic hydrocarbon.

A milestone in the art of anellation from a preparative standpoint was achieved with **kekulene** (**9**), prepared by Staab and Diederich,[13] perhaps the most popular representative of the class of cycloarenes. Not only its synthesis, but also its structure determination demanded a certain amount of inventiveness on the part of the chemists involved. NMR-spectroscopic characterization of **9** had to be carried out at 155 °C in $[D_2]$-1,2,4,5-tetrachlorobenzene because of the incredible insolubility of the compound in ordinary solvents. By the same token, suitable crystals[14] for X-ray structural analysis of **9** could be obtained only by slow cooling of a melt of **9** in pyrene from 450 °C to 150 °C over a period of 24 hours.

But this is by no means the end of the chapter on record structures based on benzene. For example, one might ask how many phenyl rings can be substituted onto a single carbon atom. This is easy to answer due to the tetravalent nature of carbon, leading to the structure of **tetraphenylmethane**. If one proceeds one sphere farther out the result is tetrakis(biphenyl)methane. The current record in this particular series is reached at the very next step with the tetrakis(terphenyl)methane derivative **10** (Fig. 4), synthesized by Griffin starting from tetraphenylmethane.[15] If the presence of heteroatoms is allowed, the winner in this category is tetrakis(diazaquaterphenyl)methane **11** from the research team of Gompper.[16]

[11] R. H. Martin, *Angew. Chem.* **1974**, *86*, 727–738; *Angew. Chem. Int. Ed. Engl.* **1974**, *13*, 649.
[12] J. R. Dias, *Can. J. Chem.* **1984**, *62*, 2914–2922.
[13] H. A. Staab, F. Diederich, *Chem. Ber.* **1983**, *116*, 3487–3503.
[14] H. A. Staab, F. Diederich, C. Krieger, D. Schweitzer, *Chem. Ber.* **1983**, *116*, 3504–3512.
[15] L. M. Wilson, A. C. Griffin, *J. Mat. Chem.* **1993**, *3*, 991–994.

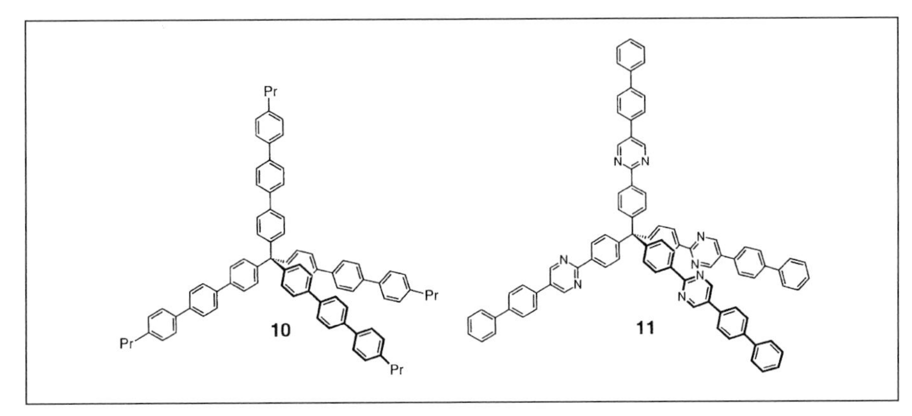

Fig. 4: The compounds with the largest number of phenyl rings as substituents.

Coupling together two benzene rings by at least two aliphatic chains leads into the realm of **cyclophanes**, where several record achievements are to be found (Fig. 5). For example, in the "superphane" **12**, synthesized in 10 steps by Boekelheide et al.,[17] every corner of the two benzene rings is subject to connection through ethylene bridges. A crystal structure determination[18] showed that in the process the two rings become fixed at what is probably a record close approach of only 2.624 Å. (For comparison, in the related [2.2]paracyclophane the separation between the two ring surfaces is 3.093 Å.[19]) If the bridging elements are elongated by one methylene group to three carbon atoms, what results is structure **13**, in which for steric reasons the central positions of the bridging units are articulated as in a molecular paddle wheel.[20] A structurally related supercyclophane based on the cyclopentadienyl anion is the so-called superferrocenophane **14**, prepared by Hisatome and coworkers,[21] in whose cage-like center an iron atom is encapsulated. But the story is far from complete with the stapling together of only two benzene units. The Japanese team led by Misumi[22] managed to synthesize **15**, a six-fold stacked cyclophane.

[16] O. Freundel, dissertation, Ludwig-Maximilians-Universität, Munich, **1996**.
[17] Y. Sekine, M. Brown, V. Boekelheide, *J. Am. Chem. Soc.* **1979**, *101*, 3126–3127.
[18] Y. Sekine, V. Boekelheide, *J. Am. Chem. Soc.* **1981**, *103*, 1777–1785.
[19] F. Vögtle, *Reizvolle Moleküle der Organischen Chemie*, Teubner, Stuttgart, **1989**, p. 255.
[20] Y. Sakamoto, N. Miyoshi, T. Shinmyozi, *Angew. Chem.* **1996**, *108*, 585–586; *Angew. Chem. Int. Ed. Engl.* **1996**, *35*, 549–550. Y. Sakamoto, N. Miyoshi, M. Hirakida, S. Kusumoto, H. Kawase, J. M. Rudzinski, T. Shinmyozu, *J. Am. Chem. Soc.* **1996**, *118*, 12 267–12 275.
[21] M. Hisatome, J. Watanabe, K. Yamakawa, Y. Iitaka, *J. Am. Chem. Soc.* **1986**, *108*, 1333–1334.
[22] T. Otsubo, S. Mizogami, I. Otsubo, Z. Tozuka, A. Sakagami, Y. Sakata, S. Misomi, *Bull Chem. Soc. Jpn.* **1973**, *46*, 3519–3530.

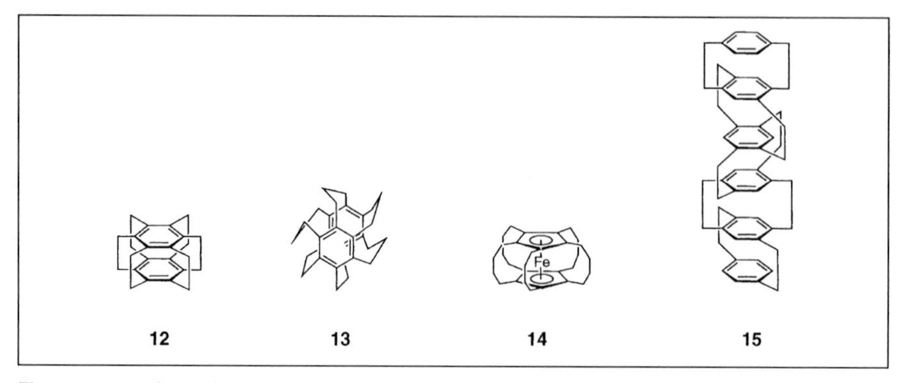

Fig. 5: Record-worthy cyclophanes

Probable Record Three-Membered Ring Systems

Three-membered ring systems have repeatedly challenged chemists to design intriguing new molecular structures. One of the greatest challenges in this area was for a long time the preparation of the highly strained molecule (→ Molecular Energy, Strain) [1.1.1]propellane (**16**, Fig. 6), first accomplished in 1982 by Wiberg[23] via a multistep process. In the meantime the technique has been improved to such an extent, especially by work emanating from the research group of Szeimies,[24] that not only the chemistry of this substance but also the properties of various derivatives have been studied. In contrast to the joined sides of three-membered rings characteristic of propellanes, ring systems connected at the vertices – so-called spiro compounds – also lend themselves to synthesis. The masterpiece in this arena has come from the team of de Meijere,[25] which assembled the perspirocyclopropanated [3]rotane **17**. Its structure can be envisioned as a fragment cut out of a carbon-atom network based on spirocyclopropane units, and despite the fact that its strain energy is ten times that of cubane, it shows surprising thermal stability. Thus, compound **17** melts without decomposition above 200 °C!

[23] K. B. Wiberg, F. H. Walker, *J. Am. Chem. Soc.* **1982**, *104*, 5239–5240.
[24] M. Werner, D. S. Stephenson, G. Szeimies, *Liebigs Ann.* **1996**, 1705–1715.
[25] S. J. Kozhushkov, T. Haumann, R. Boese, A. de Meijere, *Angew. Chem.* **1993**, *105*, 426–429; *Angew. Chem. Int. Ed. Engl.* **1993**, *32*, 401.

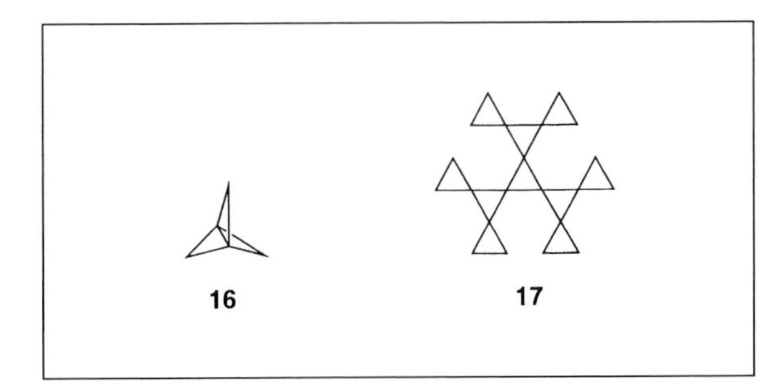

Fig. 6: Three-membered ring compounds

Olympic Connections

The molecule appropriately designated **olympiadane** (**18**, Fig. 7) opens entirely new dimensions in the search for systems with unusual "bonding." It was prepared by Stoddart's research group,[26] and consists of **five rings interlocked in the manner of the Olympic symbol**, a fitting reminder of the title of this book. The molecule was constructed according to the principle of self-organization in a remarkably straightforward two-step synthesis. Admittedly the symmetry of olympiadane is quite low based on a strict interpretation, but this example from supramolecular chemistry reveals once again that intriguing structures can spur chemists on to truly heroic accomplishments.

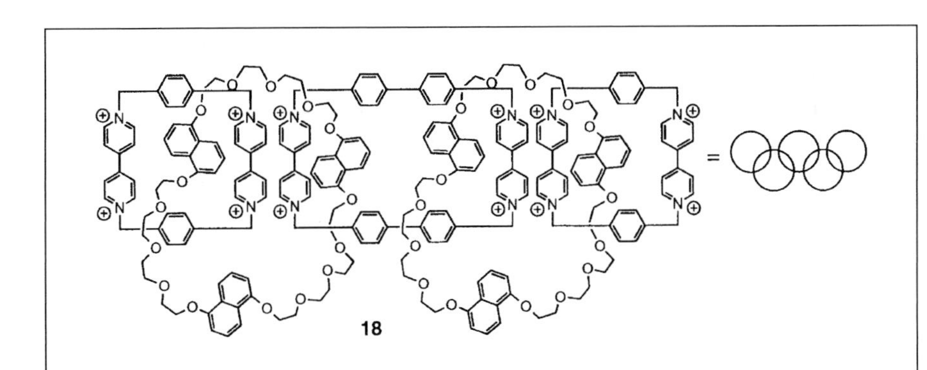

Fig. 7: Olympiadane

[26] D. B. Amabilino, P. R. Ashton, A. S. Reder, N. Spencer, J. F. Stoddart, *Angew. Chem.* **1994**, *106*, 1316–1319; *Angew. Chem. Int. Ed. Engl.* **1994**, *33*, 1286.

Unsymmetry

In the world of **inorganic chemistry** one is confronted with little evidence of any race to synthesize symmetric species. The number of highly symmetrical molecules here is so large – such as SF_6 (point group O_h), $W(NMe_2)_6$ (T_h), and SiF_4 (T_d) – that an **unexpected lack of symmetry** is more likely to lead to raised eyebrows. For instance, WF_6 has the highly symmetric octahedral structure[27] predicted by the simple VSEPR model, whereas WMe_6 takes the form of a very distorted trigonal prism.[28]

The Molecule Containing the Most Elements

The basic structure of a molecule is established by covalent bonds between its constituent atoms. In the simplest case, as in the **smallest molecule** hydrogen (H_2) or elemental sulfur (S_8), the bonds are between atoms of the same type. But even much more complex molecules such as proteins (→ Molecular Form, Giants) usually consist only of atoms of the five elements C, H, N, O, and S, perhaps joined by one or another metal atom.

In contrast to these very "simple" molecules (from the standpoint of composition), organometallic compounds often contain many more diverse elements. Compound **1**,[1] with the molecular formula $C_{30}H_{34}AuBClF_3N_6O_2P_2PtW$ with 11 (!) different types of atoms is the champion in this category. But despite its compositional variety, this compound is still not unique. The same publication[1] describes two additional, structurally related compounds with the molecular formulas $C_{30}H_{34}BClCuF_3N_6O_2P_2PtW$ and $C_{35}H_{34}AuBF_3MnN_6O_7P_2PtW$ that also include 11 different elements in their molecular makeup.

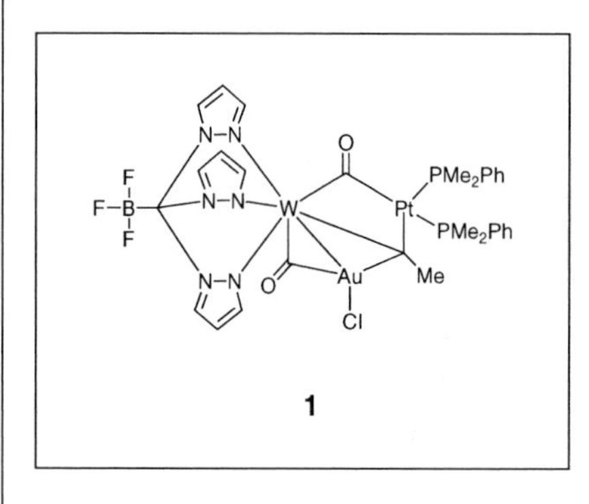

1

[1] P. K. Byers, F. G. A. Stone, *J. Chem. Soc. Dalton Trans.* **1991**, 93–100.

[27] N. N. Greenwood, A. Earnshaw, *Chemistry of the Elements*, Pergamon, Oxford, **1985**, p. 1493.
[28] V. Pfennig, K. Seppelt, *Science* **1996**, *271*, 626.

The History of the Nobel Prizes

Ever since 1910, the 10th of December has witnessed the awarding of Nobel Prizes in chemistry, physics, medicine/physiology, and (beginning in 1968) economics, as well as non-scientific prizes for literature and in the pursuit of peace, to distinguished laureates for outstanding accomplishments in their respective fields. These are indisputably the most prestigious recognitions that can be bestowed on a researcher in the course of his or her career.

Nobel Prizes are a legacy of the Swedish chemical engineer Alfred Nobel (1833–1896), who enjoyed extraordinary economic success thanks to his invention of a readily controllable and industrially accessible form of nitroglycerin, which he marketed under the trade name Dynamit (→ Molecular Energy, Explosives). In the course of his commercial activities Nobel founded ninety different factories and laboratories located in more than twenty countries, some of which (e.g., ICI) continue today to play an important role in the world market (→ Chemical Industry, Company Hits). Nobel's not inconsiderable estate provided the capital for a foundation bearing his name that manages prize funds distributed annually on the anniversary of his death (1996: 7.4 million Swedish krøner per Nobel Prize, equivalent to ca. $1.7 million). The actual selection of Nobel Laureates in the sciences is in the hands of an independent commission of experts from the Royal Swedish Academy of Science, which bases its work on nominations it receives as well as on the results of an intense scrutiny of the nominees and their activities. A complete list of prizewinners in chemistry, physics and medicine/physiology for the years 1901–1998 is provided in an appendix to this book, along with brief descriptions of the achievements for which the respective prizes were awarded.

Multiple Nobel-Prize Winners

Trying to identify "outstanding" characteristics associated with scholars already acknowledged as at the summit of their respective disciplines is a bit like carrying coals to Newcastle. Nevertheless, even among Nobel prizewinners there are a few superlatives worth calling to the reader's attention.

For example, of the 438 Nobel Laureates so far singled out in chemistry, physics, and medicine/physiology (as of 1998), only four are recipients of **multiple Nobel Prizes**. These four are **Marie Curie** (winner for chemistry in 1911 and physics in 1903), **John Bardeen** (physics in both 1972 and 1956), **Linus Pauling** (for peace in 1962 and chemistry in 1954), and **Frederick Sanger** (chemistry in both 1980 and 1958). No triple Nobel Laureates have yet been designated. Of these four, Pauling is the only one who was the *exclusive* recipient of both his honors, whereas Bardeen and Sanger are the only ones to be recognized twice in the same discipline. Mme. Curie is not only the lone woman with two Nobel Prizes, she is also by a wide margin the **most popular Nobel Laureate**. In the course of collecting a set of internet statistics for the Nobel

Prize internet archives (www.almaz.com/nobel) it was noted that Marie Curie's biographical data have been requested three times as often as those of the "next most popular" recipient, Albert Einstein (Nobel Laureate in physics, 1921).

The Youngest and Oldest Nobel Laureates

The fact that a summons to Stockholm can come at almost any stage in a scholar's career is quickly established by a scrutiny of the ages of the various Nobel Laureates. Thus, Sir **William Lawrence Bragg** (born in 1890, physics prize in 1915 jointly with his father, Sir William Henry Bragg) became the **youngest of all Nobel Laureates** at the early age of 25. Others have been forced to wait until their twilight years for the crowning of their life's work: **Pyotr Leonidovich Kapitsa** (born in 1894, physics prize in 1978), **Charles J. Pedersen** (born in 1904, chemistry prize in 1987), and **Georg Wittig** (born in 1897, chemistry prize in 1979) were the **oldest laureates**, all of them having passed the age of 80 before receiving their awards.

Dissolved Nobel Prize Medals

The great symbolic value of the gold medals conferred in conjunction with the Nobel Prize, as seen through the eyes of the prizewinners themselves, can be inferred from two separate events involving the Dane Niels Bohr[1] (Nobel Prize in physics, 1922).

During the stormy years of the Third Reich, two Nobel Laureates living at the time in Germany, Max von Laue (physics, 1914) and James Franck (physics, 1925), sent their precious medals for safe keeping to Niels Bohr in Copenhagen, a city assumed to be immune to any threat from the Nazis to confiscate the trophies for their valuable gold content (a value enhanced by the circumstances of war). When German troops also occupied Denmark in 1940, Bohr and his friend Georg von Hevesy (a physical chemist who had won the chemistry prize in 1943) decided to "protect" the sacred medals from capture by dissolving them in aqua regia (HCl/HNO_3 3:1). After the war the dissolved gold was recovered, and the Nobel Foundation was able to present von Laue and Frank with newly struck medals.

Bohr, like Schack August Steenberger Krogh (Nobel Prize in medicine, 1920), placed the proceeds from sale of his medal in January, 1940, at the disposal of the Finnish people, who had suffered tremendously during the so-called "Winter War" with Russia (November 1939 to March 1940).

---

[1] A. Pais, *Niels Bohr's Times*, Clarendon, Oxford **1991**, chap. 21.

Hit List of Countries

Global frontrunners can also be singled out from the list of the **native lands of Nobel prizewinners** in science. Nearly one-third of all awardees in the sciences since the inception of the prize were born in the United States. Only when winners from the three closest "runner-up" countries (Germany, Great Britain, and France) are combined does one reach a tally equal to that of the United States alone (Fig. 1). If the focus is directed not on the place of birth of the winners but rather on the countries in which they were living at the time of the award, then the **dominance by the United States** becomes even more apparent: 44% of the awardees. Based on this measure, Great Britain (16%) also moves ahead of Germany (13%) into second place.

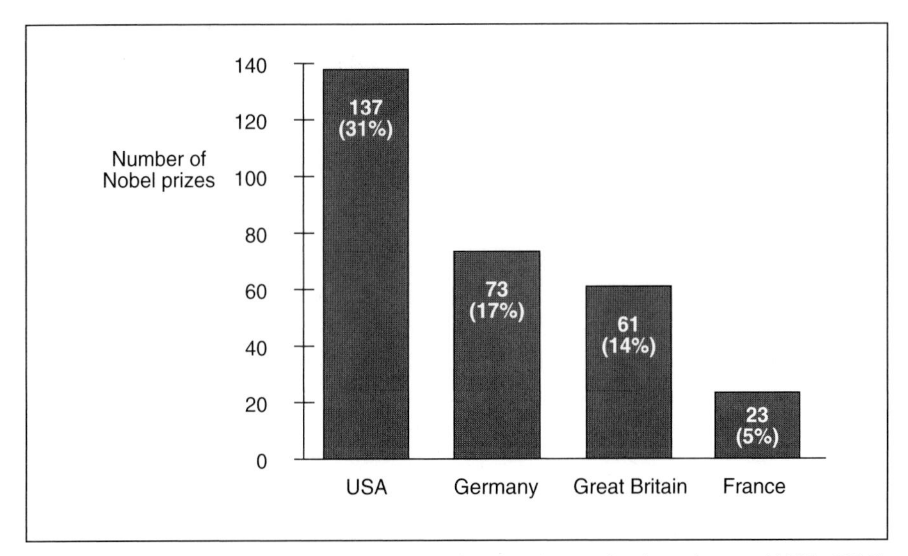

Fig. 1: The countries where the most Nobel prizewinners in the sciences (1901–1996) were born

Nobel Prizes in Chemistry: The USA Sailing Before the Wind

A survey of the number of Nobel Prizes in chemistry awarded to scholars from Germany, Great Britain, France, and the United States in each decade since 1900 reveals a clear trend. Early in this century the field of chemistry was dominated by the Germans. As late as 1950 no other country was responsible for as many Nobel Prizes in chemistry as Germany. In the subsequent decade, however, Great Britain and the United States moved to the forefront. Great Britain's high-altitude flight came to an end in the 1960s, but that of the United States continues unabated.

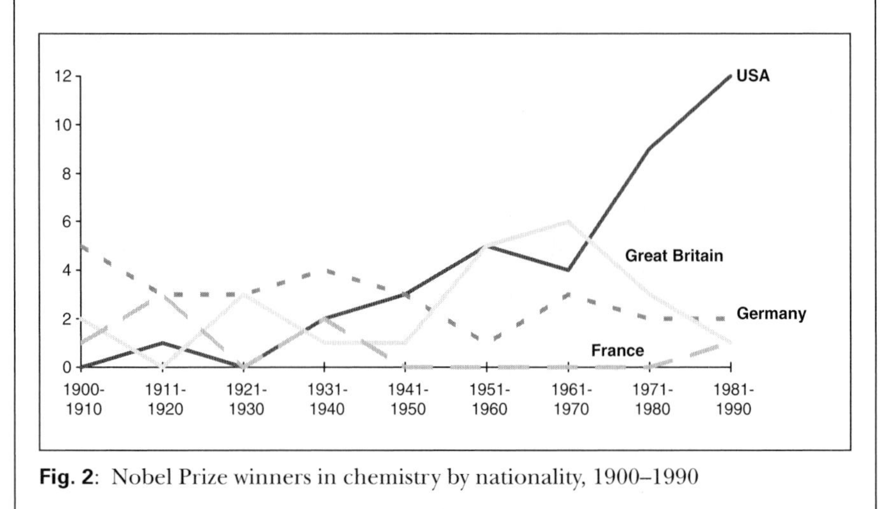

Fig. 2: Nobel Prize winners in chemistry by nationality, 1900–1990

---

[1] data from: *dtv-Atlas zur Chemie*, Vol. 2, **1994**, appendix.

The Earliest Patent

Patents are designed to provide protection for inventions, and they are issued upon application provided the applicant can document some specific technical innovation that holds the prospect of being exploited economically. The idea of protecting inventions reaches back to the late Middle Ages in Venice, where the **first patent law**, the "Parte Veneziana," was enacted in 1474.[1] Although various forms of imperial and princely privileges related to inventions are known from the early years of the German Empire, the second oldest codification of patent law is the English "Statute of Monopolies" of 1623/1624. The United States of America followed with an organized code of patents in 1790, shortly after that country won its independence from England.

The First Chemical Patent

The **first patent** in the United States for an invention in the field of chemistry, shown in Fig. 1, belonged to Samuel Hopkins and was issued on July 31, 1790. It provided legal protection for an improved process for the preparation of potash (K_2CO_3).[2]

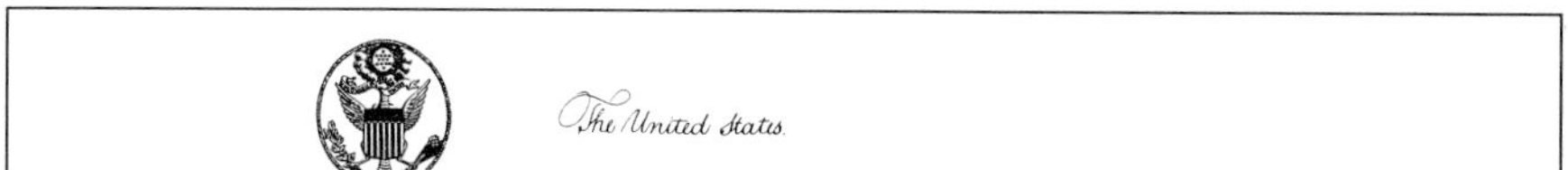
The United States.

To all to whom these Presents shall come. Greeting.

Whereas Samuel Hopkins of the City of Philadelphia and State of Pensylvania hath discovered an Improvement, not known or used before such Discovery, in the making of Pot ash and Pearl ash by a new Apparatus and Procefs; that is to say, in the making of Pearl ash 1st by burning the raw Ashes in a Furnace, 2d by difsolving and boiling them when so burnt in Water, 3d by drawing off and settling the Ley, and 4th by boiling the Ley into Salts which then are the true Pearl ash, and also in the making of Pot ash by fluxing the Pearl ash so made as aforesaid; which Operation of burning the raw Ashes in a Furnace, preparatory to their Difsolution and boiling in Water, is new, leaves little Residuum, and produces a much greater Quantity of Salt: These are therefore in pursuance of the Act, entituled "An Act to promote the Progrefs of useful Arts", to grant to the said Samuel Hopkins, his Heirs, Administrators and Afsigns, for the Term of fourteen Years, the sole and exclusive Right and Liberty of using, and vending to others the said Discovery, of burning the raw Ashes previous to their being difsolved and boiled in Water, according to the true Intent and Meaning of the Act aforesaid. In Testimony whereof I have caused these Letters to be made patent, and the Seal of the United States to be hereunto affixed. Given under my Hand at the City of New York this thirty first Day of July in the Year of our Lord one thousand seven hundred & Ninety.

G Washington

City of New York July 31st 1790. –

I do hereby Certify that the foregoing Letters patent were delivered to me in pursuance of the Act, entituled "An Act to promote the Progrefs of useful Arts"; that I have examined the same and find them conformable to the said Act.

Edm: Randolph Attorney General for the United States. –

(Endorsement on back of grant)

Delivered to the within named Samuel Hopkins this fourth day of August 1790.

Th: Jefferson

First United States Patent Grant
July 31, 1790
(Reproduced from the original in the collection of the Chicago Historical Society)

Fig. 1: The first chemical patent issued by the United States

[1] P. Kurz, *Mitteilungen der deutschen Patentanwälte* **1996**, *87* (3), 65–75.
[2] US-Patent X000001, United States Patent & Trademark Office, Washington, D. C.

The First German Patent

Considerable time passed before uniform German patent legislation was adopted. Only with the establishment of the Imperial Patent Office ("Kaiserliches Patentamt") in Berlin in 1877 were the nearly thirty different patent laws of the individual German principalities brought into conformity. The day the office opened – July 1, 1877 – Johannes Zeltner of the Nürnberger Ultramarinfabrik (Nuremberg Ultramarine Factory) submitted an application for a patent covering a "technique for the production of a red ultramarine dye." It was advertised on July 7, and the resulting Patent Number 1 took effect within three and one-half months (!) on November 29, 1877 (Fig. 2).[3] It is probably significant that the very **first German patent** was in the field of chemistry, and perhaps also no coincidence that this patent had to do with an invention in dye chemistry (→ Pigments, Records), which proved so important in the development of the chemical industry in Germany.

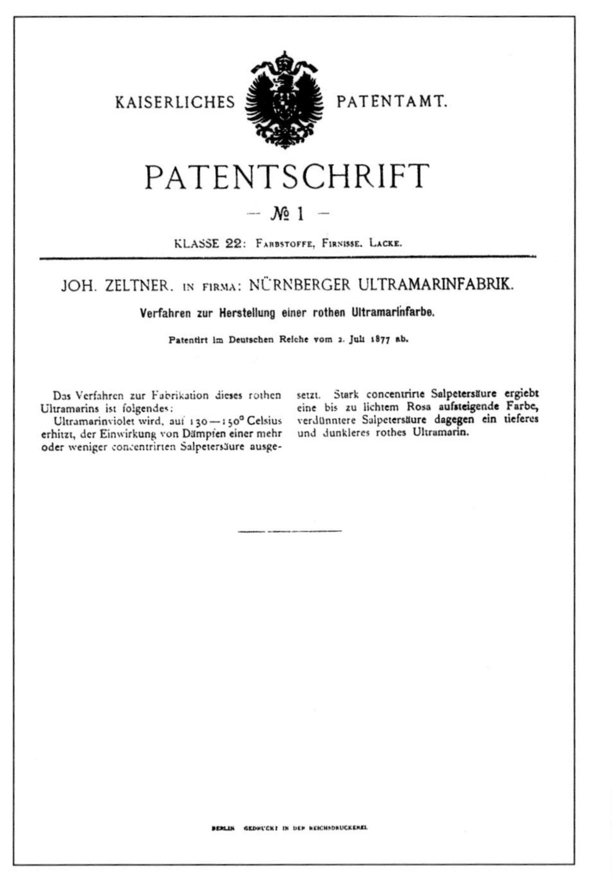

KAISERLICHES PATENTAMT.

PATENTSCHRIFT

— № 1 —

KLASSE 22: FARBSTOFFE, FIRNISSE, LACKE.

JOH. ZELTNER, IN FIRMA: NÜRNBERGER ULTRAMARINFABRIK.

Verfahren zur Herstellung einer rothen Ultramarinfarbe.

Patentirt im Deutschen Reiche vom 2. Juli 1877 ab.

Das Verfahren zur Fabrikation dieses rothen Ultramarins ist folgendes:

Ultramarinviolet wird, auf 130—150° Celsius erhitzt, der Einwirkung von Dämpfen einer mehr oder weniger concentrirten Salpetersäure ausgesetzt. Stark concentrirte Salpetersäure ergiebt eine bis zu lichtem Rosa aufsteigende Farbe, verdünntere Salpetersäure dagegen ein tieferes und dunkleres rothes Ultramarin.

BERLIN. GEDRUCKT IN DER REICHSDRUCKEREI.

Fig. 2: The first patent issued by the German patent office

[3] D. R. Schneider, *Mitteilungen des deutschen Patentanwälte* **1994**, *81* (10), 192–193.

National Parade of Hits

Since its origins, the patent literature has consistently grown at a rapid rate, in step with the increasing globalization of markets. At the end of 1994, over 3.9 million patents were active worldwide, covering a wide range of scientific and technical fields. More than 82% of these active patents originated in Japan, the United States, and the 17 countries making up the European Patent Convention (EPC; Austria, Belgium, Denmark, France, Germany, Greece, Ireland, Italy, Liechtenstein, Luxembourg, Monaco, Portugal, the Netherlands, Spain, Sweden, Switzerland, and the United Kingdom; Finland became the 18th member on March 1, 1996). From a trilateral comparison (Fig. 3)[4] it is apparent that considerably more patents are active in Europe (36.9%) than in either Japan (16.7%) or the USA (28.9%). Patents are at best an imprecise indication of the technical and scientific creativity of a nation.[5] The large majority of patents are economically inactive and fail to lead to third-party licensing agreements. For this reason, populist patent statistics designed to provide confirmation of the growth or decline of national progress make little sense. Somewhat more meaningful is a trilateral comparison of **mutual patent applications** within the three geographical blocks represented by the EPC countries, Japan, and the United States (Fig. 4).[1] Here one sees that, despite the often overworked claim of an "innovative advantage" in Japan, Europe still maintains a leadership position. Over 4000 more patents from Europe were registered in Japan in 1994 than Japanese patents in Europe. Still, one cannot overlook the quantitative imbalance evident in a comparison of Europe with the United States. In 1994 more than twice as many patents from the United States were registered in Europe as the reverse. From a purely numerical standpoint the United States was trumped by Japan, however, with 50% more Japanese applications filed in the United States than American applications in Japan.

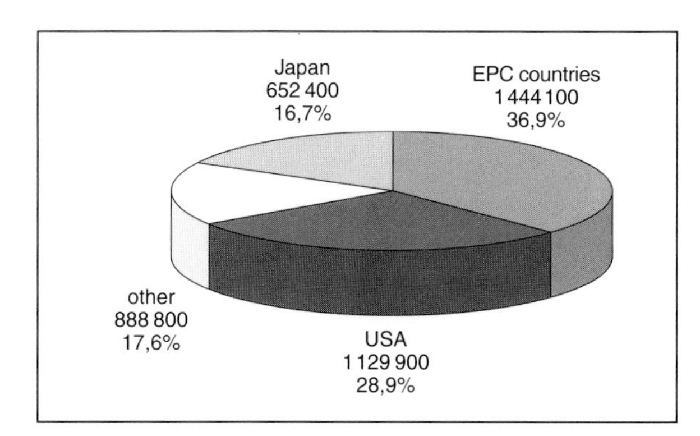

Fig. 3: Number of active patents, 1994

[4] *Trilateral Statistical Report, 1995*, European Patent Office, Munich.
[5] See for example W.-D. Wirth, *Nachr. Chem. Techn. Lab.* **1994**, *42*, 884.

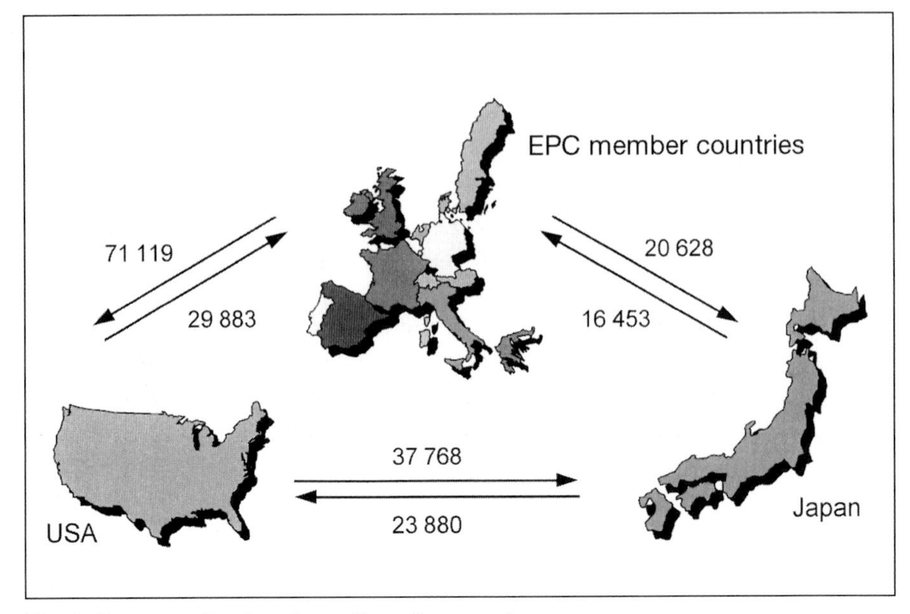

Fig. 4: Patent applications in a trilateral comparison

Corporate Hit Parade

Which company is the current **leader with respect to patent applications in the field of chemistry**? Figure 5 covers the time period 1986–1995 and provides a comparison of "chemical" patent applications filed by the chemical companies with the highest sales figures (→ Chemical Industry, Company Hits).[6] It is clear that the German "big three" maintain strong competitive positions behind the Japanese firm Mitsubishi Chemicals. The two American companies Du Pont and Dow are next, but some distance down the line.

The chemistry sector fares well in a comparison including German companies in other industrial branches as well (Table 1). Among the ten most active seekers of patents, four are chemical companies, the "big three" being joined this time by seventh-place Henkel.

[6] Derwent Patent Database study conducted in March, 1997; rights retained by BASF.

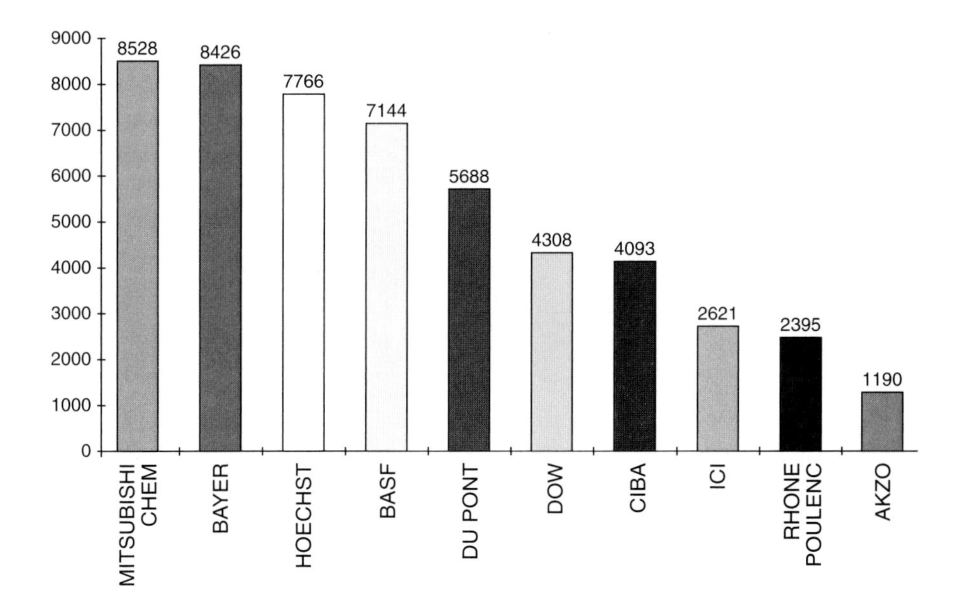

Fig. 5: Patent applications in the field of chemistry from selected companies, 1986–1995

Table 1: The ten most active German patent applicants

Rank	Company	Patents, 1991–1995
1	Siemens AG	4682
2	Robert Bosch GmbH	3830
3	BASF AG	3404
4	Bayer AG	2920
5	Daimler–Benz AG	2256
6	Hoechst AG	1895
7	Henkel KGaA	1550
8	BMW	1349
9	Volkswagen AG	962
10	Fraunhofer Gesellschaft	898

Source: German Patent Office, *Jahresberichte 1991–1995* (only since 1991 have patent applications been indexed according to the company filing the application).

Countries Responsible for the Most Patent Applications

In 1995, 60 078 patent applications were registered at the European Patent Office (EPA) in Munich (1994: 57 800). Of these, 49% originated in European countries. The **leading European country of origin** was Germany. Nevertheless, just as in the previous year, even more applications originated in the United States, which was the source of 29.3% of all the filings. Japan was in third place with a share of only 17%. Thus, now as in the past, Germany classes as one of the five most "inventive nations," although it is important once again to emphasize the limited predictive value of a purely numerical comparison (Fig. 6).

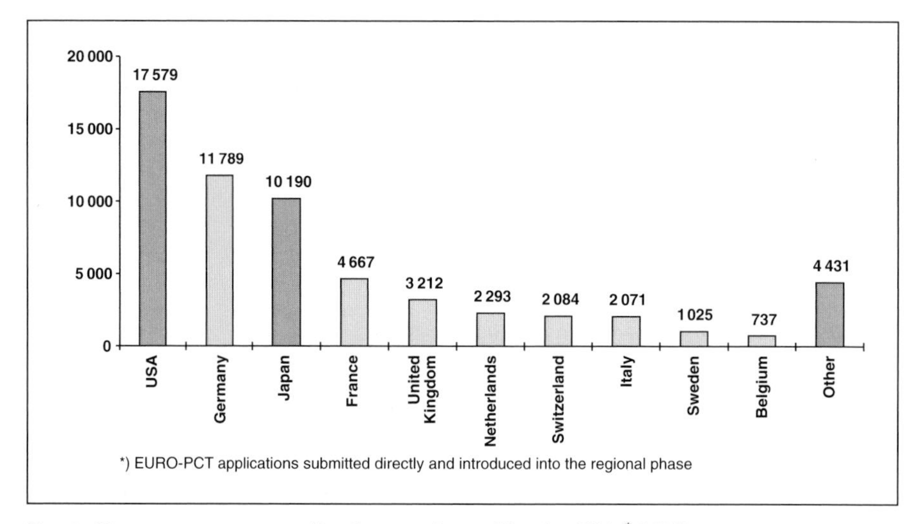

Fig. 6: European patent applications registered by the EPA,* 1995

Of the patent applications in question, nearly 60% were associated with just seven technical fields. Organic chemistry claimed 6.6%, 5.3% were related to health care, and 4.1% covered macromolecular organic compounds.[7]

[7] *Nachr. Chem. Tech. Lab.* **1996**, *44*(9): EPA.

The Largest Pharmaceutical Firms

If one takes market value (current price per share × number of shares outstanding) as a measure of the size of a corporation, then at the beginning of February, 1998, Merck & Co. could call itself **the largest pharmaceutical company in the world**. Its market value on that date amounted to an impressive $140.6 billion (Fig. 1).

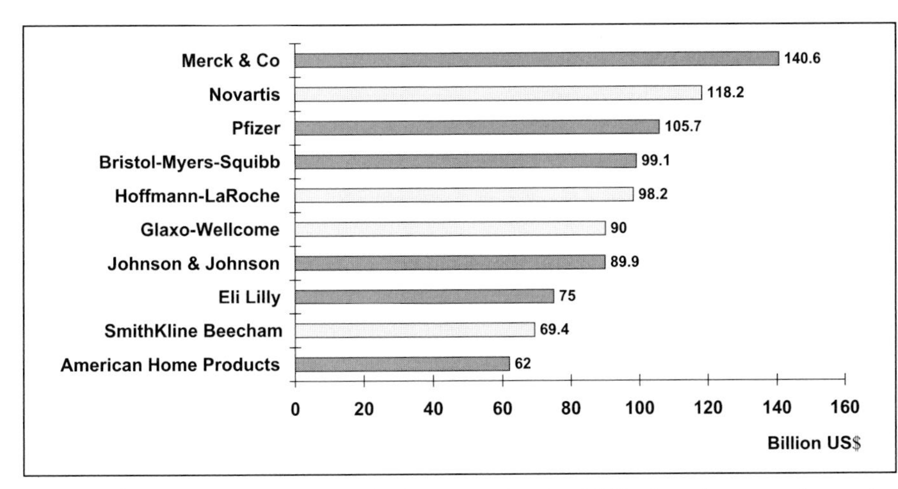

Fig. 1: The top ten pharmaceutical companies in 1998 (based on stock market valuations, February 1998)[1]

Novartis, the largest European pharmaceutical company, followed in second place. The total market value of the top ten pharmaceutical companies amounted to $948 billion on the date cited above, an increase of 76% since September 30, 1996.

In 1997 Merck was also the company with the highest level of pharmaceutical sales. The corresponding list of rankings from IMS Health includes once again almost all the "market-value top ten," albeit in somewhat altered sequence (Fig. 2). The only exception is Hoffman–LaRoche, which has been replaced in the sales list by Hoechst.

[1] data from: *Handelsblatt*, February 3, **1998**, p. 29.

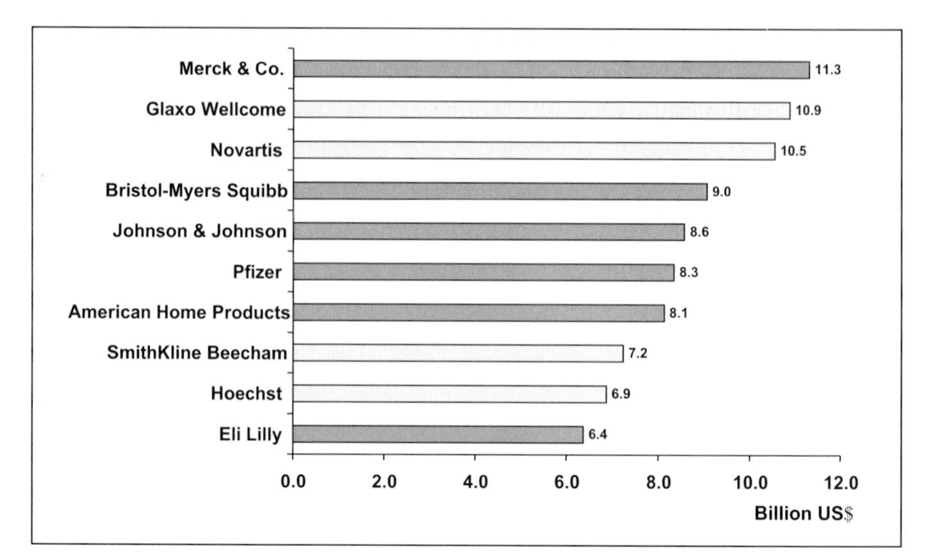

Fig. 2: The top ten companies in 1997 based on pharmaceutical sales[2]

The Largest Manufacturers of Generic Drugs

Copycat products based on unpatented active ingredients are playing an ever-increasing role in the pharmaceuticals business. Their share of the world drug market was only 12.3% as recently as 1994, but this is predicted to rise by the year 2000 to slightly above 20%. Roland Berger & Partners estimates that by then the market volume will exceed $30 billion. The driving force behind this anticipated high growth rate of 14% a year (in an overall market expected to grow by only 6% a year) is the necessity to reduce costs in the healthcare system coupled with increasing consolidation on the purchasing side and a corresponding increase in the market power wielded by the remaining purchasers.

Despite impressive growth rates, this market is not exerting an irresistible force of attraction on all pharmaceutical suppliers. On the list of important producers of generic drugs one searches in vain for such famous names as Glaxo–Wellcome, Hoffman–LaRoche, or Pfizer, who have so far shown virtually no interest in this side of the business.

[2] data from: *CHEManager* 10, **1998**, p. 9.

Leading the list of **largest suppliers of generics** in 1995 was Ivax, with a market share of 6.7%. Hoechst Marion Roussell occupied second place some way behind, followed by Ratiopharm (3.7%), another manufacturer dealing almost exclusively in generic drugs (Fig. 3). In 1995 the ten leading firms, with a combined volume of roughly $4.4 billion, probably controlled about 30% of the market.

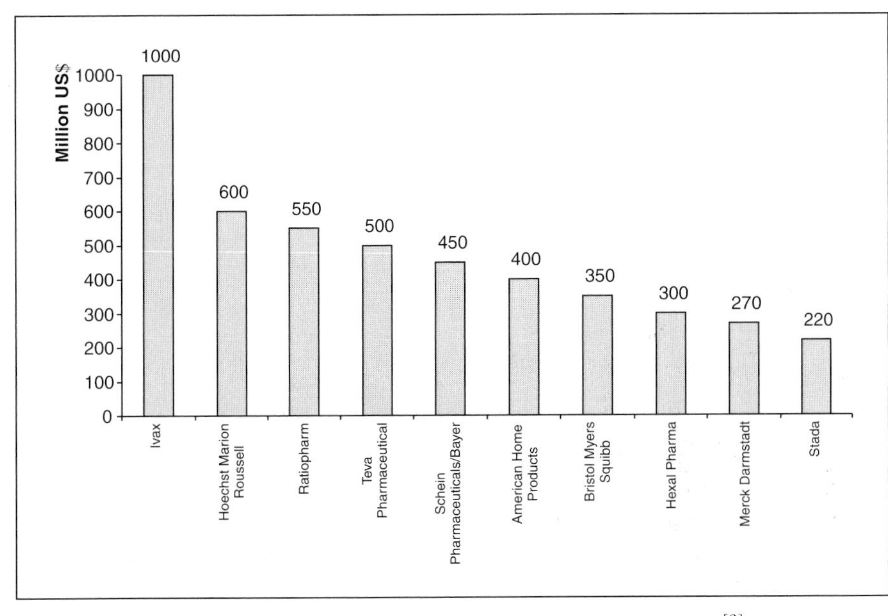

Fig. 3: Sales volumes for the leading suppliers of generic drugs (1995)[3]

[3] data from: *Handelsblatt*, September 13, **1996**, p. 28.

Healthcare Costs in Europe, Japan, and the United States

In this "triad" of regions, the Americans appear to "invest the greatest interest" in their healthcare. Thus, in 1995 the United States spent 14.5% of its gross domestic product for this purpose, which constitutes an annual increase of 3.1% based on 1985 (Fig. 5). Only the Spanish reported a similarly large annual rate of increase, 2.9%. Spain thus overtook Great Britain, but in 1995 still was near the end of the list of Western European countries. The lowest growth rates among the latter were registered by Italy and Germany (0.96% and 0.99%, respectively).

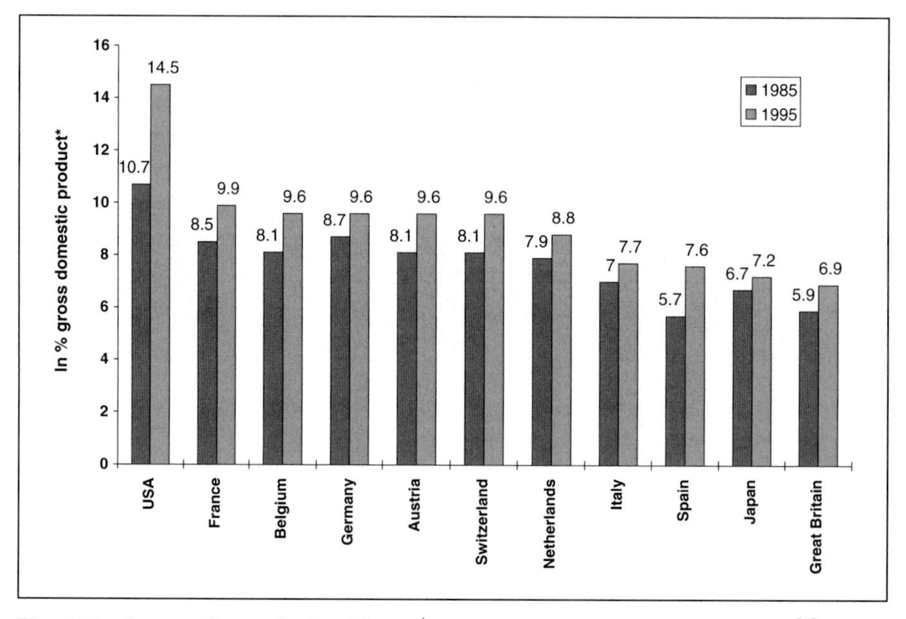

Fig. 4: Total expenditures for healthcare* as a percentage of economic output[4]

The rate of growth over the period 1985–1995 as a fraction of gross domestic product was even smaller for Japan: 0.72%. In 1995 the Japanese spent only 7.2% of their gross domestic product on healthcare, roughly half the relative outlay of the United States.

* (excluding such income benefits as sick pay or income protection)

[4] data from: Verband Forschender Arzneimittelhersteller, Statistics 97, p. 47.

Expenditures for Pharmaceuticals in the "Triad"

The cost of medication represents only a small fraction of overall healthcare costs. In Japan in 1995 this fraction amounted to 17%, significantly higher than that in Europe (Germany: 12.7%) and somewhat more than twice that in the United States. The order is thus the reverse of that for total healthcare costs (Fig. 6).

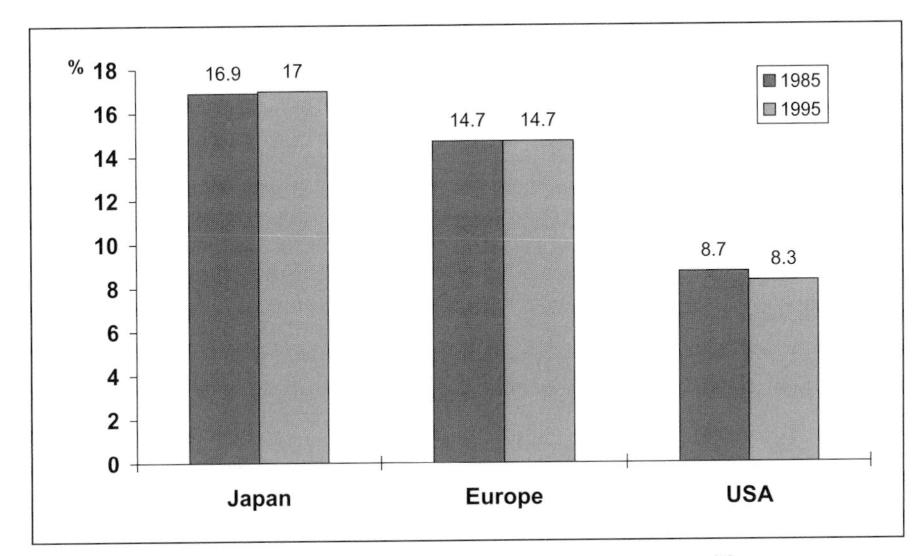

Fig. 5: Medication expenses as a fraction of overall healthcare costs.[5]

The share of the total healthcare expenses associated directly with medication changed very little between 1985 and 1995.

[5] data from: Verband Forschender Arzneimittelhersteller, Statistics 97, p. 48.

The World's Ten Largest Pharmaceutical Markets

The ten **most important pharmaceutical markets in the world** (in the sense of drug sales in pharmacies) generated a total volume in 1997 of about $165.54 billion.[1]

As would be expected, the single most important market (40.2%) was the United States, followed by Japan (25.2%) (Fig. 1). In third place was Germany, slightly ahead of France (8.9 and 8.3%, respectively).

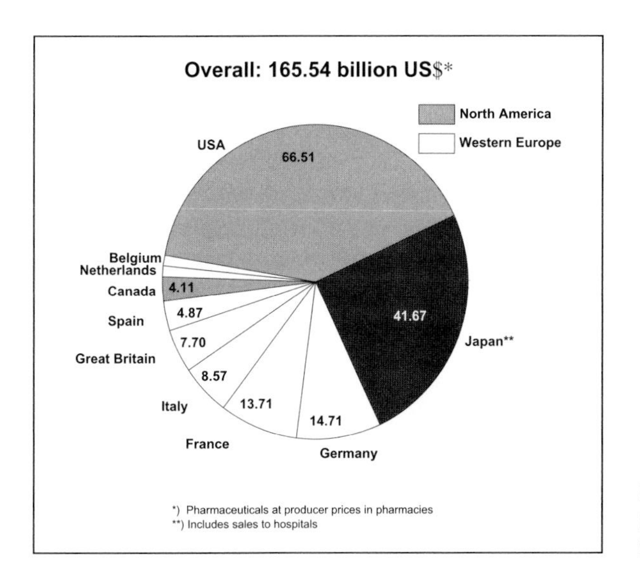

Fig. 1: The top ten pharmaceutical markets in 1997[1]

The average growth for 1996/1997 (disregarding fluctuations in exchange rates) was 6%. The U. S. market played an especially major role, with a double-digit growth rate (10%), just as in the preceding year.

The Major Growth Markets

During the period 1994–1997 the world pharmaceutical market grew steadily at an average annual rate of ca. 6.4% from $244 billion to $294 billion.

The **highest annual growth rates** of 33 and 30% were registered in North and South America. The region Africa/Asia/Australia, which increased greatly in significance in 1991–1995, surpassing first Europe and then in 1995 even North America, declined again in the two subsequent years (Fig. 2a, 2b).

[1] data from: Verband Forschender Arzneimittelhersteller, Statistics 98, p. 60.

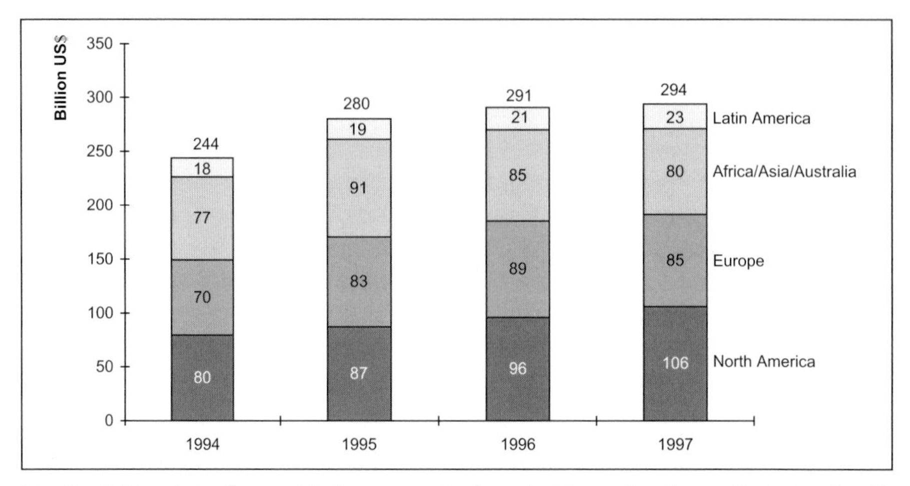

Fig. 2: a) Trends in the world pharmaceutical market by region (manufacturing level)

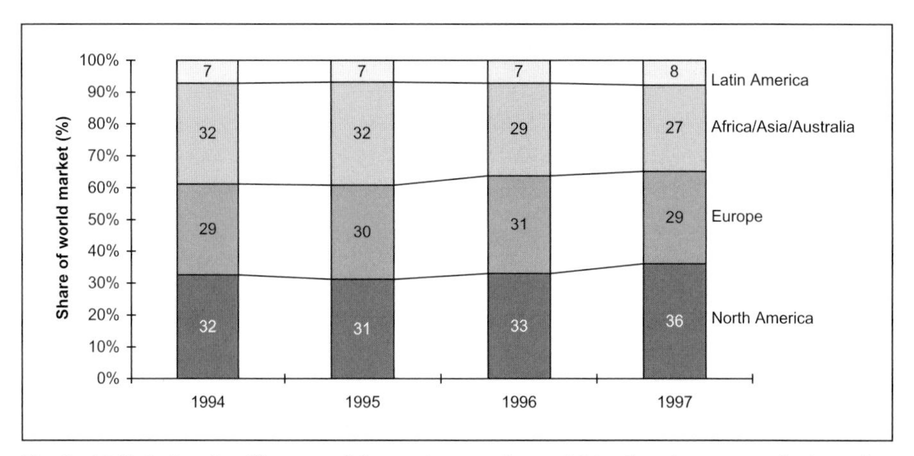

Fig. 2: b) Relative significance of the various regions within the pharmaceutical market

Export/Import Hit List

With a value of $10.92 billion, Germany was again the **world champion exporter of pharmaceuticals** in 1996 (excluding pharmaceutical raw materials). Its lead over second-place Switzerland amounted to $2.49 billion, larger than the total pharmaceutical export figure for Japan (Fig. 3).

On the other side of the coin, the United States, which had $7.49 billion in pharmaceutical exports, was in 1996 the largest pharmaceutical importer ($7.64 billion). Germany was in second place, so that the export champion achieved a surplus of $3.66 billion, enough to secure second place in the list of export surpluses (a list headed for years by the Swiss, with a growing lead).

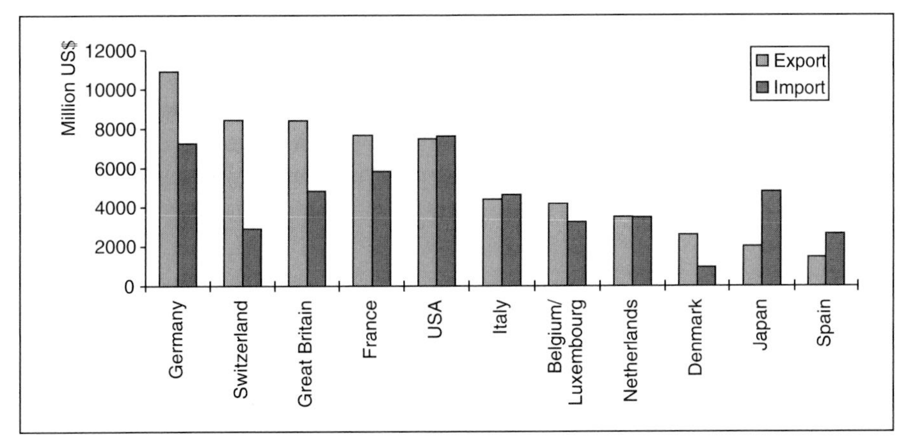

Fig. 3: Drug exports/imports of leading industrial nations, 1996 (excluding pharmaceutical raw materials) [2]

At the bottom of the latter list was Japan, with a striking negative balance (Fig. 4).

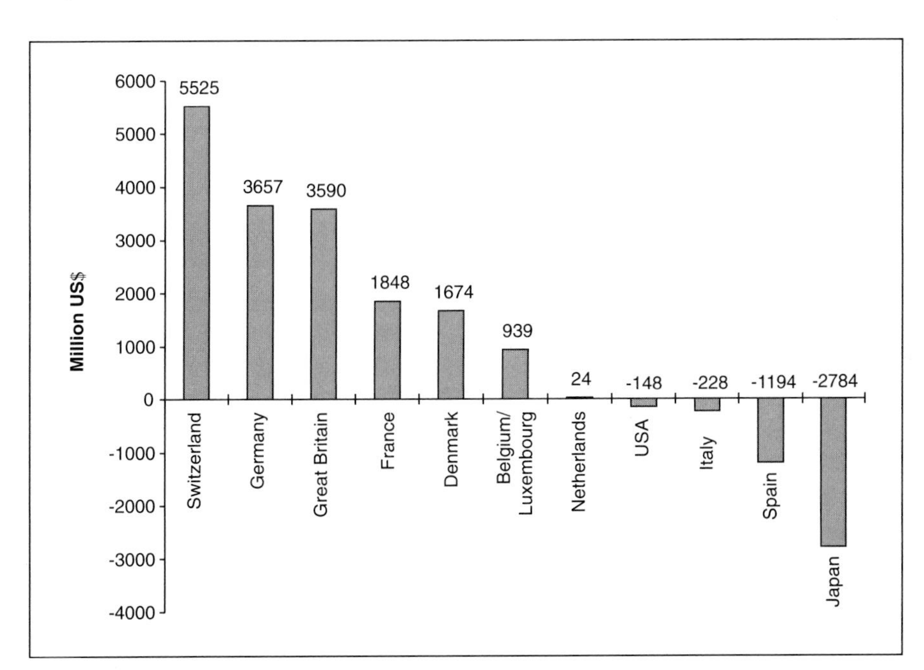

Fig. 4: Pharmaceutical export surpluses for leading industrial nations in 1996 (excluding pharmaceutical raw materials) [2]

[2] data from: Verband Forschender Arzneimittelhersteller, Statistics 98, p. 17.

The Leading Producing Region for Pharmaceuticals

Again in 1996 the EU was the **world champion producer of pharmaceuticals**, with $105.15 billion (Japan: $56.14 billion; USA in 1995: $91.04 billion). Of the total, 87.3% (i.e., $91.82 billion) could be ascribed to the five largest producing countries. Just as in 1995, this list of the top five was led by France, which was responsible for pharmaceuticals worth nearly twice as much as those from fifth-place Switzerland (Fig. 5).[3]

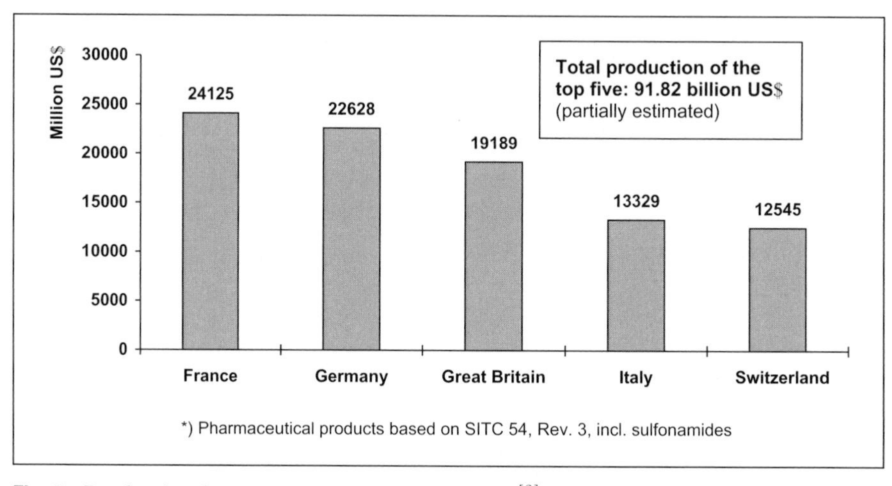

Fig. 5: Production by country within the EU, 1996[3]

The Major Suppliers of Pharmaceuticals to Germany

Germany draws a large fraction of its pharmaceutical imports from the very region to which it preferentially sends its exports: Western Europe. Only the USA and Japan were able again in 1995 to break into the phalanx of European countries (Fig. 6). The United States (1 238 million German marks) captured second place among the top ten **suppliers of pharmaceutical products to Germany**, although well behind the leader, Switzerland (2 434 million German marks), which has maintained that position for years (Fig. 7). France and Great Britain, surpassed for the first time by the USA in 1993, followed with figures of slightly more and slightly less than 1 billion German marks, respectively. Japan was the caboose on this hit-list train, with an export value of 422 million German marks.[3]

[3] data from: Verband Forschender Arzneimittelhersteller, Statistics 98, p. 10.

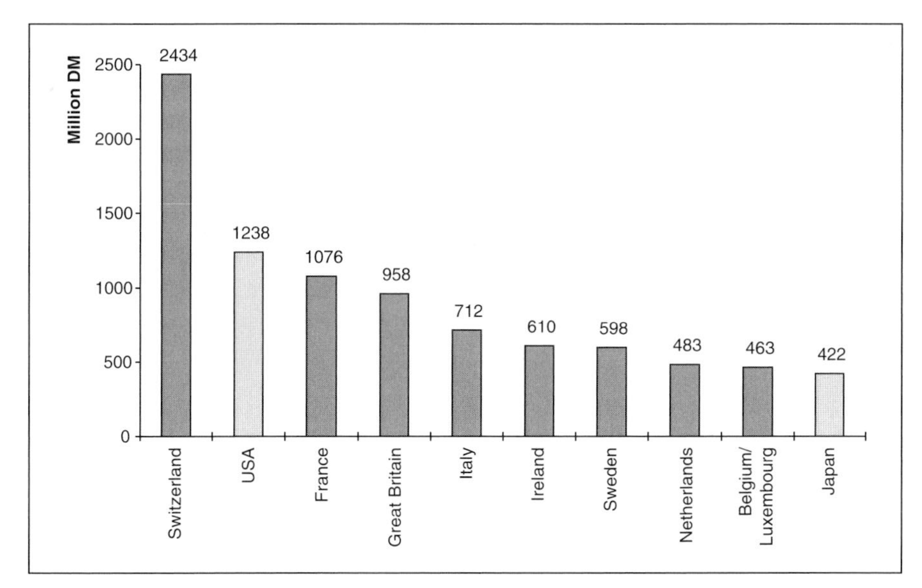

Fig. 6: Principal sources of pharmaceutical products for Germany, 1995[4]

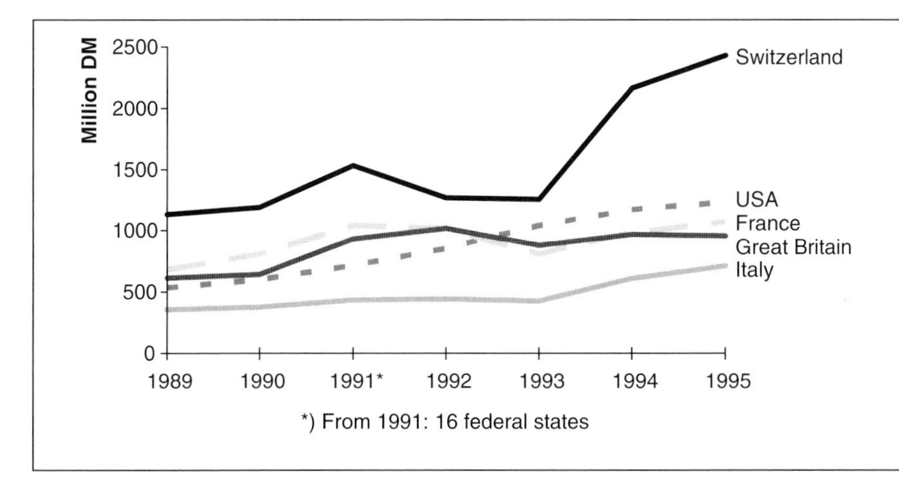

Fig. 7: Import trends from Germany's most important source countries[4]

[4] data from: Federal Association of the Pharmaceutical Industry e.V. (Germany), *Pharma Daten 1996.*

The Drugs with the Highest Sales Figures, 1997

Astra produced in 1997 the **drug with the year's highest sales figures**: Losec®. Merck followed in second place at a considerable distance with Zocor®, but was also able to place a second drug in the top ten list: Renitec®. Equally well represented were Pfizer and SmithKline Beecham.

Over the period 1996–1997 the product with the highest rate of growth was Seroxat®, ahead of Zocor® and Norvasc®. The greatest decrease in sales was suffered by Zantac®, which still led the list in 1995 (Table 1).

Table 1: Top ten drugs, 1997

Trade name®	Active ingredient	Company	Sales, 1997 (in $billion)	Indication
Losec	omeprazole	Astra	3.8	stomach ulcers
Zocor	simvastatin	Merck & Co	2.8	hypercholesterolemia
Prozac	fluoxetine	Lilly	2.4	depression
Zantac	ranitidine	Glaxo–Wellcome	2.2	stomach ulcers
Norvasc	amplodipin	Pfizer	2.0	high blood pressure
Renitec	enalapril	Merck & Co.	1.9	high blood pressure
Augmentin	amoxicillin	SmithKline Beecham	1.4	infections
Zoloft	sertralin	Pfizer	1.4	depression
Seroxat	paroxetin	SmithKline Beecham	1.4	depression
Ciproxin	ciprofloxacin	Bayer	1.4	bacterial infections

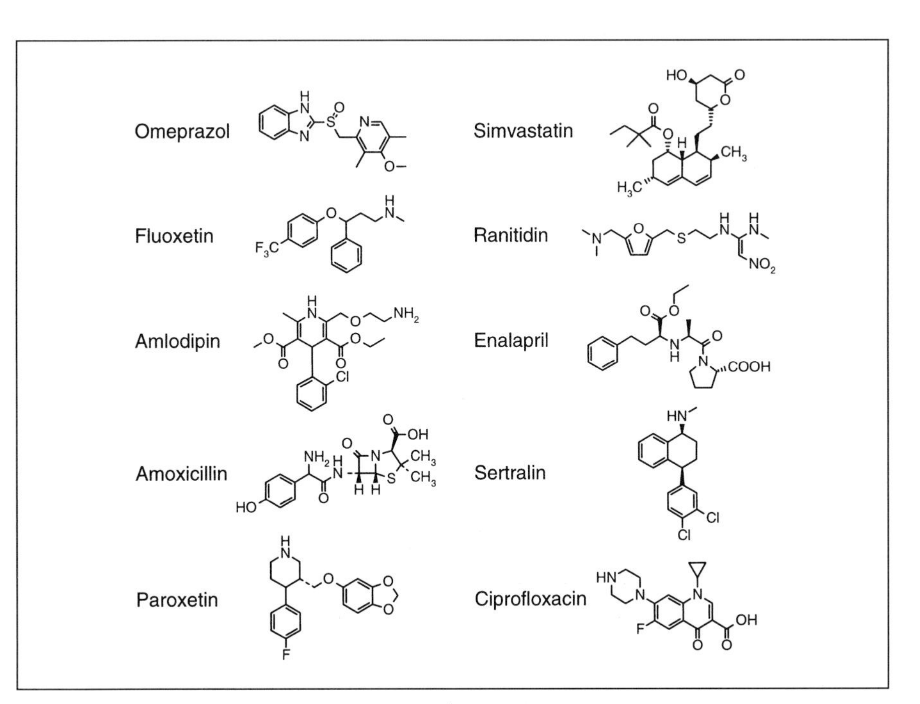

Fig. 1: Structures of the top ten drugs, 1997

Aspirin Breaks All Records

The best known and most frequently used medication in the world is Aspirin®. Headaches and toothaches, fever and colds, rheumatism, heart attack, stroke, and perhaps even certain types of cancer – the list of illnesses for which aspirin is taken gets longer and longer. Acetylsalicylic acid, Aspirin's active ingredient, was discovered by Dr. Felix Hoffmann at Farbenfabriken Bayer, and in 1997 it celebrated its 100th birthday.

The Most Important Groups of Drugs in Germany

The German pharmacy market in 1995 dispensed drugs with a total value of $16.5 billion, where the four **leading groups of pharmaceuticals** accounted for 62% of the volume. The top positions were held by medications for the treatment of "civilization diseases" and degenerative processes (Fig. 2), which are of course taken disproportionately by people over 50.

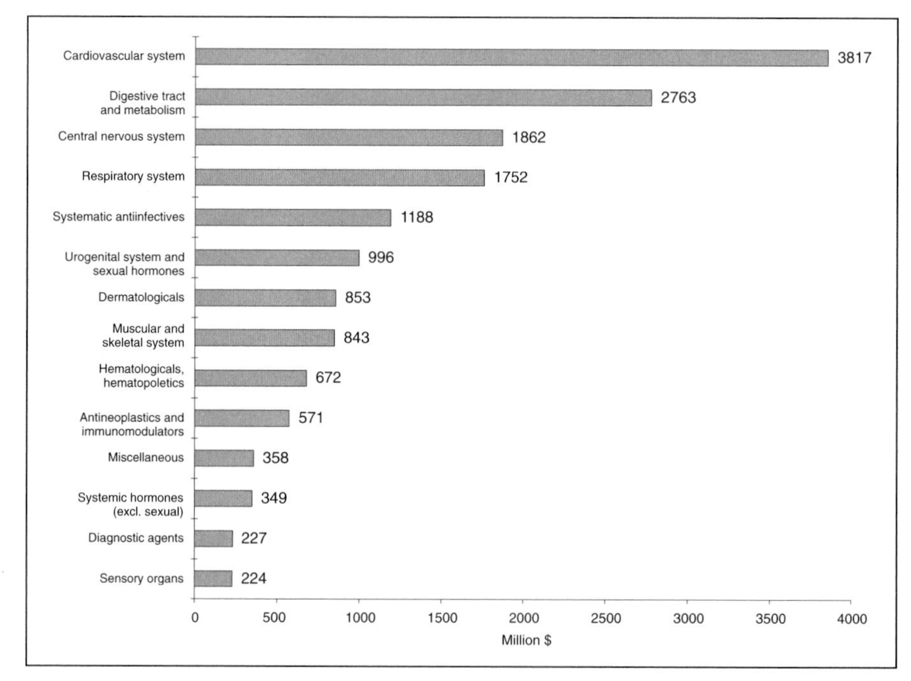

Fig. 2: Sales by pharmaceutical category, 1995 (producer prices)[1]

[1] data from: Federal Association of the Pharmaceutical Industry e.V. (Germany), *Pharma Daten 1996*, p. 54.

Tablet consumption in this age group is significantly higher than for 30- to 49-year-olds. Whereas 22% of the latter take pills daily or almost daily, the figure climbs to 43% for people 50–59 and to 66% for those over 60. Nevertheless, the same Emnid poll (1995) showed that 40% of the general population rarely takes pills, and 10% even claim never to do so.

Influence on Average Life Expectancy

The **average life expectancy** of a boy born today in Germany is estimated to be 72.8 years, that of a girl, 79.3 years. These are roughly double the life expectancies in 1880.

The dramatic increases are due in part to improvements in nutrition (→ Fertilizers) as well as living conditions and hygiene (→ Chemical Industry, Sector Records), but also to far better medical care for the population. The availability of more effective medications plays a major role in this context. Thus it has proven possible not only to suppress once dreaded infectious diseases, but also significantly to reduce the mortality rate with respect to diseases related to civilization and to aging (Fig. 3).

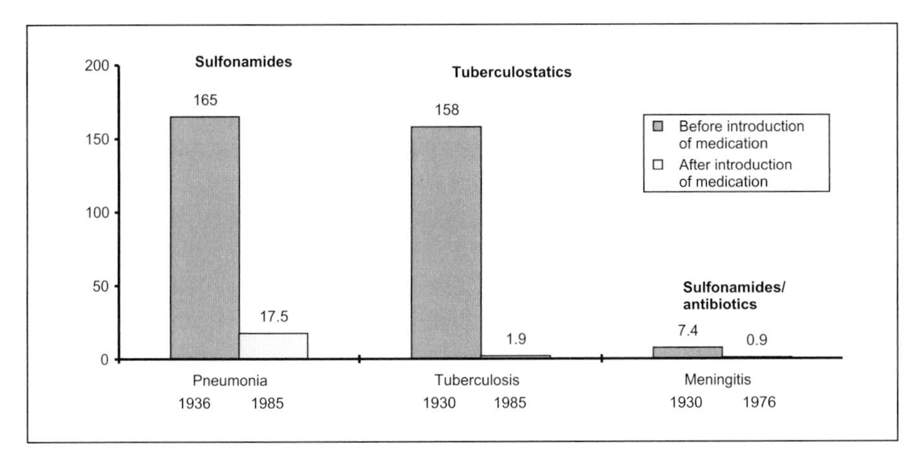

Fig. 3: a) Mortality rates per 100 000 population (Germany)[2]

[2] data from: Monograph series from the Chemical Industry Fund (Germany), Vol. 5, Federal Statistics Office.

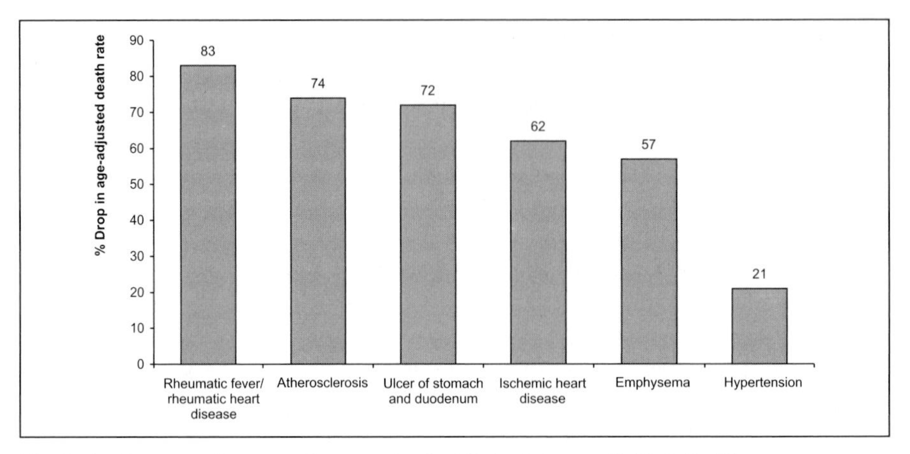

Fig. 3: b) Decrease in mortality rates in the United States, 1965–1996[3]

[3] data from: Pharmaceutical Research and Manufacturers of America (PhRMA), *Facts& Figures 1998, 8.*

Infectious Diseases of the Past

With the aid of modern drugs it is now possible to prevent or cure many diseases that in the past were fatal.

One can only speculate on the number of important works that might have been produced by the various artists listed in Table 2 if such medications had been available when they were still alive.[4]

Fortunately, not all important personages of the past died so young. Isaac Newton lived to be 74, for example, and Tizian was 87 when he died.

Table 2: Selected creative individuals of the past: causes and ages of death

Identity	Profession	Year of birth	Age at death	Cause of death
Masaccio	painter	1401	27	plague
Giorgione	painter	1477	33	plague
Raffael	painter	1483	37	sudden fever
W. A. Mozart	composer	1756	35	inflammatory fever
John Keats	poet	1795	26	tuberculosis
Heinrich Heine	poet	1797	59	tuberculosis
Franz Schubert	composer	1797	31	typhus
Robert Schumann	composer	1810	39	syphilis
Frederic Chopin	composer	1810	39	tuberculosis
Emily Brontë	author	1818	22	tuberculosis
Ann Brontë	author	1820	29	tuberculosis
Charles Baudelaire	author	1821	46	syphilis
Friedrich Nietzsche	poet, philosopher	1844	56	syphilis
Paul Gauguin	painter	1848	55	syphilis
Guy de Maupassant	author	1850	43	syphilis
Georges Seurat	painter	1859	31	throat infection
Hugo Wolf	composer	1860	43	syphilis
D. H. Lawrence	author	1885	45	tuberculosis
George Orwell	author	1903	47	tuberculosis

[4] Max F. Perutz, *Is Science Necessary?*, E. P. Dutton, New York, **1991**.

New Active Agents

Over the time period 1975–1997, 1163 **new active pharmaceutical agents** (new chemical entities, NCEs) were introduced into the marketplace. The largest share of these successes can be credited to European firms, the source of nearly half the newly developed compounds (Fig. 4). Does it follow that Europe is by far the most important breeding ground for innovation in the pharmaceutical industry?

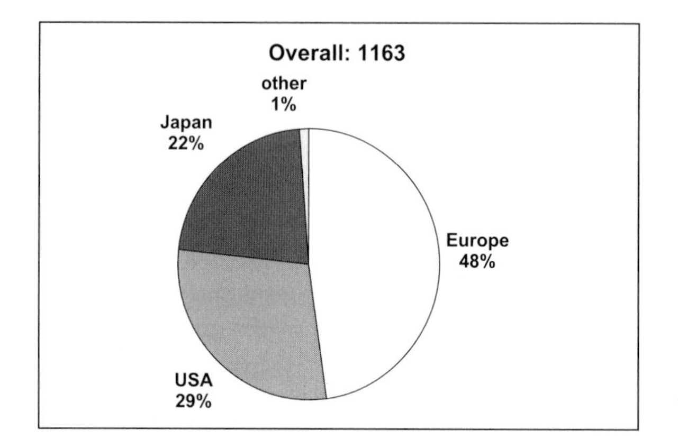

Fig. 4: Regional origins of new pharmaceutical agents (NCEs), 1975–1997[5]

The question can no longer be answered by an unqualified "yes." A careful look at historical trends (Fig. 5) shows that in the recent past Europe was indeed still the leading region, but the United States and especially Japan have taken major strides forward. One can no longer point to a market dominance like that in the second half of the 1970s, when 60.3% of the newly developed NCEs came from Europe (USA: 26.7%; Japan: 11.3%). Corresponding figures for the 1990s are: Europe, 39.1%; USA, 33.0%; and Japan, 26.1%.

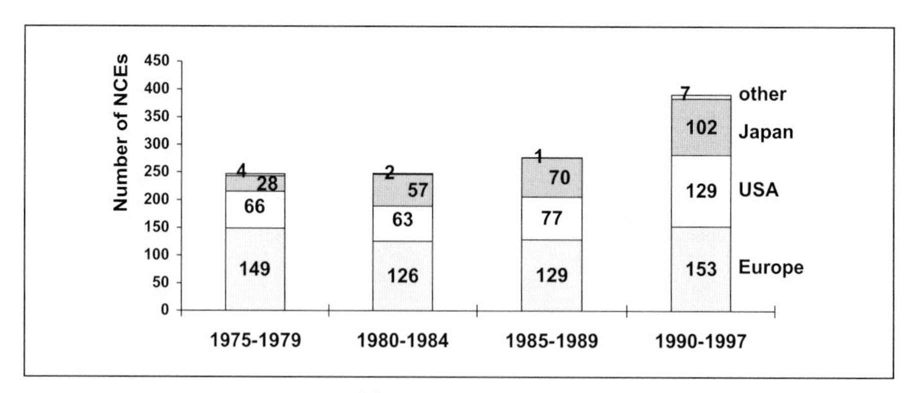

Fig. 5: Historical trends for NCEs[5]

[5] data from: Pharmaceutical Research and Manufacturers Association (Germany), *Statistics 96, 98,* and calculations by the authors.

The Smallest Pharmaceutical Agent

In general one suspects that an effective synthetic active agent will consist of a complex array of atoms custom-tailored to perform a particular task and thus act in a highly specific way. This is indeed often the case, but certainly not always, as illustrated by lithium. Lithium is administered in the form of one of its salts – such as the carbonate (Hypnorex®), the acetate (Quilonum®), or the sulfate (Duriles®) – as a psychotic agent for the treatment of manic depression. The mechanism by which Li^+ acts has not yet been fully clarified, but presumably it is related to the cation's exceptionally small size. The effective ionic radius of Li^+ is only 76 pm (= 76×10^{-12} m), which gives this ion the right to claim first place as the smallest active pharmaceutical agent.

Active Agents Under Development

The World Health Organization (WHO), a unit of the United Nations, has concluded that worldwide there are ca. 30 000 discrete diseases. So far, only about one-third of these can be cured by medication. Every disease is subjected to some sort of treatment, but there remain a great many therapeutic gaps. Convincing therapeutic strategies are lacking precisely in the cases of such serious and widespread diseases as cancer, rheumatoid conditions, viral infections, and central nervous system complaints. The pharmaceutical industry is striving worldwide to record further advances (Fig. 6).

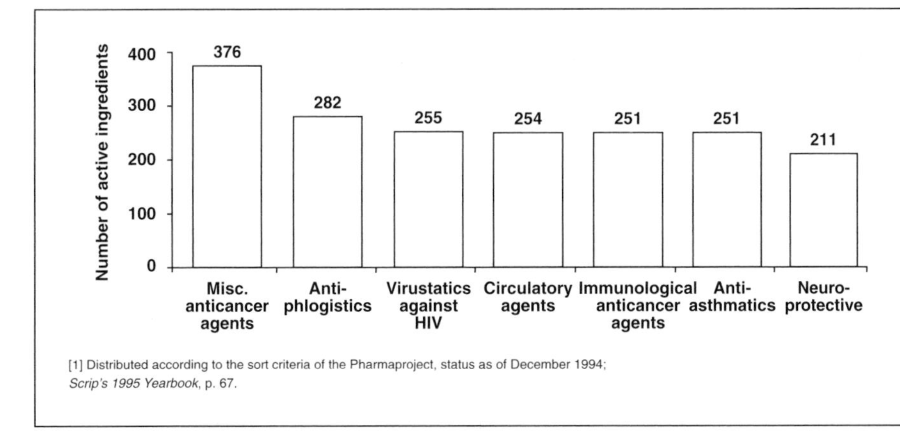

Fig. 6: Over 5500 active drug ingredients are currently under development worldwide. Important therapeutic sectors are the subject of active research[1]

Pharmaceuticals Hit List

Expenditures for Pharmaceutical R&D at a Record Level

Research and development play a large and increasing role in the pharmaceutical industry. For this reason, expenditures for such activities – relative to sales – are significantly higher for the innovative pharmaceutical manufacturers than for purely chemical firms. In general they represent a share of between 15 and 18%.

Research expenditures worldwide in the pharmaceutical industry have climbed recently from $19.5 billion in 1990 to $35.2 billion in 1996, an increase of 81% (Fig. 7).

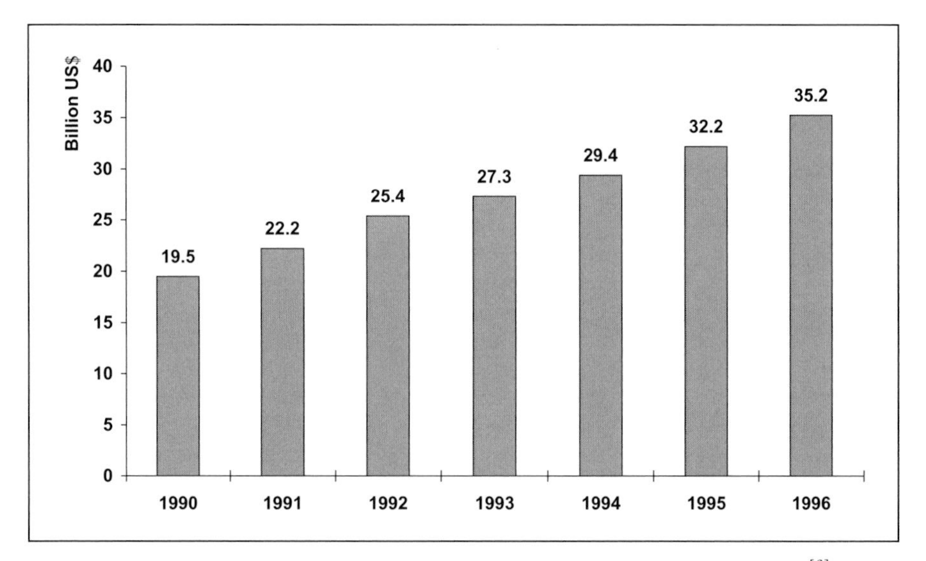

Fig. 7: Historical course of R&D expenditures in the pharmaceutical industry[6]

In order to increase the effectiveness and efficiency of pharmaceutical research, use is increasingly being made of the most modern techniques, such as high-efficiency screening (with which 100 000 compounds per day can be tested for activity) and combinatorial chemistry. Very high expectations are associated with decoding of the human genome, which is expected to provide new points of attack for pharmaceutical agents. So far only ca. 400 so-called "pharmaceutical genes" are known, and for this reason many drug companies have recently entered into alliances with genetic engineering firms. The most comprehensive cooperative agreement[7] so far developed is that of September 1998 between Bayer and Millennium Pharmaceuticals, one of the leading

[6] data from: *Handelsblatt*, January 9, **1998**, p. 18.
[7] Press release, Bayer, September **1998**.

genetic research companies in the world. For $465 million – including a capital participation of 14% – Bayer acquired access to the genetic research activities of Millenium. The hope is that more than 200 new genes will result in the next 5 years.

Pharmaceutical Research is Tiresome and Expensive

- Only about 1 of every 10 000 investigated substances ever reaches the market in the form of a drug
- Drug development requires 8–15 years
- The associated costs amount to about $300–500 million

The Oldest Pigment

The world's perhaps oldest pigment came into use shortly after the discovery of fire: **Soot** has been employed as a coloring agent since about 20 000 B.C., and is there to be admired in impressive stone-age cave paintings, including those at Lascaux (France) where it was used in conjunction with reddish-brown tones.[1] Even today soot is the basis for black printing inks, as one recognizes from traces left on the fingers after reading the daily newspaper.

Indigo: The "Bluest" Dye

The color blue has played a special role in the history of dyes since the most ancient of times. This ubiquitous color (the sky, water) proved very difficult to reproduce, and was long regarded as a great luxury ("royal blue"). The first use of blue mineral pigments as a way of achieving artistic effects can be dated before 3000 B.C. thanks to findings from excavations of the Sumerian city Ur in Chaldea. Here were unearthed animal figures of gold and silver decorated with blue *lapis lazuli*, a semiprecious stone that is the source of the pigment **ultramarine**.[2] The molecular basis of the blue color of *lapis lazuli* first became intelligible with the help of modern science, which showed its origin to lie in the radical anion $S_3^{\cdot -}$, stabilized here by an aluminosilicate lattice (→ Reactive Intermediates, Radicals).

Yet another blue colorant has also made history, this time an organic one.[3] **Indigo** (**1**) began its triumphal journey around the world from the Indian subcontinent sometime before 2000 B.C. A colorless precursor of indigo (3-hydroxyindole) is present there in plants of the genus *Indigofera*, and in Europe the compound is found in the plant *Isatis tinctoria*, commonly known as "dyer's woad." The dye itself is obtained in a tedious manual process by fermentation of the appropriate vegetable matter with urine (!) and subsequent oxidation in sunlight and air. From these humble beginnings there developed a successful handicraft tradition of dying, albeit one constantly beset with problems of variability in the quality of the dyes themselves and of the colors produced in the dyed fabrics. As chemistry gradually matured into a science, 19th century chemists began to take an interest in the study of indigo, especially Adolf von Baeyer (→ Molecular Energy, Strain). In 1883 (about 18 years after he began his investigations and ca. 5000 years after indigo's discovery), Baeyer succeeded in deducing the dye's structure. On March 19, 1880, he patented (→ Patents) a technically feasible synthesis for indigo – though with little practical potential on an industrial scale – starting from phenylacetic acid. BASF found a way to prepare commercial quantities of indigo in 1887 starting from the inexpensive

[1] G. Pfaff, *Chem. Unserer Zeit* **1997**, *31*, 6–16.
[2] F. Seel, G. Schäfer, H.-J. Güttler, G. Simon, *Chem. Unserer Zeit* **1974**, *8*, 65–71.
[3] M. Seefelder, *Indigo – Kultur, Wissenschaft und Technik*, Ecomed, Landsberg, 2nd ed. **1994**.

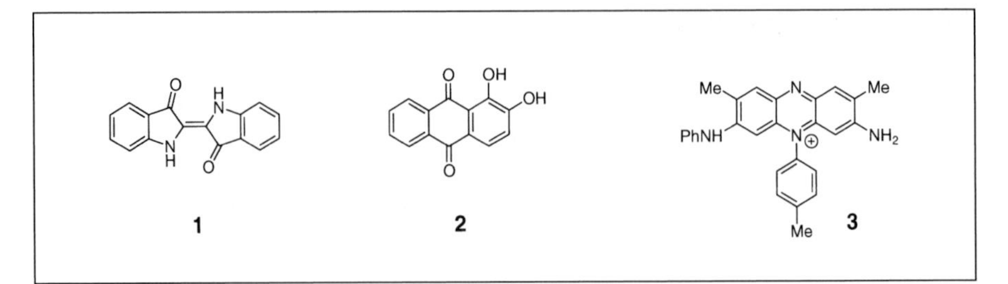

Fig. 1: Historically significant dyes

compound phenylglycine and based on a procedure devised by the Swiss chemist Karl Heumann. The Ludwigshafen entrepreneurs invested far more money in the process than their entire company was at that time worth – a record that will almost certainly never be, equaled. This courageous decision proved to be a wise one, and it still would be, given the legion of jeans wearers in these closing years of the 20th century.

The First Synthetic Dye

Several other records were achieved along the path leading to synthetic indigo. Thus, Graebe and Liebermann, two students of Adolf von Baeyer, elucidated the structure of alizarin (**2**), the red dye of the madder root.[3, 4] On the basis of a production method they developed, Heinrich Caro at BASF perfected the **first industrial synthesis of a natural product**. Nine years earlier, **production of the first synthetic dye** was accomplished by an 18-year-old Englishman, William Perkin.[3,4] His oxidation of aniline to mauveine (**3**) marked the beginning of the era of the coal-tar dyes.

[4] J. R. Partington, *A Short History of Chemistry*, Dover, New York, **1989**, reprint of the 3rd ed. **1957**, p. 318.

Absorption Records

The interaction of light with a chromophore in a dye molecule depends sometimes to a dramatic extent on the medium in which the measurement is conducted, and in some cases also on that to which the dye is applied. Solvents can have a significant influence on the **absorption maxima** of organic compounds; the phenomenon of changes in dye absorption in various solvents is called solvatochromism.[5] The frontrunner in this discipline is the pyridinium compound **4** (Fig. 2), which has its absorption maximum in water at 452.9 nm but in diphenyl ether at 809.7 nm. The corresponding difference is an impressive 356.8 nm, greater than the wavelength span of the visible region of the spectrum (violet, 420 nm; red, 700 nm). "Reverse" solvatochromism is also known. Compound **5** absorbs in nonpolar solvents at a lower wavelength (462.1 nm, hexane) than it does in polar solvents (597.0 nm, formamide/H_2O).

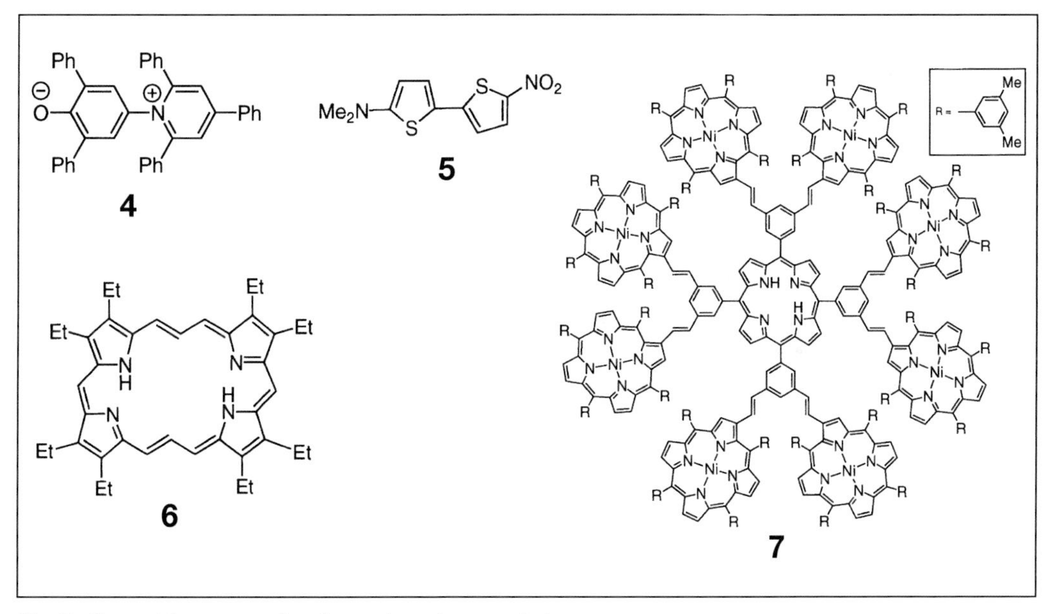

Fig. 2: Dyes with noteworthy absorption characteristics

Apart from its absorption maximum, the extinction coefficient also plays a major role in the intensity of a dye, albeit not an absolute one. For a long time, octaethyl-[22]porphyrin **6**[6] was the most intensely absorbing compound known, with an extinction coefficient ε = 1 120 000 $M^{-1}cm^{-1}$ [λ_{max} ($CHCl_3$/1% TFA) = 460 nm]. Recently, however, this compound was knocked from its

[5] C. Reichardt, *Chem. Rev.* **1994**, *94*, 2319–2358.
[6] B. Franck, A. Nonn, *Angew. Chem.* **1995**, *107*, 1941–1957; *Angew. Chem. Int. Ed. Engl.* **1995**, *34*, 1795–1811.

pedestal by the porphyrin nonamer **7**.[7] The latter not only broke the record for highest extinction coefficient (ε = 1 150 000 $M^{-1}cm^{-1}$, λ_{max} = 620 nm), but is also the **(nonpolymeric) compound with the largest number of π electrons.** This feat is achieved in **7** with 8 double bonds, 36 benzene rings, and 9 porphyrin units, for a total of 430 π electrons!

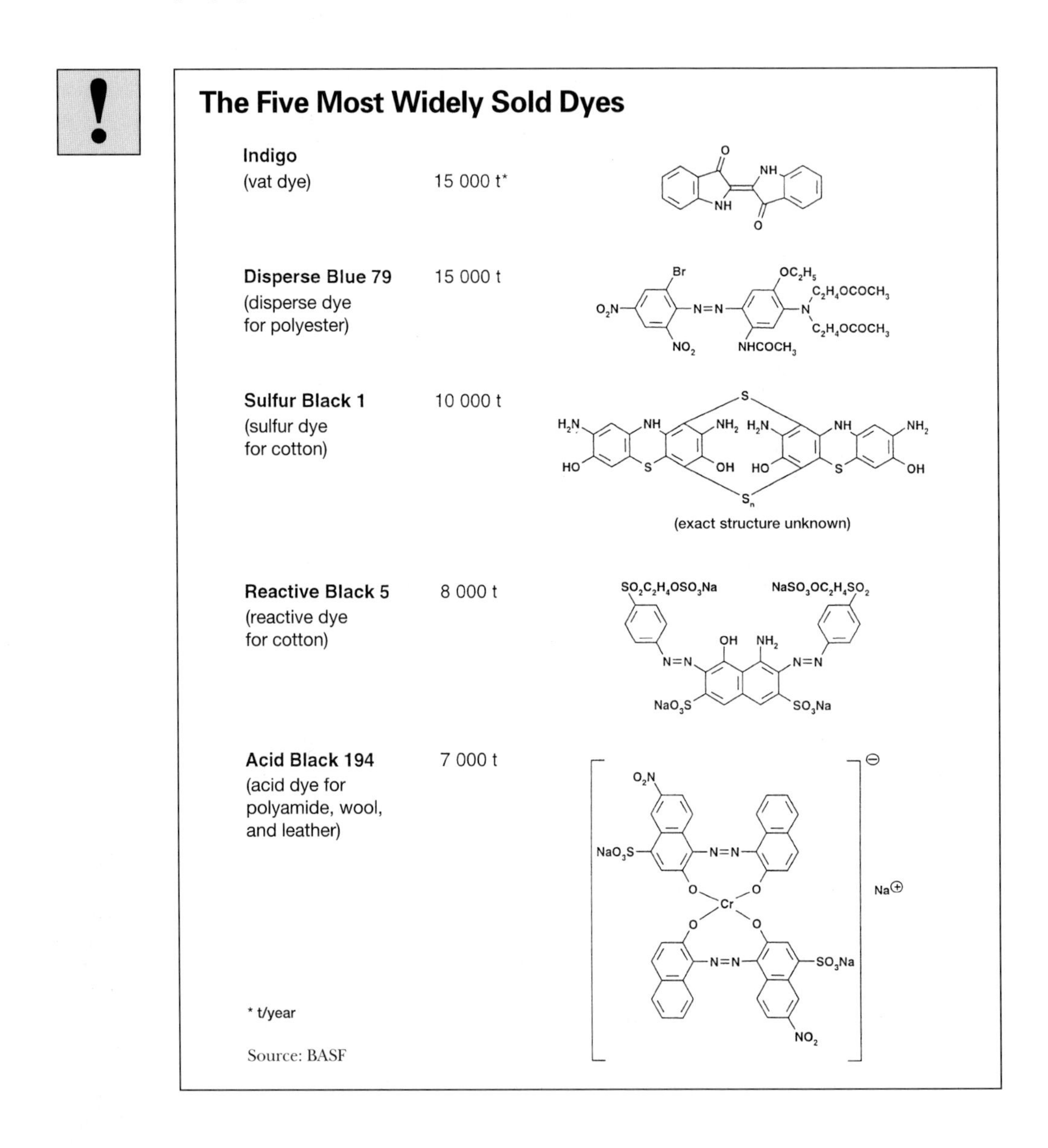

[7] D. L. Officer, A. K. Burrell, D. C. W. Reid, *Chem. Commun.* **1996**, 1657–1658.

The Largest Paint Markets

The world market for paints and coatings in 1997 grew by 1.7% to ca. 25 million metric tons, which corresponds to a value of $58–60 billion. Whereas the economic problems in Asia resulted in a growth rate there of only 0.5%, Latin America (2.8%) and North America (2.0%) achieved above average growths.

The **leading sales region** was North America, with 27.2% of the world market. Nevertheless, its lead over Western Europe and Asia, which currently are very significant markets, was only 2.4% (Fig. 1).[1]

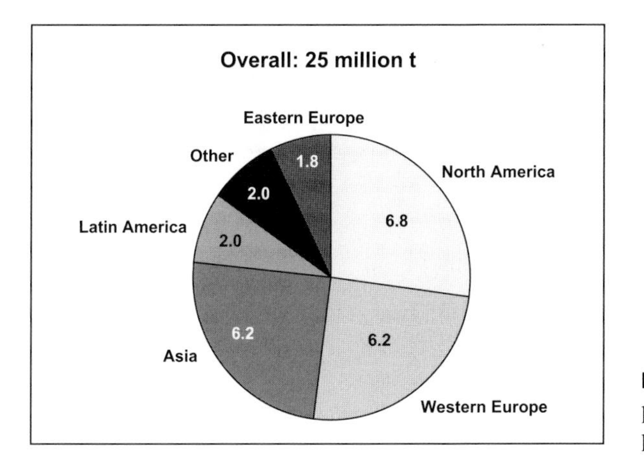

Fig. 1: World market for paint and coatings pigments, 1997[1]

Accelerated growth is anticipated in the years ahead, especially since demand in developing nations is expected to increase. Here the annual per capita use of paints and coatings is well below the typical Western European and North American levels of 15–20 kg.

Leading Paint Producers

A series of takeovers in recent years has advanced ICI to the forefront in the international list of **leading producers** of paint and coating pigments. Thus, ICI was the only company in 1996 to hold a greater than 10% share of the world market. This lead was maintained in 1997 as ICI continued to surpass second-place Akzo Nobel by a small margin (Fig. 2). Taken together, the top ten firms recorded sales of $24 billion. In 1998 Akzo Nobel succeeded in a takeover of Courtauld's for $3 billion (the third largest takeover in Dutch economic

[1] data from: Christoph Maier, *Farbe und Lack*, **7/1998**, *104*, p. 98–100.

history), thereby ascending to the rank of world's leading paint and coatings producer.

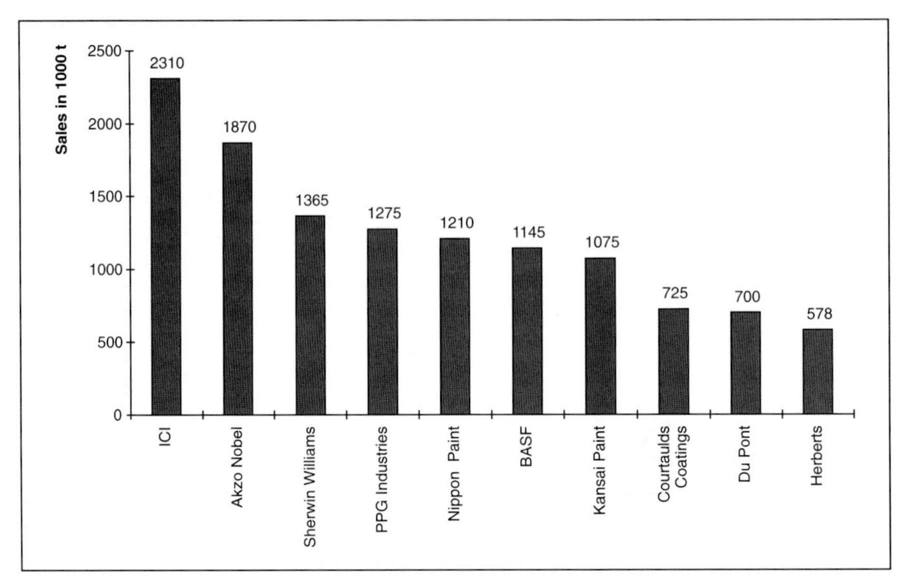

Fig. 2: a) Leading producers of paints and coatings, 1996 (in thousands of metric tons)[2]

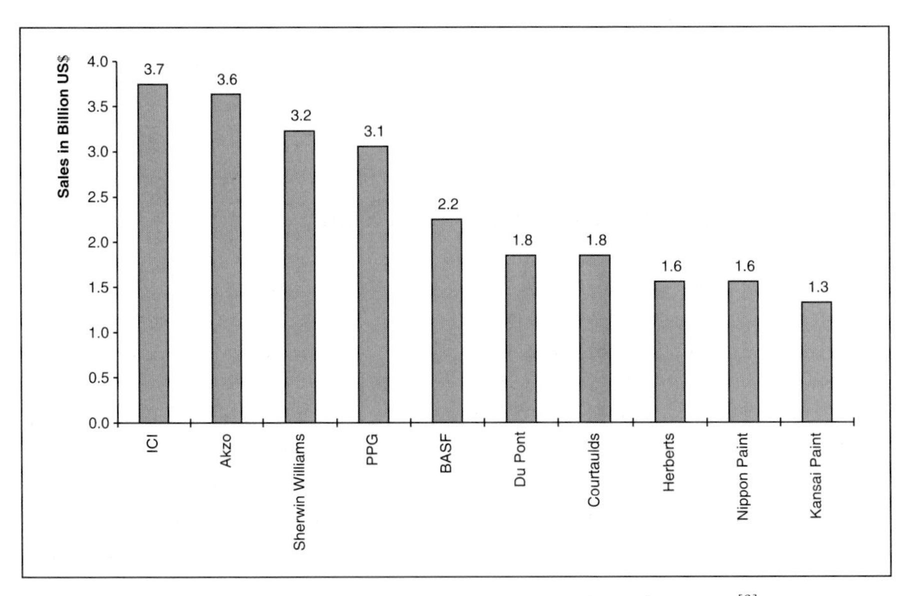

Fig. 2: b) Leading producers of paints and coatings, 1997 (in $ billion)[3]

[2] data from: *Farbe und Lack*, **2/1997**, *103*, p. 25.
[3] BASF Coatings.

Areas of Application

The largest quantities of paints and coatings in 1997 were consumed in the housing and construction sectors. This application represented as much as 70% of total demand in certain countries.

Other areas of application – apart from general industry – all consumed market shares of less than 8%, although the relative consumptions varied from region to region. Thus, in Europe, unlike North America, corrosion protection was more important than wood preservation.

Automotive repair paints and sheet-metal packing lacquer are used in relatively small quantity, but from a revenue standpoint they are very significant. (Fig. 3).

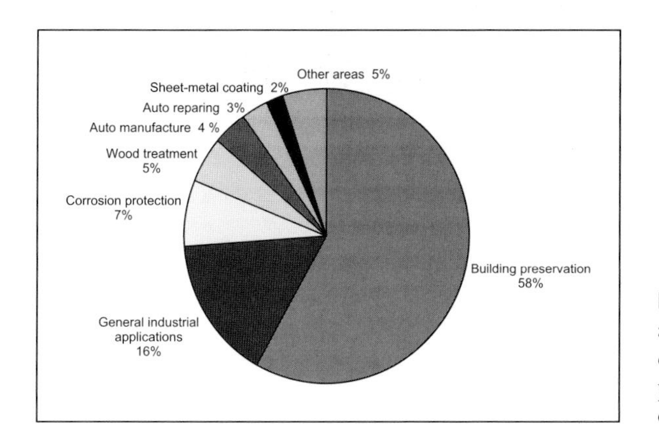

Fig. 3: Fields of application for paints and coatings, 1997 (worldwide percentages by quantity) [1]

The Most Rapidly Growing Types of Paint

The paint market was long dominated by solvent-based paints. However, for environmental reasons consumers are increasingly demanding paints that contain less organic solvents, or none at all (→ Atoms and Molecules, Solvent Records). Such "clean" technologies include water-based paints, powdered coatings, radiation-crosslinked paints, and solvent-based systems with high solids contents.

In Japan until at least 1994 these technologies acquired little attention from industry, where they constituted only 17% of the market. In Western Europe and the United States, by contrast, they held market shares of 43% and 45%, respectively (Fig. 4).

Water-based solvents were rather significant in all the regions reported, but high-solids products played a significantly larger role in the U.S. compared to Western Europe. In the latter case the powdered systems achieved a larger market share than elsewhere.

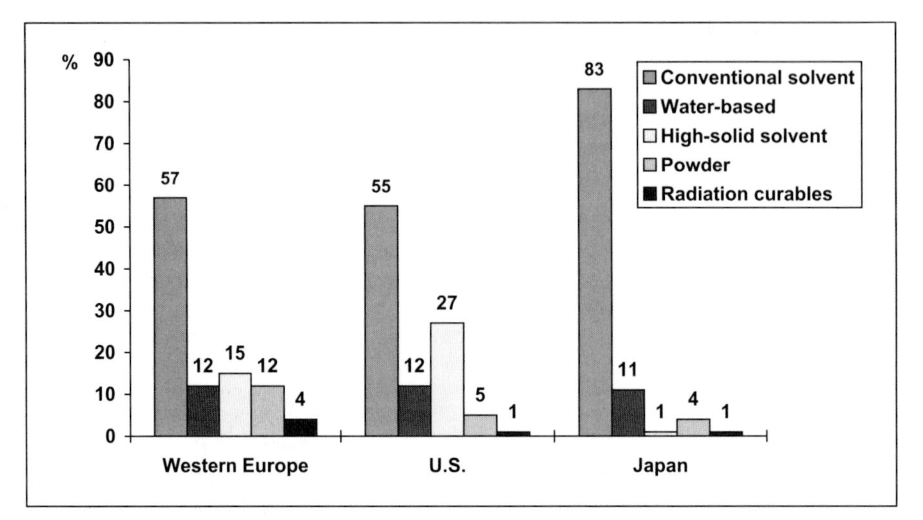

Fig. 4: a) Market shares for various paint technologies in industrial applications, 1996 (in % of total consumption)[4]

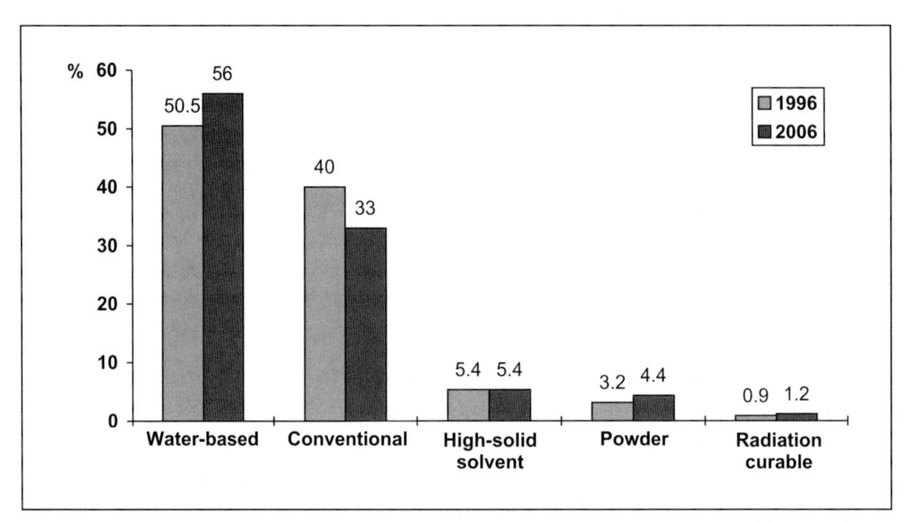

Fig. 4: b) Trends in paint technology in Western Europe, 1996–2006 (in % of total consumption)[4]

What are the technologies of the future? In Western Europe it is assumed that water-based paints will increase their market share for construction and industry at the expense of conventional paints. Nevertheless, the conventional products are still expected to enjoy considerable use at least until the middle of the next decade.

---

[4] data from: Franco Busato, *Farbe und Lack,* **9/1997**, *103*, p. 114–116.

The Leading Producer Regions

During 1997, 126 million metric tons of plastics were produced worldwide. This excludes fibers (→ Chemical Products), dispersions, adhesives, paints (→ Pigments), and similar materials. The **largest plastics-producing region** was North America, followed by Western Europe, which had a significant lead over South and East Asia (Fig. 1).

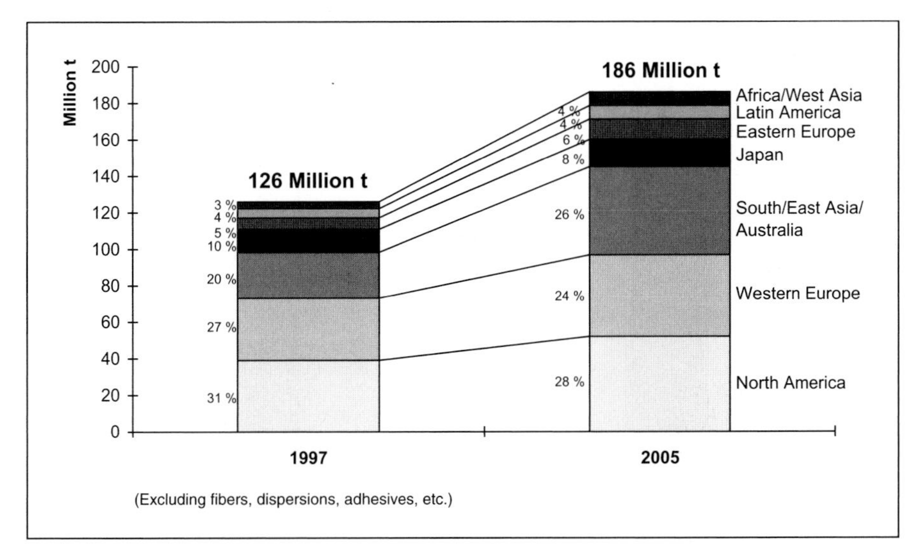

Fig. 1: Worldwide plastics production, 1997–2005[1]

The latter region is expected to surpass Western Europe in the next few years, however. Its projected growth rate of at least 8% per year is more than double that anticipated for Western Europe and North America. Nevertheless, North America will continue to be the leading producer region through 2005.

Overall, the expected average growth in world plastics production is 5% per year to 186 million tons in 2005.

[1] BASF.

The Largest Markets

As in the case of production, North America is at the present time also the **leading plastics consuming region**, accounting in 1997 for 28% of the world consumption of 126 million metric tons (Fig. 2). Its lead over second-place South and East Asia (which as recently as 1985 consumed only 17%) was reduced to a slim 2%. The South and East Asian/Australian market thus gained markedly in importance between 1985 and 1997, having first displaced Western Europe from its runner-up position. The end is not yet in sight, however: by 2000 this region is expected to draw even with North America.

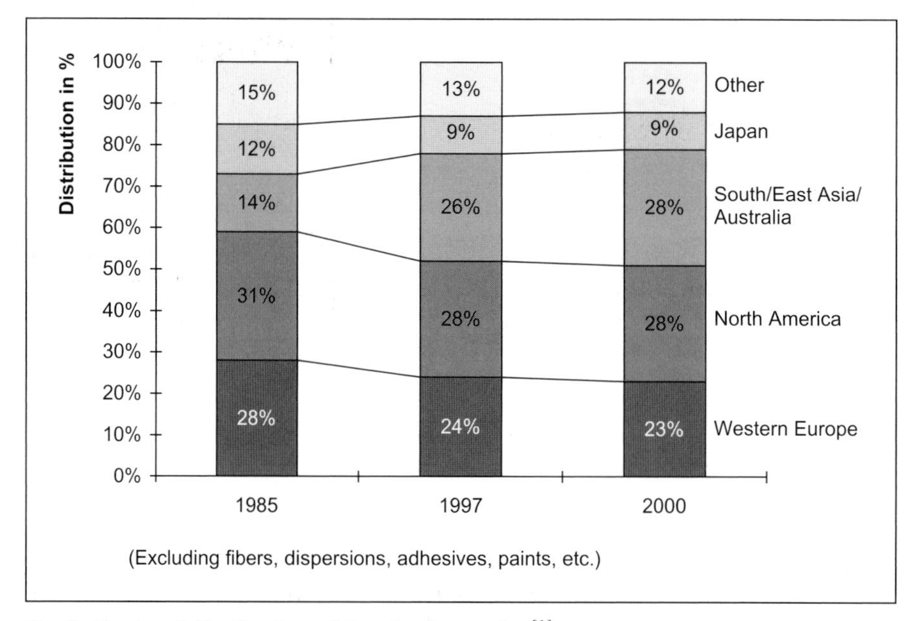

Fig. 2: Regional distribution of the plastics market[1]

The Most Important Consumer Sectors in Western Europe

Plastics consumption in Western Europe has climbed steadily in recent years, reaching a record of 31 million metric tons in 1997.

Who were the main consumers? Above all, the packaging industry should be cited, along with construction. These two sectors dominated with a combined market share of 60%. Demand in the electrical and automotive areas appears modest by comparison. Plastics demand in each of the other market sectors represented a share well under 5% of the total (Fig. 3).

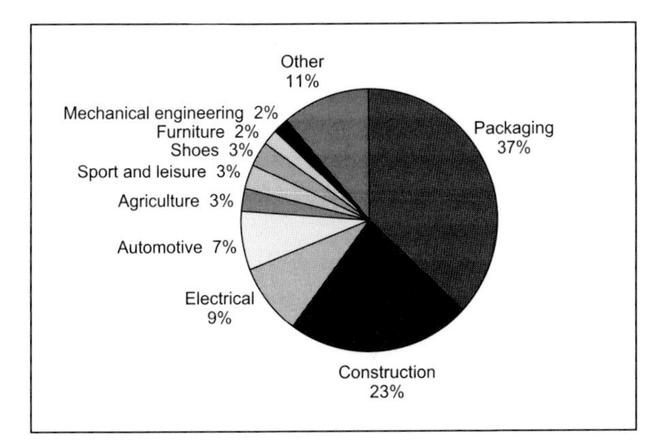

Fig. 3: Industrial sectors with the highest plastics consumptions in Western Europe, 1997

The Plastic Showing the Most Dynamic Growth

Over the period 1990–1997, demand for plastics increased by 6% a year to 126 million tons. Two of the three **leading plastics**, polyethylene and poly(vinyl chloride), showed below average rates of growth, whereas the third contender, polypropylene, turned in a stellar performance of 9% per year (Fig. 4). Polypropylene is likely to remain the **most rapidly growing plastic** in the next few years, and it may well acquire sole possession of second place in the list of most utilized plastics.

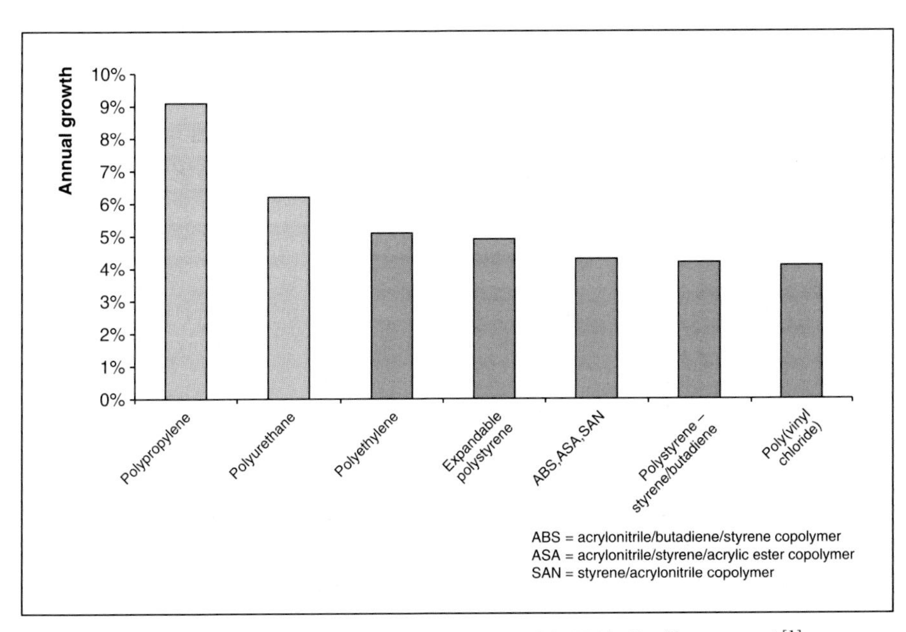

Fig. 4: Growth rates for various types of plastics, 1990–1997 (in % per year) [1]

The Most Widely Used Plastic

Of the total 1997 plastics consumption of 126 million metric tons (excluding fibers, dispersions, adhesives, paints, etc.), about two-thirds was accounted for by three classes of plastics. By far the **leading material** was polyethylene (PE) at 33% of consumption, with roughly comparable figures for the next two types, poly(vinyl chloride) (PVC) and polypropylene (PP). The only other plastics with market shares above 5% were polystyrene (PS) and polyurethane (PU), as shown in Fig. 5.

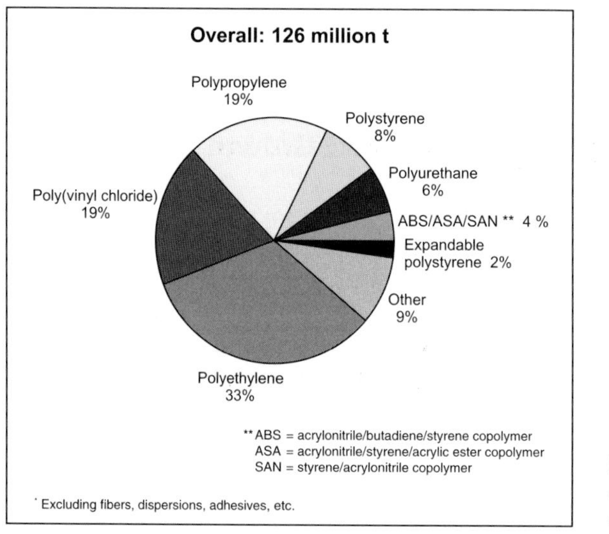

Fig. 5: Plastics consumption by category,* 1997[1]

Polyethylene: Leading Producing Regions and Producers of LDPE

LDPE (low-density polyethylene) is a partially crystalline, thermoplastic polymer prepared by the high-pressure polymerization of ethylene. Its lower density (ca. 0.91–0.94 g/cm^3) relative to HDPE is a result of more branching in its molecular chains. Most LDPE is processed into film.

In contrast to HDPE, the **leading center of production capacity for LDPE worldwide in 1998 was Western Europe**, with 5.76 million metric tons (total capacity: 18.39 million tons, somewhat less than for HDPE). North America followed, some distance behind. Eastern Europe's share of world capacity was significantly greater than in the case of the low-pressure variants of PE (Fig. 6).

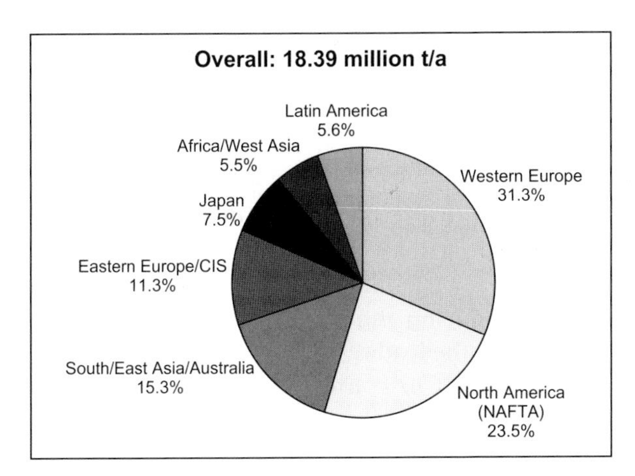

Fig. 6: Capacities for LDPE production by region, 1998[1]

The total capacity for the ten leading LDPE producers amounted to 8.19 million tons per year (44.6% of world capacity). At the forefront was Dow, the company that also laid claim to the world's largest capacity for ethylene production (Fig. 7).

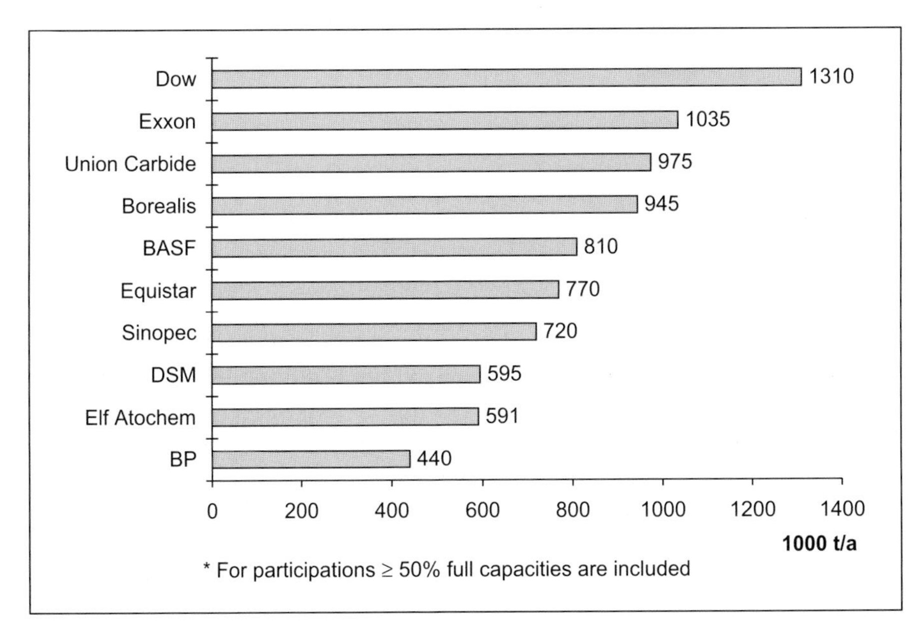

Fig. 7: Capacities* of the top ten LDPE producers in 1998[1]

Plastics Facts and Figures

Polyethylene: Leading Producing Regions and Producers of LLDPE

LLDPE (linear low-density polyethylene) is a special polyethylene of low density that is prepared by low-pressure polymerization of ethylene in the presence of 1-alkenes. It has molecular chains with defined, short branches.

World capacity for LLDPE in 1998 was 14.71 million metric tons per year. The rankings of the **most important producing regions** most nearly resembled those of HDPE; i.e., North America was the leading producing region by a wide margin, and Western Europe was forced to content itself with third place. Interestingly, the role of Eastern Europe, which is not insignificant in the case of LDPE, is strictly marginal with respect to LLDPE (Fig. 8).

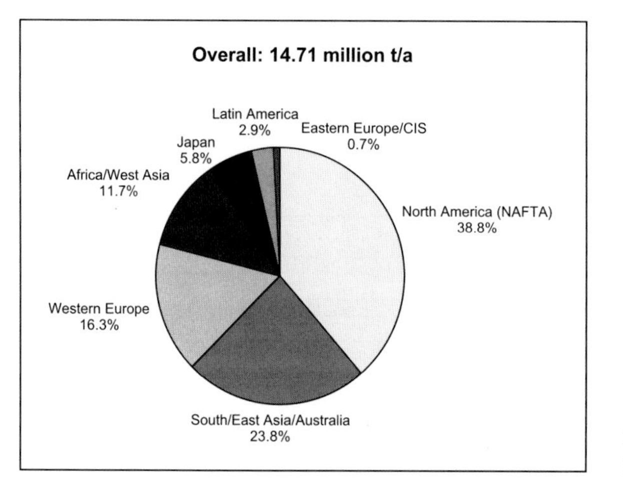

Fig. 8: LLDPE capacities by region, 1998[1]

Dow and Exxon, the **leading companies in the LDPE top ten**, also appeared at the top for LLDPE in 1998, albeit with a significantly larger fraction of world capacity. Taken together, the top ten firms could lay claim to 57% of LLDPE capacity worldwide (Fig. 9).

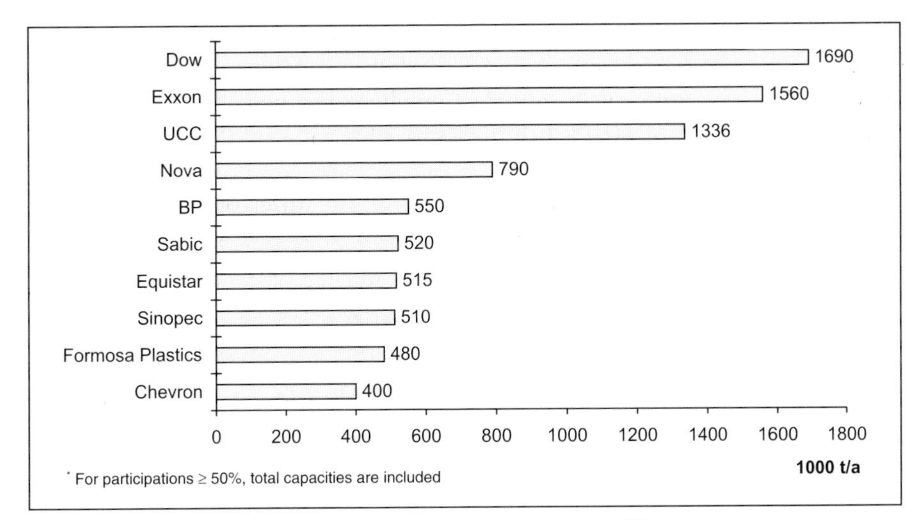

Fig. 9: Capacities* of the top ten LLDPE producers in 1998[1]

Polyethylene: Leading Producing Regions and Producers of HDPE

HDPE (high-density polyethylene) is a partially crystalline, thermoplastic polymer prepared by low-pressure polymerization of ethylene and with a density of 0.94–0.97 g/cm^3. It consists largely of unbranched molecular chains. HDPE is a mass-consumption commodity familiar in a variety of forms, including bottles, canisters, flowerpots, and also film.

As in the case of ethylene monomer, North America was the **region with the greatest production capacity for HDPE** in 1998. The importance of the South Asia/East Asia/Australia producing region (excluding Japan) was significantly greater for HDPE than for ethylene, and exceeded that of Western Europe (Fig. 10).[1]

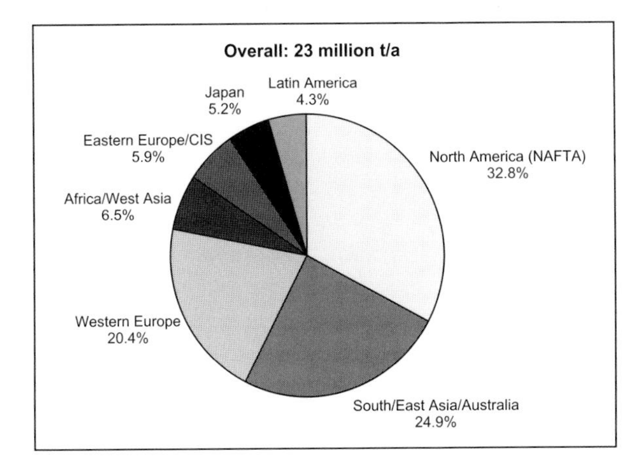

Fig. 10: HDPE capacities by region, 1998[1]

The formation in 1997 of Equistar Chemicals, a joint venture consisting of the olefin and polymer activities of Lyondell and Millenium, signaled the arrival of a new significant producer of ethylene and polyethylene. This new concern led the list of HDPE producers in 1998 with an 8% share of world capacity.

Together, the top ten accounted for 47.2% of the world production capacity of 23.0 million metric tons per year (Fig. 11).

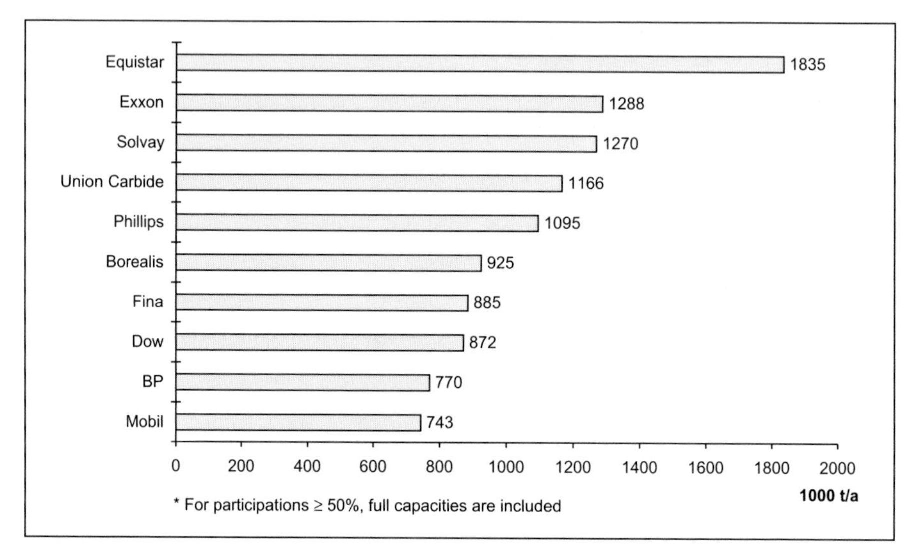

Fig. 11: Capacities* of the top ten HDPE producers in 1998

Polypropylene: Leading Producing Regions and Producers

Polypropylene is a thermoplastic material prepared by the polymerization of propylene, one that has recently made major gains in importance. It is encountered not only in the form of films and fibers, but also as a solid material suited to a wide range of applications, such as in automobiles and household appliances. In contrast to polyethylene, relative capacities for polypropylene were almost equivalent in the **three leading producer regions**: Western Europe, South Asia/East Asia/Australia, and North America (Fig. 12). Total worldwide capacity in 1998 was 29.80 million metric tons per year.

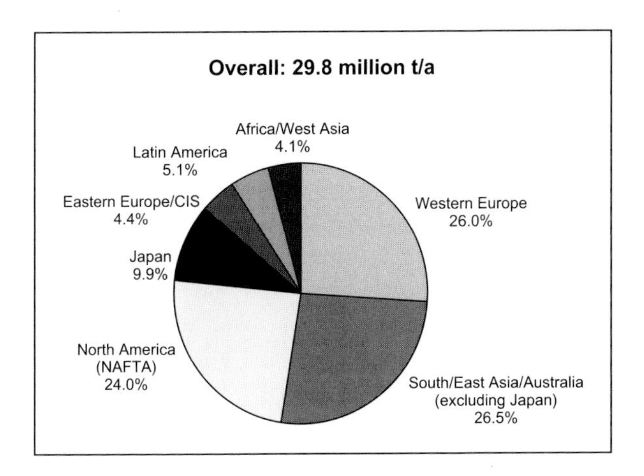

Fig. 12: Capacities for polypropylene production by region, 1998[1]

Among the top ten firms in 1998 there was one clear **market leader**: Shell, with 12.7% of world capacity, more than twice that of second-place BASF (5.7%). The top ten producers collectively accounted for 45.7% of world capacity, comparable to the situations with HDPE and LDPE (Fig. 13).

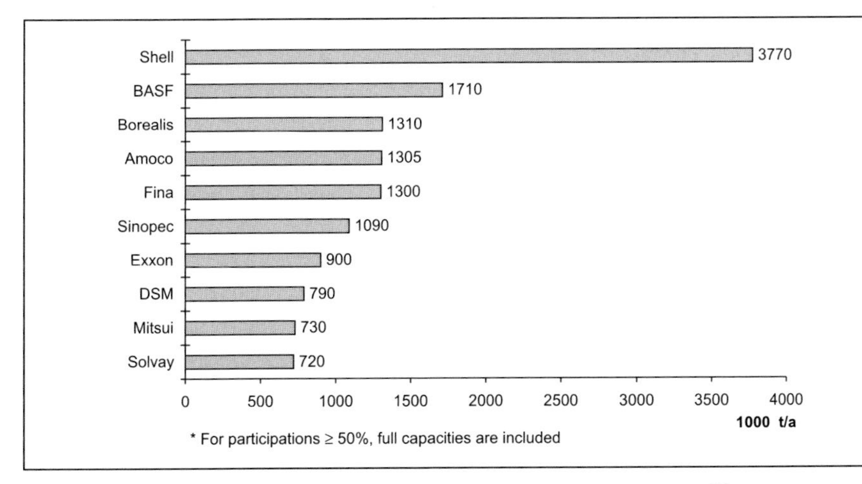

Fig. 13: Capacities* of the top ten polypropylene producers in 1998[1]

The Most Toxic Compounds

Very few classes of substances are associated with such a vast treasure of myths, half-truths, and exaggerations as poisons. Looked at more soberly, studies undertaken on the physiological activity of substances have the power of enriching important subcategories of chemistry (→ Chemical Products, Hit List), such as pharmaceutical research (→ Pharmaceuticals, Hit List) and crop protection, but also workplace safety and environmental protection (→ Environmental Protection).

From a toxicological standpoint, the so-called lethal dose (LD_{50}) is a common measure of the toxicity of a compound. This is a dosage level that would lead to death in half the cases of the research animals tested. It is a somewhat problematic measure, however, since very rarely is an LD_{50} directly transferable from one species to another, and also because biological availability and physiological reaction with respect to human subjects can differ radically from predictions based on experimental results. Moreover, nonlethal organ damage and long-term effects are ignored altogether by the parameter. On the other hand, even Paracelsus (1493–1541) recognized that the beneficial or detrimental effects of an ingested substance are a function of dose, so it is not surprising that at appropriately low concentrations many of the compounds discussed below are also exploited for therapeutic purposes.

The **known compounds that are by far the most toxic** (Table 1) are high molecular weight proteins from bacteria, which begin to exhibit toxic effects on an organism at the subnanogram level. The champion, botulinum toxin (botulin), is produced by a bacterium that reproduces preferentially on spoiled or inadequately preserved foods. Fortunately, this toxin cannot survive high temperatures (as in cooking, for example), so botulin-induced food poisoning is relatively rare. A sharp contrast is presented by tetanus toxin, which nearly everyone has encountered: It is administered in an inactivated form through innoculation as protection against lockjaw and induces the formation of protective antibodies in the human immune system. The combination of high molecular weight and a very low activity threshold makes it possible almost to supply a count of the number of physiologically active molecules present. In the case of botulin, ca. 120 million molecules suffice to provoke a deadly response, whereas with tetanus toxin the number is ca. 400 million, and for the much smaller snake poison β-bungarotoxin 5.7×10^{11} molecules are required. Given that a mole of a substance consists of 6.023×10^{23} molecules, these are all impressively small numbers.

Table 1: The most toxic compounds known

Poison	Occurrence	Compound type	Mol. wt. [g mol^{-1}]	LD_{50} [µg kg^{-1}][a]	Ref.
botulinum toxin	bacterium *Clostridium botulinum*	protein	150 000	0.0003–0.00003 MLD, mouse, i.p.	[1, 2]
tetanus toxin	bacterium *Clostridium tetani*	protein	150 000	0.001–0.0001 MLD, mouse, i.p.	[1, 2]
β-bungarotoxin	Southeast Asian snake species *Bungarus multicinctus*	protein	20 000	0.019, mouse, i.p.	[3]
maitotoxin (**1**)	dinoflagellum *Gambierdiscus toxicus*	polyketide	3 422	0.050, mouse, i.p.	[4]
ciguatoxin	dinoflagellum *Gambierdiscus toxicus*	polyketide	1 061	0.35, mouse, i.p.	[5]
palytoxin (**2**)	coral species *Palythoa toxica*	polyketide	2 679	0.45, mouse, i.v.	[6]
taipoxin	Australian taipan snake *Oxyuranus scutellatus*	glycoprotein	45 600	2, mouse, i.v.	[7]
batrachotoxin (**3**)	Columbian arrow-poison frog *Phyllobates aurotaenia*	steroid alcohol	539	2, mouse, s.c.	[8]
tetrodotoxin (**4**)	pufferfish *Spheroides rubripes*	saccharide derivative	319	10, mouse, i.p.	[9]

[a] i.p. = intraperitoneal, i.v. = intravenous, s.c. = subcutaneous, MLD = minimum lethal dose.

[1] E. J. Schantz, E. A. Johnson, *Microbiol. Rev.* **1992**, *56*, 80–99.
[2] J. L. Middlebrook, J. *Toxicol. Toxin Rev.* **1986**, *5*, 177–190.
[3] K. Kondo, K. Narita, C.-H. Lee, *J. Biochem. (Tokyo)* **1978**, *83*, 101–115.
[4] M. Murata, H. Naoki, S. Matsunaga, M. Satake, T. Yasumoto, *J. Am. Chem. Soc.* **1994**, *116*, 7098–7107.
[5] M. Murata, A. M. Legrand, Y. Ishibashi, M. Fukui, T. Yasumoto, *J. Am. Chem. Soc.* **1990**, *112*, 4380–4386.
[6] J. S. Wiles, J. A. Vick, M. K. Christensen, *Toxicon* **1974**, *12*, 427–433.
[7] J. Fohlman, D. Eaker, E. Karlsson, S. Thesleff, *Eur. J. Biochem.* **1976**, *68*, 457–469.
[8] T. Tokuyama, J. Daly, B. Witkop, *J. Am. Chem. Soc.* **1969**, *91*, 3931–3938.
[9] C. Y. Kao, F. A. Fuhrmann, *J. Pharmacol. Exp. Ther.* **1963**, *140*, 31–40.

The Most Toxic Low Molecular Weight Compounds

The most poisonous low molecular weight compounds come from the sea. One dinoflagellum, *Gambierdiscus toxicus*, produces two highly effective poisons, perhaps earning it the distinction of being the **most toxic of all organisms**. These toxins can enter the human body during the consumption of coral fish and are the source of a fish poisoning known as ciguatera that affects ca. 20 000 people a year. The formidably complex natural products maitotoxin (**1**) and palytoxin (**2**), shown in Fig. 1, together with ciguatoxin itself, are all similarly constructed (→ Synthesis, Masterful Achievements). Maitotoxin classes not only as the **most poisonous peptide** but also as the **highest molecular weight natural product** (→ Molecular Form, Giants) that is not a biopolymer.

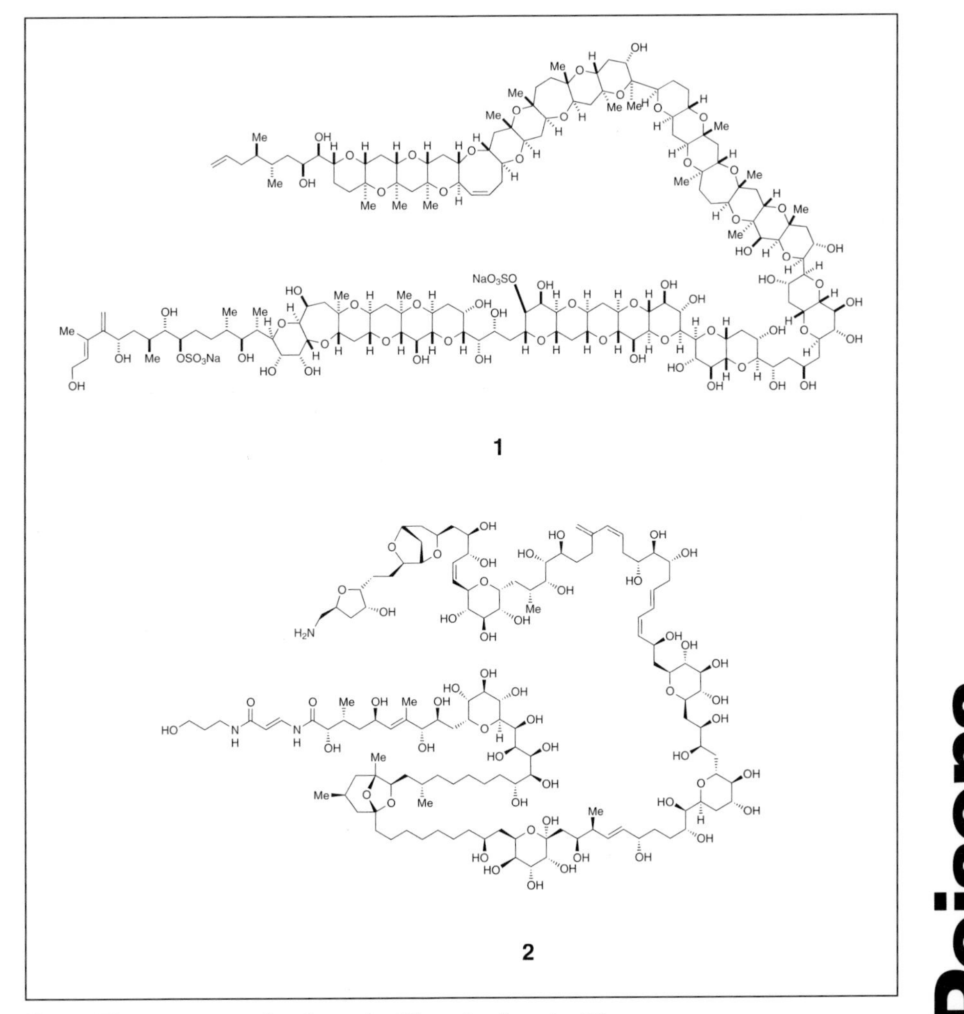

Fig. 1: The structures of maitotoxin (**1**) and palytoxin (**2**)

Among highly toxic compounds, the two with the **lowest molecular weights** are batrachotoxin (**3**) and tetrodotoxin (**4**), shown in Fig. 2. Inadvertent contact with **3**, the poison of the Columbian arrow-poison frog, can be regarded as quite unlikely. In order to isolate enough of the material to characterize it, four expeditions fought their way through the Columbian jungle over the course of eight years. The extract from 5 000 frog skins finally permitted isolation of 11 mg of pure batrachotoxin.[1] On the other hand, every year enough connoisseurs eat the fugu pufferfish, which classes as a delicacy, to cause several poisonings attributable to tetrodotoxin.

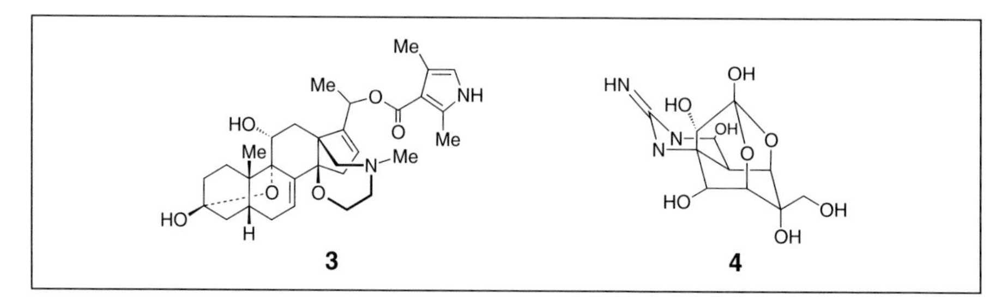

Fig. 2: Highly toxic compounds of low molecular weight

The Most Toxic Plant Poisons

Among the poisonous compounds found in plants (Table 2), the glycoprotein ricin, from the castor bean, displays the greatest toxicity. Two well-known compounds shown in Fig. 3, nicotine (**5**) and strychnine (**6**), both from the large and structurally very diverse alkaloid family, are three orders of magnitude less toxic. Worth special mention in the case of nicotine is the long-term damage to human organs caused by repeated inhalation of tobacco smoke. The physiological potency of nicotine becomes strikingly clear from the fact that injection of the amount present in the tobacco of a single cigar is enough to kill an adult.[2] By comparison, the arrow poison curare (**8**), used by South American Indian tribes, seems almost harmless, because it is two orders of magnitude less toxic than the two plant alkaloids just cited. Oral toxin intake is the **most common cause of poisoning by atropine** (**9**), the poisonous substance from deadly nightshade. This especially poses a threat to small children, since they readily confuse the sweet black nightshade berries with cherries, and may die from eating as few as three or four.[2]

[1] T. Tokuyama, J. Daly, B. Witkop, *J. Am. Chem. Soc.* **1969**, *91*, 3931–3938.
[2] L. Roth, M. Daunderer, K. Kormann, *Giftpflanzen – Pflanzengifte*, Ecomed, Landsberg/Lech, 4th ed., **1994**.

Table 2: The most toxic plant poisons

Poison	Occurrence	Compound type	Mol. wt. [g mol^{-1}]	LD_{50} [µg kg^{-1}][a]	Ref.
ricin	castor bean *Ricinus communis*	glycoprotein (lectin)	62 400	0.10, mouse, i.p.	[1]
nicotine (**5**)	tobacco plant *Nicotiana tabacum*	alkaloid	162	300, mouse, i.v.	[2]
strychnine (**6**)	poison nut *Strychnos nux-vomica*	alkaloid	334	750, cat, oral	[1]
cymarine (**7**)	*Strophantus* species of tropical creeping shrub	digitalis glycoside	549	25 000, rat, i.v.	[3]
tubocurarine chloride (**8**), "curare"	*Chondrodendron tomentosum*	alkaloid	682	33 200, mouse, oral	[4]
atropine (**9**)	deadly nightshade *Atropa belladonna*	alkaloid	289	400 000, mouse, oral	[1]

[a] i.p. = intraperitoneal, i.v. = intravenous.
[1] L. Roth, M. Daunderer, K. Kormann, *Giftpflanzen–Pflanzengifte*, Ecomed, Landsberg/Lech, 4th ed., **1994**.
[2] R. B. Barlow, L. J. McCleod, *Brit. J. Pharmacol.* **1969**, *35*, 161–174.
[3] V. G. Vogel, E. Kluge, *Arzneimittel-Forsch.* **1961**, *11*, 848–850.
[4] R. D. Sofia, L. C. Knobloch, *Toxicol. Appl. Pharmacol.* **1974**, *28*, 227–233.

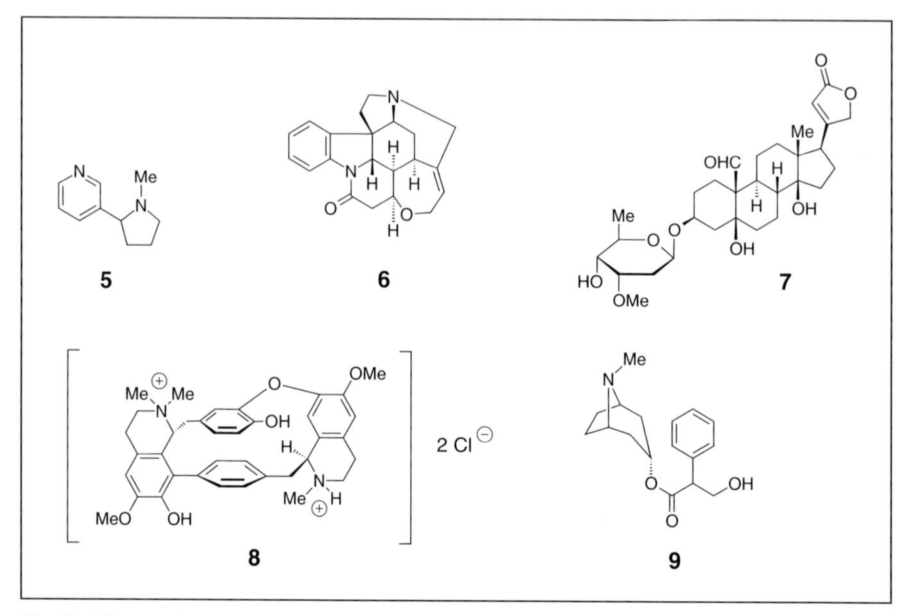

Fig. 3: Plant poisons

The Most Toxic Fungal Poisons

The fungal poison data presented in Table 3 dramatize the high toxicity of the compound muscarine (**10**), whose structure is shown in Fig. 4 and which is found in toadstools, as well as α-amanitine from the deadly amanita ("death cup"). Whereas the toadstool with its characteristic red cap is seldom confused with other fungi, extra caution is warranted with respect to the death cup, because its very faint warning color makes it easy to confuse with certain edible species. In 100 g of fresh death-cup mushrooms there can be ca. 8 mg of α-amanitine, which explains why this is the source of up to 90% of fatal mushroom poisonings.[3]

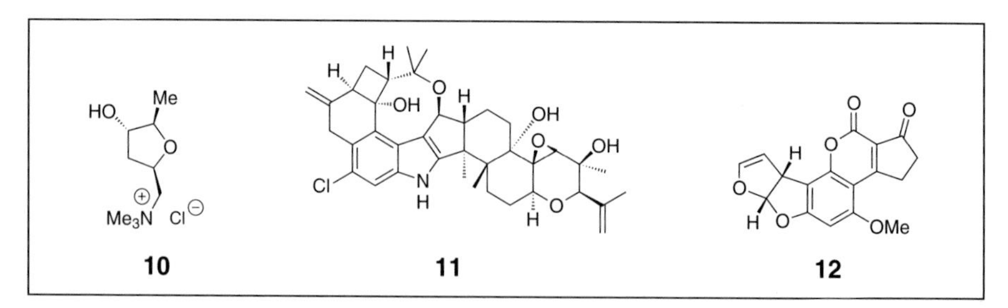

Fig. 4: Fungal poisons

[3] L. Roth, H. Frank, K. Kormann, *Giftpilze, Pilzgifte, Schimmelpilze, Mycotoxine*, Ecomed, Landsberg/Lech, **1990**.

Table 3: The most toxic fungal poisons

Poison	Occurrence	Compound type	Mol. wt. [g mol^{-1}]	LD_{50} [µg kg^{-1}][a]	Ref.
L-(+)-muscarine (**10**)	toadstool *Amanita muscaria*	alkaloid	174	230, mouse, i.v.	[1]
α-amanitine	deadly amanita (death cup), *Amanita phalloides*	bicyclic octapeptide	919	300, mouse	[1]
penitrem A (**11**)	mold *Penicillium crustosum*	polycyclic indole derivative	634	1 050, mouse, i.p.	[1]
aflatoxin B_1 (**12**)	mold *Aspergillus flavus*	difuran coumarin derivative	312	1 700, mouse, oral	[1]

[a] i.p. = intraperitoneal, i.v. = intravenous.

[1] L. Roth, H. Frank, K. Kormann, *Giftpilze, Pilzgifte, Schimmelpilze, Mykotoxine*, Ecomed, Landsberg/Lech, **1990**.

The mycotoxins penitrem A (**11**) and aflatoxin B_1 (**12**) from the molds *Penicillium crustosum* and *Aspergillus flavus*, respectively, are toxic metabolites that have lethal effects in animal models at a level slightly above one milligram per kilogram. The starting point for mycotoxin research is related to a mass poisoning of English turkeys initiated by moldy peanuts,[4] the molecular origins of which could be traced to aflatoxin B_1. Also, the so-called "curse of the pharaohs," which led to the mysterious deaths of archeologists after the opening of the pyramid of Tutankhamen (1347–1339 B.C.), has been attributed to the mold-generated coumarin derivative **12** that was present inside the tomb.[3] Apart from its acute toxic effects, aflatoxin B_1 is the compound that shows the **highest, orally active liver carcinogenicity** yet detected, with an activity threshold of only 10 $\mu g\ kg^{-1}$ in rats.[3] Fortunately, instinct and nausea usually prevent humans from ingesting large amounts of these dangerous toxins.

The Most Toxic Synthetic Poisons

The list of poisons is by no means limited to natural products like those described above; certain organic compounds of anthropogenic origin are regarded as typical poisons as well, as are several inorganic substances (Table 4). 2,3,7,8-Tetrachlorodibenzodioxin (dioxin, TCDD) **13**, shown in Fig. 5 and sometimes given the sensational name "ultrapoison", has been "on everyone's tongue" since the tragic accident at Seveso, Italy, on July 10, 1976, when 2 kg of the substance escaped into the environment. From a toxicological standpoint dioxin certainly deserves to be classed as a powerful poison, leading to chloracne in humans and shown to be carcinogenic in animal studies. One of the fundamental differences between this and toxic natural products is the disconcerting persistence of synthetic TCDD, which is not metabolized by living organisms or decomposed in sediments. Its half-life in soil is 2–9 years, and in the air up to 32 days.[5] It therefore is urgently necessary that introduction of this compound into the environment be subject to ongoing stringent control and further reduction.

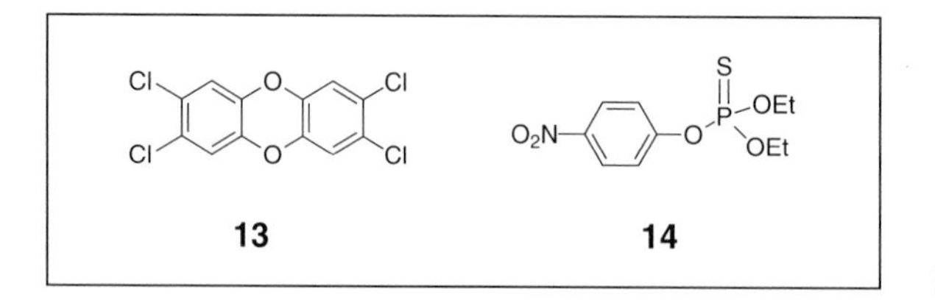

Fig. 5: Synthetic poisons

[4] B. Franck, *Angew. Chem.* **1984**, *96*, 462–474; *Angew. Chem. Int. Ed. Engl.* **1984**, *23*, 493.
[5] D. Lenoir, S. Leichsenring, *Chem. Unserer Zeit* **1996**, *30*, 182–191.

Table 4: Inorganic and synthetic poisons

Poison	Compound	LD_{50} [µg kg^{-1}]	Ref.
2,3,7,8-TCDD (dioxin)	**13**	22, rat, oral	[1]
parathion (E605)	**14**	3 600, rat, oral	[2]
potassium cyanide	KCN	10 000, rat, oral	[3]
arsenic oxide	As_2O_3	15 100, rat, oral	[4]

[1] B. A. Schwetz, J. M. Norris, G. L. Sparschu, V. K. Rowe, P. J. Gehring, J. L. Emerson, C. G. Gerbig, *Chorodioxins – Origin and Fate*, ACS Symp. Ser. **1973**, *120*, 55–69.
[2] T. B. Gaines, *Toxicol. Appl. Pharmacol.* **1969**, *14*, 515–534.
[3] W. J. Hayes, Jr., *Toxicol. Appl. Pharmacol.* **1967**, *11*, 327–335.
[4] J. Harrison et al., *Arch. Ind. Health* **1958**, *17*, 118.

The crop-protection agent parathion, **14** ("E 605"; see Fig. 5; → Crop Protection), which has acquired a sad reputation as a suicide agent, is roughly 100 times less toxic than dioxin. Another difference between the two is the relatively rapid environmental decomposition to which the phosphate esters are subject. "Traditional" poisons like potassium cyanide and arsenic oxide, the chemical lead actors in countless detective tales, must be taken in relatively large doses if they are to have a lethal effect. It is remarkable in any event that the powerful natural toxins exceed the potency of synthetic poisons by orders of magnitude.

World Energy Consumption

The steadily increasing world population creates an ongoing need for additional energy. The **worldwide consumption of energy** in 1970 was 206.7 quadrillion (10^{15}) Btu (British thermal units), but by 1990 it had risen at an annual rate of 2.6% to 345.6 quadrillion Btu. There is no end to this process in sight, although based on current estimates the **energy growth rate** over the period 1990–2010 will fall to 1.6% per year as a consequence of more energy-efficient technologies, which should result in a worldwide demand of 472 quadrillion Btu in 2010.

Nothing is expected to change in the **ranking of energy sources** over the time period in question. The most important source of energy now and for the foreseeable future is oil (Fig. 1). Oil's relative importance declined significantly during the period 1970–1990, but it is likely to remain stable at the present level until 2010 due to rapid expansion in the field of transportation in the developing countries. On the other hand, a major increase was registered in use of the more environmentally friendly energy source natural gas. With annual growth rates of 3.5% (1970–1990) and 2.0% (1990–2010), gas should become nearly as important as coal by 2010 (Fig. 2).

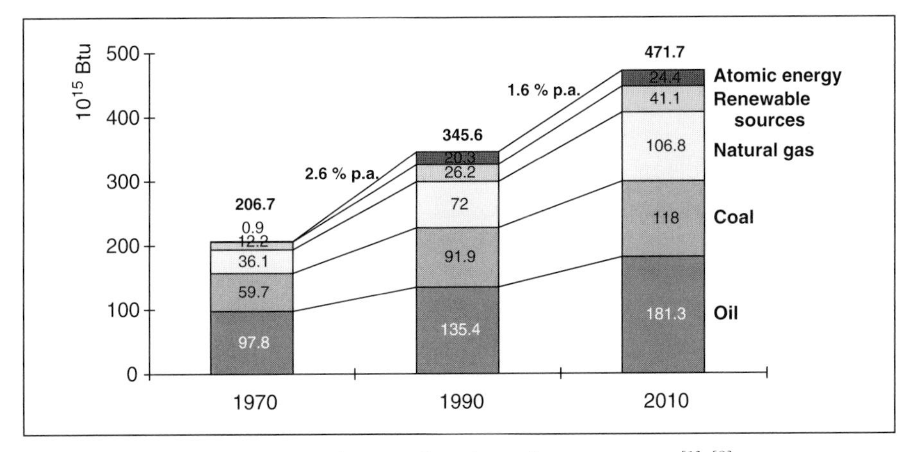

Fig. 1: World energy consumption as a function of energy source[1], [2]

Lagging noticeably behind these three principal sources in fourth place are renewable energy sources, for which high annual growth rates are projected in the future. The opposite applies to atomic energy, which is trailing the pack. The meteoric rise in this particular energy source prior to 1990 is expected to be followed by much quieter times.

---

[1] historical data from: Energy Information Administration (EIA), Office of Energy Markets and End Use, International Statistics Database.

[2] projection from: EIA, *World Energy Projection System* (**1995**), *Reference Case.*

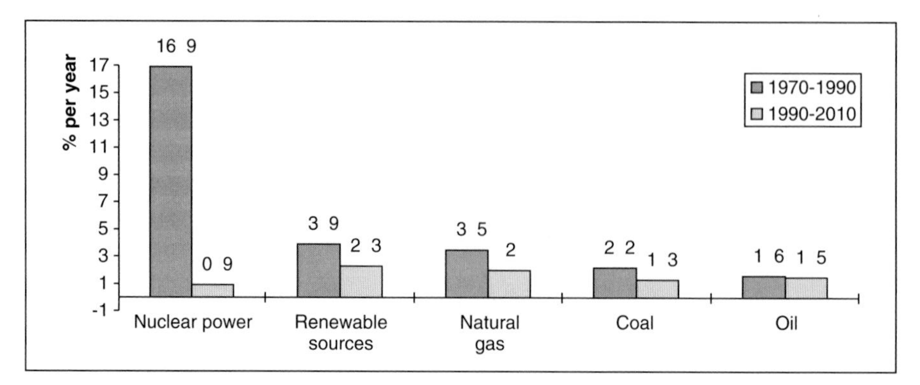

Fig. 2: Growth rates for the various major sources of energy[1], [2]

Worldwide Reserves of Fossil Fuels

Most of the known and economically recoverable worldwide **reserves of fossil energy** are in the form of bituminous coal. Coal reserves are currently estimated at 494 billion bituminous coal units, greater than the reserves of petroleum, natural gas, and lignite taken together (Fig. 3).

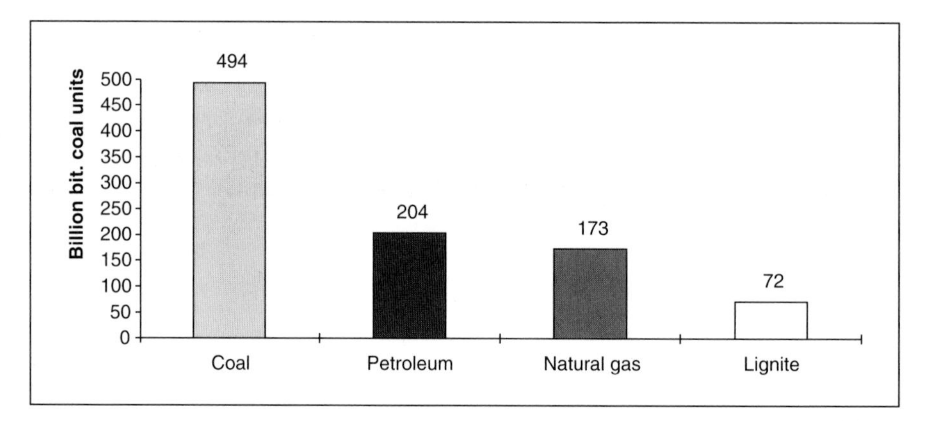

Fig. 3: World reserves of fossil fuels by type (in billions of bituminous coal units)[3]

Assuming a constant level of utilization, the world coal reserves should last another 180 years. Only the reserves of lignite are projected to have a longer lifetime (Fig. 4).

[3] data from: *Wirtschaftsmagazin Aktiv*, January 4, **1997**, p. 3.

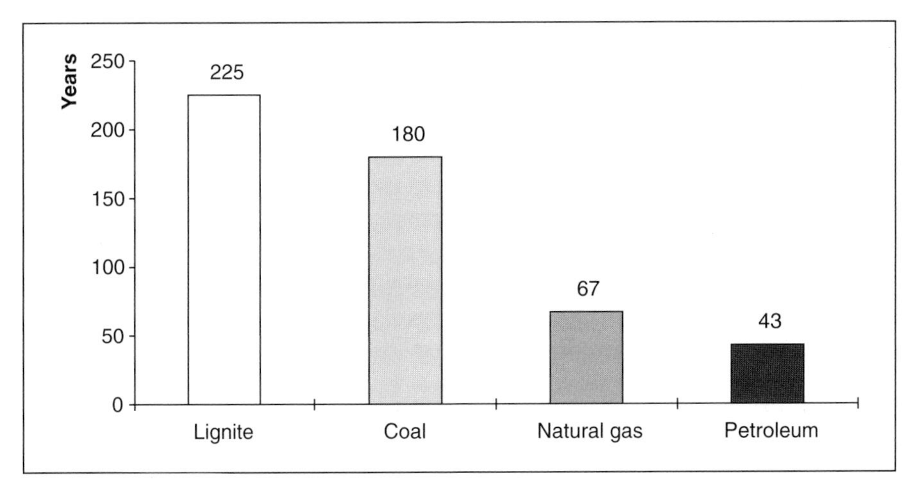

Fig. 4: Life expectancies of the known reserves of fossil fuels

Petroleum, extremely important as a raw material for the chemical industry, is expected to have a much shorter lifespan for humanity, as is natural gas, which has recently increased in importance as an energy source. Nevertheless, past experience has taught us that in 40 years there will still be oil reserves available. This seems probable due both to the great likelihood that additional oil resources remain to be discovered as well as the fact that use of energy-conserving technologies is expected to increase. To what extent these hopeful prognoses will be counteracted by rapidly increasing appetites for energy in the industrializing nations remains to be seen.

The Significance of OPEC with Respect to World Energy Supplies

Since 1985, the impact of the Organization of Petroleum Exporting Countries (OPEC) on world oil and gas supplies has increased dramatically. OPEC's **share of world production** (Fig. 5) has increased in the case of petroleum from 29.7% (818.6 million metric tons) in 1985 to 40.9% (1 436.0 million tons) in 1997. For gas the corresponding figures are 8.8% (154 billion m^3) and 13.5% (312 billion m^3).

In 1997, OPEC controlled 78.2% of the world's known oil reserves (108 304 million tons).

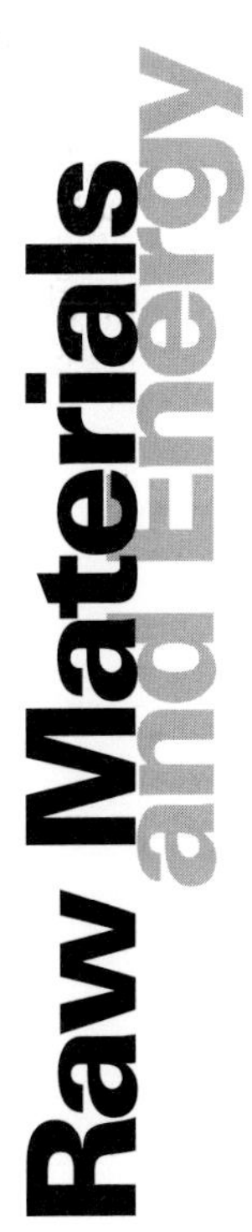

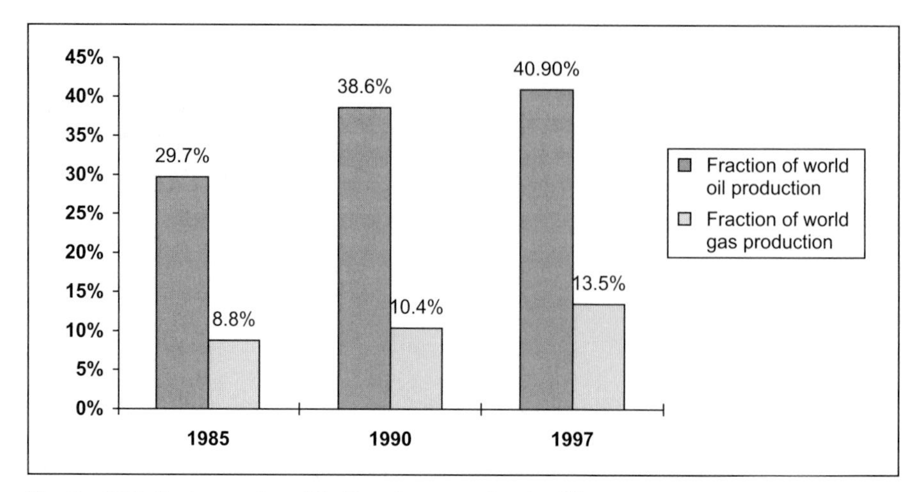

Fig. 5: OPEC's share of world oil and gas production[4]

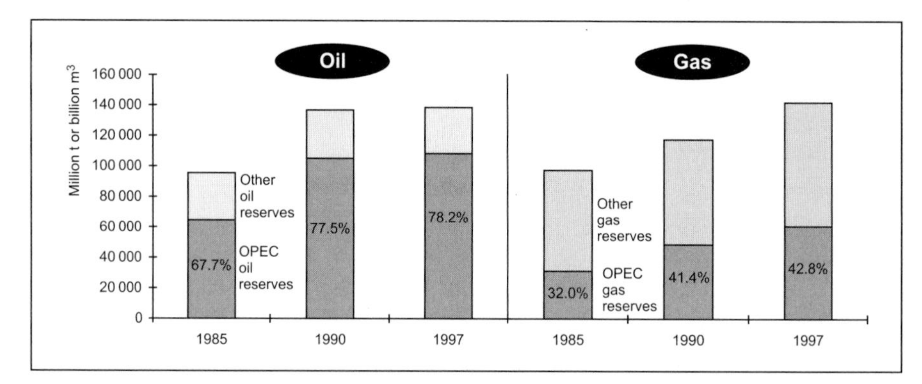

Fig. 6: OPEC's share of known world oil and gas reserves[4]

OPEC's dominance in natural gas is not quite so complete. Even so, the known OPEC gas reserves in 1997 amounted to 312 billion m^3, 42.8% of world reserves.

[4] data from: *Esso Oeldorado 1998*; data for 1997 are preliminary.

Principal Energy Sources in Germany

No changes are expected in the **ranking of primary energy sources** in Germany between now and 2010 (Fig. 7). Thus, petroleum is expected to have no problem maintaining its leading position over this period despite declines in absolute consumption and significant increases in the demand for natural gas. Apart from natural gas, only renewable energy sources can anticipate increasing demand, which will in no way affect their last-place standing.

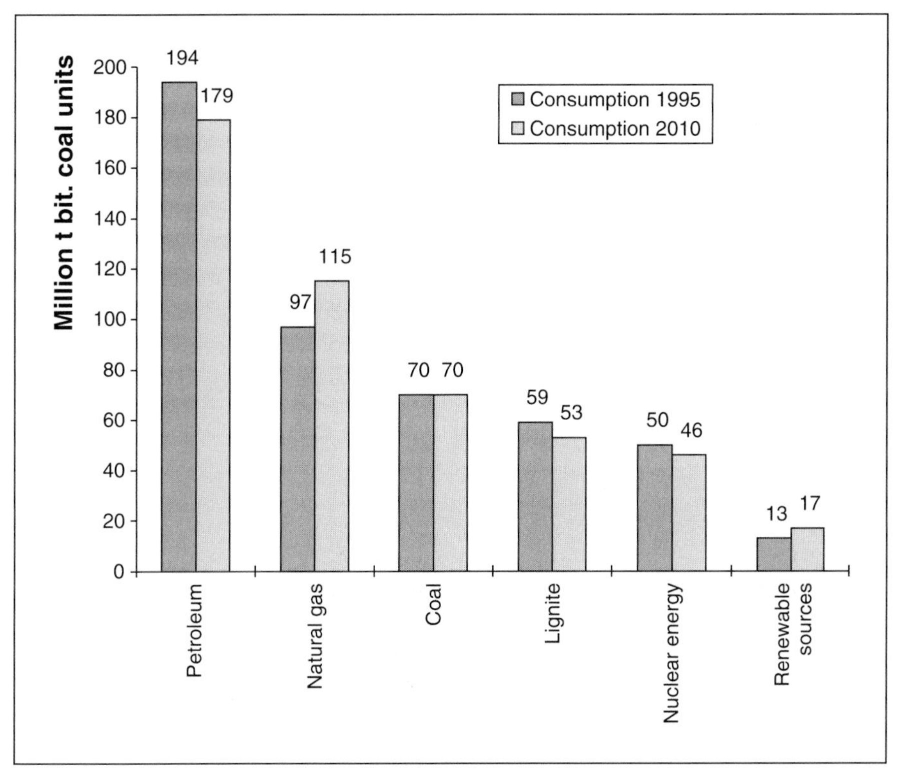

Fig. 7: Energy consumption as a function of source[5]

[5] data from: *Handelsblatt*, January 22, **1997**, p. 32.

The Countries with the Largest Natural Gas Reserves

World reserves of natural gas (Fig. 8) increased in 1997 by 2.9% to 143 942 billion m^3, which represented a 45.7% increase over 1985. The ten **countries richest in natural gas** increased their share in world reserves from 77.7% to 81.5% (117 311 billion m^3) over this period.

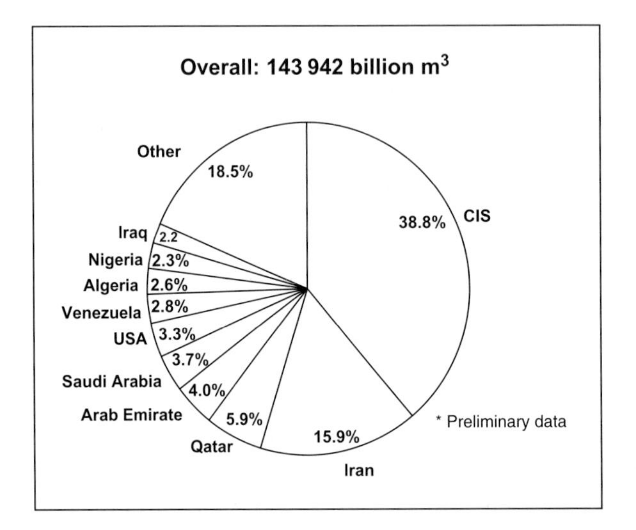

Fig. 8: Natural gas reserves as a function of country, 1997* [4]

Two countries continued to dominate the list in 1997, just as in 1985: The Commonwealth of Independent States (CIS, the new name for the loose confederation between Russia and its neighbors, formed after the disintegration of the Soviet Union) followed at some distance by Iran. The CIS alone possessed 55 949 billion m^3 of natural gas, 39% of the worldwide reserves, while Iran accounted for another 15.9% (22 923 billion m^3). Qatar was in third place with 8490 billion m^3, considerably behind Iran.

The other members of the top ten, ending with Iraq (2.2% of the world gas reserves), are all in North and South America, the Near East, and Africa. Western Europe is as completely unrepresented as Asia.

With the exception of the United States, proven natural gas reserves in all the top-ten countries climbed between 1985 and 1997 (Fig. 9). Especially impressive were the rates of increase in the United Arab Emirates and Iraq. The dwindling gas reserves in the United States, the world's second largest producer, are an indicator – against a background of an increasingly critical oil situation – of the growing dependence of the USA on foreign sources of energy.

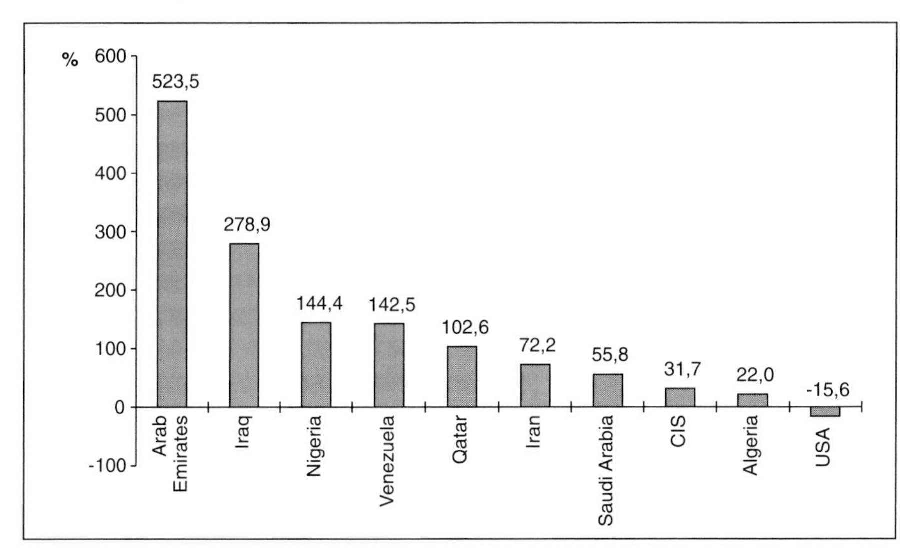

Fig. 9: Growth rates of natural gas reserves by country, 1985–1997[4]

The World's Most Important Producers of Natural Gas in 1997

Net production of natural gas and petroleum gas has grown steadily in recent years, increasing from 1 747 billion m^3 in 1985 to 2 312 billion m^3 in 1996 (growth rate: 2.6% per year). For the first time, however, this trend has now been broken. In 1997 natural gas production declined by about 5 billion m^3 to 2 307 billion m^3, attributable largely to the difficult circumstances in the CIS and Eastern Europe (Fig. 10).

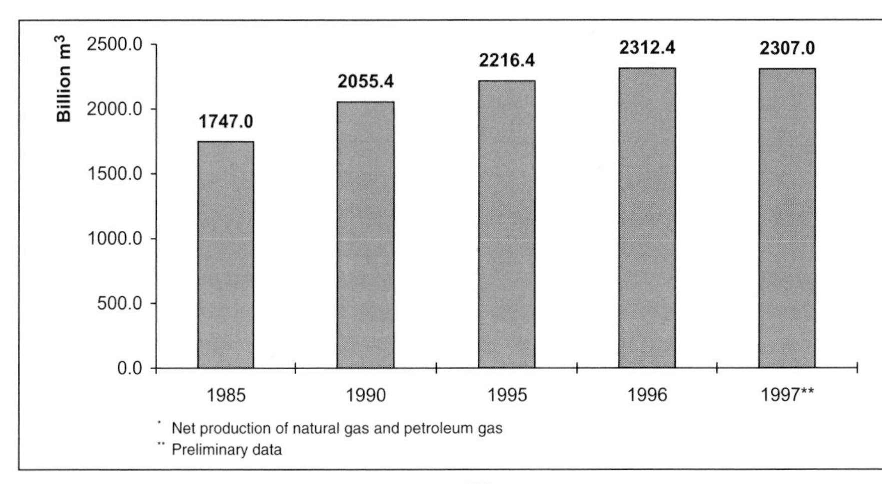

Fig. 10: Natural gas production*, 1985–1997[4]

Nevertheless, the CIS remained in 1997 the champion among the top ten gas producers. The CIS and the United States together dominated the field, representing 54% of total world natural gas production. Canada occupied third place, but a considerable distance behind (Fig. 11).

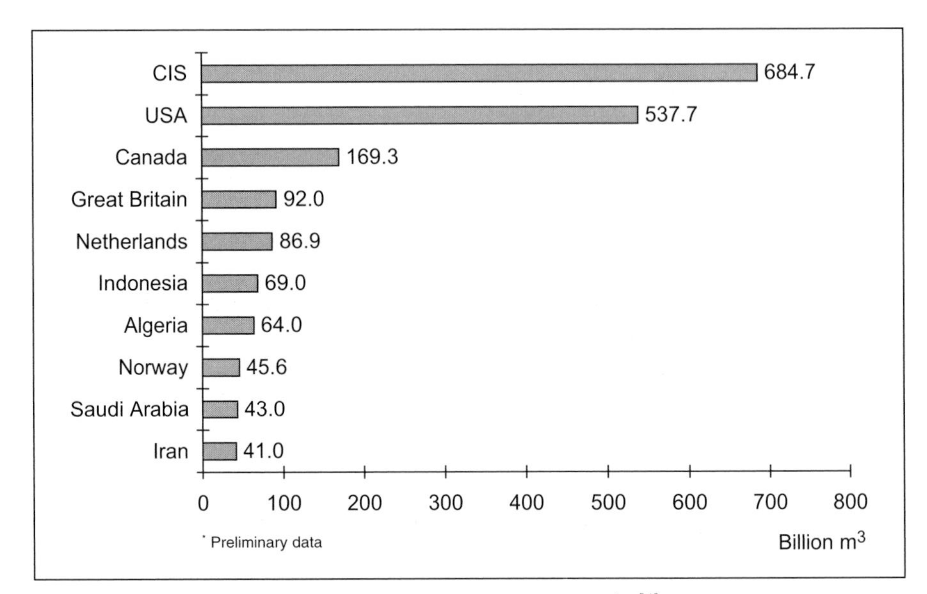

Fig. 11: The top ten natural gas producing countries, 1997*[4]

The only Western European countries represented on this list are Great Britain and the Netherlands, in fourth and fifth place, respectively. Their production quotas (92 billion and 86.9 billion m^3) appear quite modest in comparison with that of the CIS.

Among the countries in the Near East, which plays such an important role in petroleum production, only Saudi Arabia is in the list of the top ten natural gas producers (in ninth place). China, which has accomplished an impressive increase in oil production, is nowhere to be found among the leading natural gas producers.

All of the **top-ten producing countries**, which were responsible for 79.5% of world natural gas production in 1997, increased their production quotas between 1985 and 1997, most at double-digit rates. The most spectacular rate increase, a hefty 181%, was recorded by tenth-place Iran (Fig. 12).

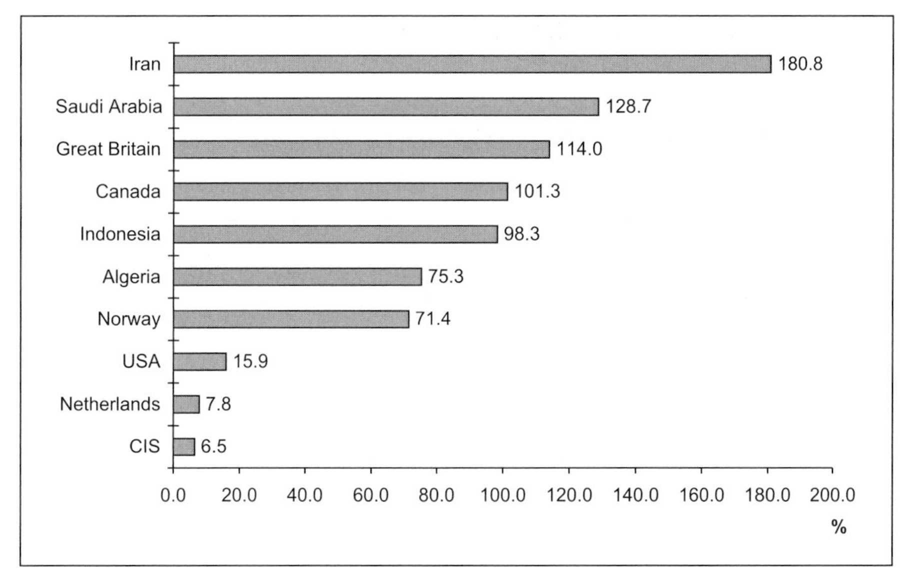

Fig. 12: Increases in natural gas production by country, 1985–1997[4]

World Oil Reserves at Record Levels

After a dramatic increase in the second half of the 1980s and several years unchanged at a high level, **world oil reserves** began to climb again during the past three years. They reached a record level in 1997 of 138 533 million metric tons. These proven reserves include only oil supplies that have been verified by drilling and that are economically accessible with current technology at current oil prices (Fig. 13).

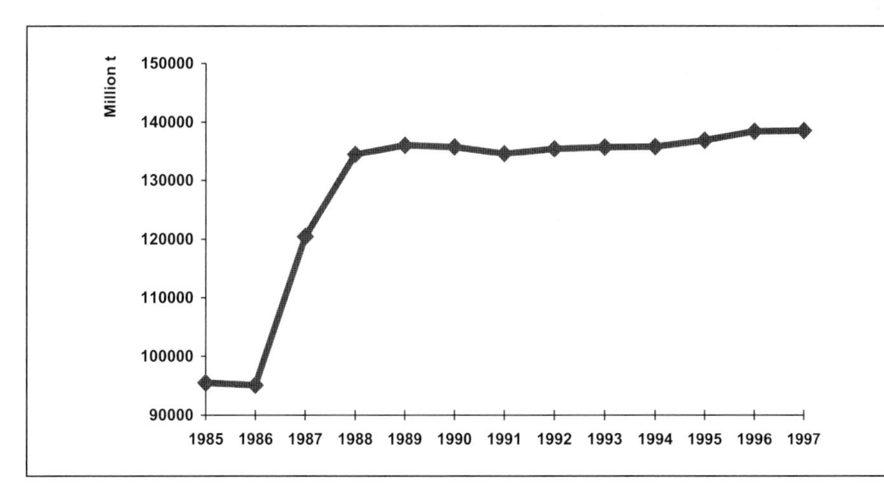

Fig. 13: Historical development of world oil reserves[4]

The Countries with the Most Petroleum

The fraction of world oil reserves associated with the **ten countries with the largest reserves** increased over the period 1985–1987 from 80% to 86.2% (119 410 million metric tons), as shown in Fig. 14. This can be attributed largely to strong growth in the oil reserves of the Near East (Fig. 15), the home of the five countries with the highest reserves. Saudi Arabia in 1997 reported 25.5% of the known worldwide reserves of petroleum, and it was also the **world's largest producer of oil**. The Saudis held a substantial lead over Iraq. Kuwait, the United Arab Emirates, and Iran, all with roughly comparable petroleum reserves, round out the list of leaders. Venezuela and Mexico, both of which are in the list of top ten oil producers, are also among the ten countries with the largest reserves (in sixth and eighth place, respectively). This applies also to China, as well as to the CIS, which was the third largest oil producer in the world. The CIS and Mexico are the only ones among the top ten oil producers whose known oil reserves fell between 1985 and 1997.

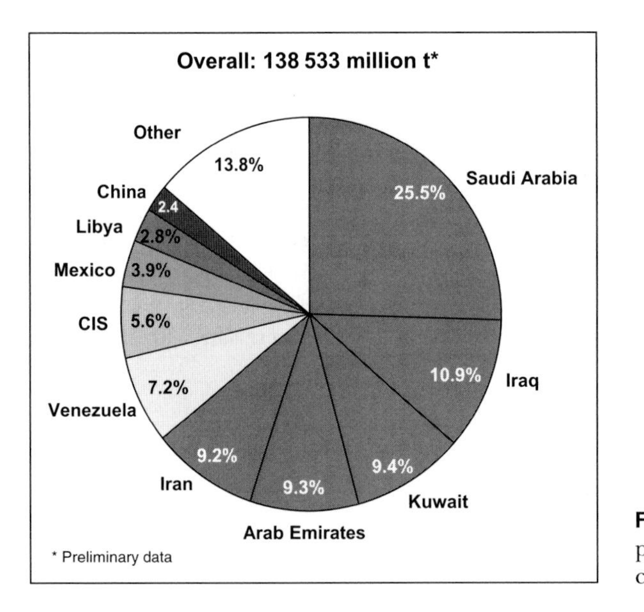

Fig. 14: Distribution of proven oil reserves by country, 1997*[4]

The largest consumer of petroleum by far, and the world's second largest oil producer – the United States – interestingly is not among the top ten countries in the world with respect to oil reserves. For several years the USA has been forced to import more oil than it produces, and the trend is an increasing one.

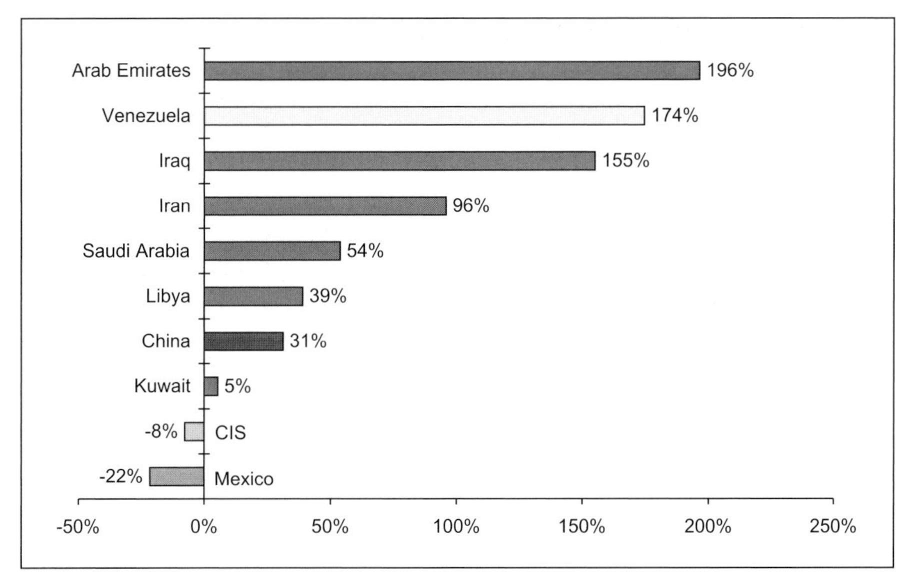

Fig. 15: Changes in known reserves in the principal petroleum-producing nations, 1985–1997[4]

The Leading Oil Producers

In 1997 the ten **leading oil-producing countries** generated 2251.8 million metric tons of petroleum, which represented 64.2% of total world output (3508.8 million tons). More than half the production of the top ten was attributable to a leading trio of Saudi Arabia, the United States, and the CIS (Fig. 16).

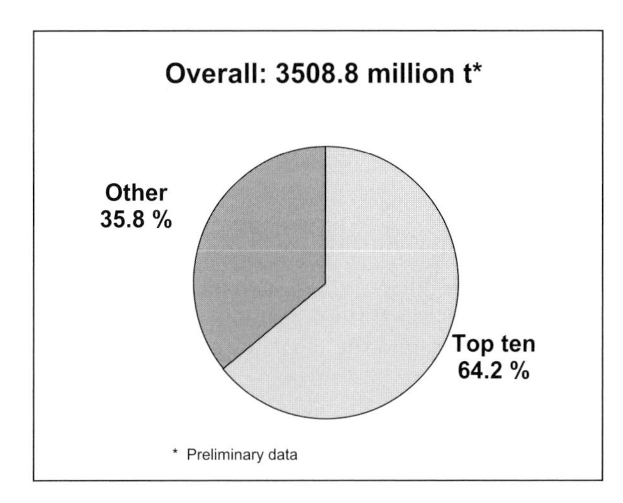

Fig. 16: Petroleum production, 1997*[4]

It is intriguing to compare the ranking of oil-producing nations in 1997 with that in 1985. In contrast to the ranking for consuming nations, dramatic changes within the leading countries themselves are clearly reflected in the production data. The leader in 1985, then known as the Soviet Union, dropped from 595 to 357.7 million tons, so that the newly named CIS ranked only third by 1997. This decline is readily understandable against the background of internal political and economic changes that occurred there, but it is still surprising that over the same period the United States suffered a production decline from 491.3 to 398.1 million tons while remaining in second place. Increased consumption with decreased production of course means that the United States became even more dependent upon oil imports between 1985 and 1997. Over that same period, Saudi Arabia, a crucial source of petroleum supplies for the West, increased its production from 158.2 to 415.9 million tons, thus becoming the largest oil producer in the world (Fig. 17).

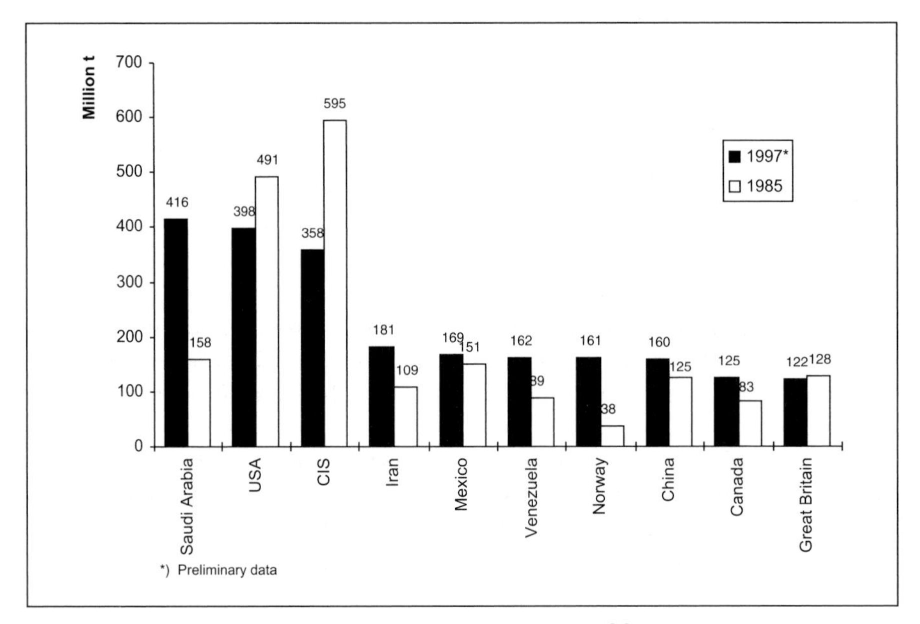

Fig. 17: The top ten petroleum producers, 1985 and 1997[4]

The only European countries appearing on the list are Norway (160.8 million tons) and Great Britain (122.3 million tons), in seventh and tenth place, respectively. Nevertheless, the Norwegian production figures were nearly comparable to those of China, a country displaying a dramatic **rate of increase** in oil consumption. Even so, China's increase in domestic production was still insufficient to keep pace with demand. Production exceeded consumption in 1995 by only 19.3 million tons (13.8%), whereas the surplus in 1985 was 47.7 million tons (61.9%).

The Leading Oil Consumers

In 1997, 61.8% of the world oil consumption was attributable to the **ten most important consuming nations** (Fig. 18), although their share of the demand for this valuable energy source actually declined from 66.2% in 1985 to 61.8%. Still, their absolute consumption in tons increased significantly: from 1846 million metric tons in 1985 to 2059 million tons in 1997.[4]

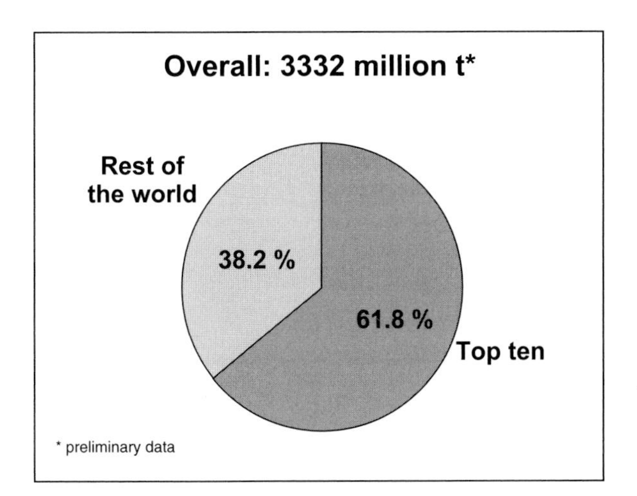

Fig. 18: The share of world oil consumption consigned to the top ten consuming nations, 1997[4]

All the countries in the list registered higher petroleum consumptions in 1997 than in 1985 with the single exception of the CIS, where a dramatic economic retreat caused the use of oil to fall from 397.9 million tons in 1985 to 220.3 million tons in 1997.

The rank-order list of leading oil consumers was headed by the United States in 1997 just as in 1985. The U.S. was by far the largest oil consumer in the world, accounting for 25.1% of total world oil consumption. Moreover, America's "thirst for oil" has increased by 15% since 1985 (Fig. 19).

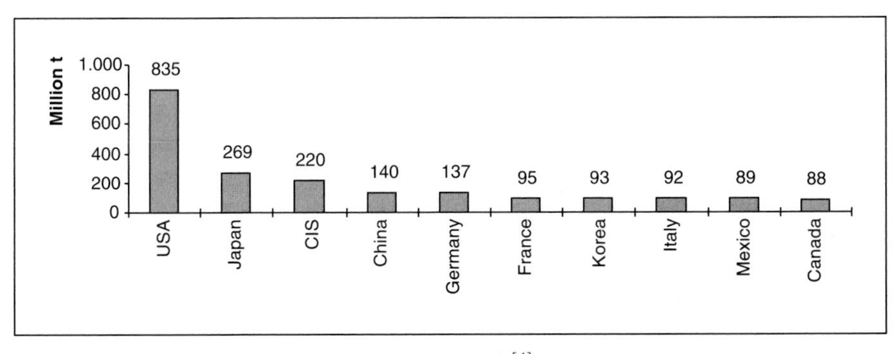

Fig. 19: The leading oil-consuming nations, 1997[4]

Much higher rates of increase (Fig. 20) were reported, as anticipated, by the rapidly developing South Korea (246%) and China (82%). In fact, China has overtaken the European countries one by one, and by 1996 it surpassed Germany for fourth place among oil-consuming nations.

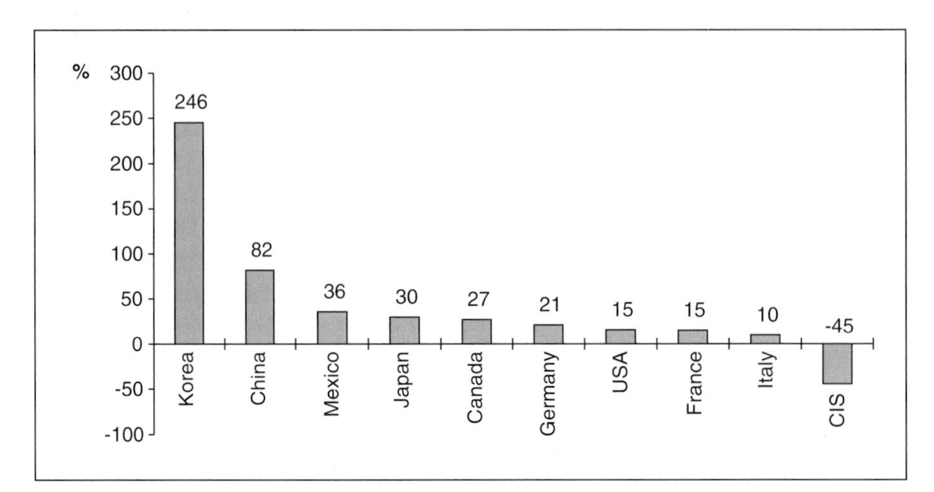

Fig. 20: Changes in oil consumption, 1985–1997[4]

Petroleum as a Raw Material

The entire output of the chemical industry can be traced back to ca. 300 basic chemicals, many of which are molecules ultimately acquired most readily and economically from petroleum and natural gas. In 1991 these two raw materials accounted for 90% of the raw material input into the German chemical industry, and **petroleum played the dominant role** (Fig. 21). The dependence upon this raw material is not expected to change in the foreseeable future.

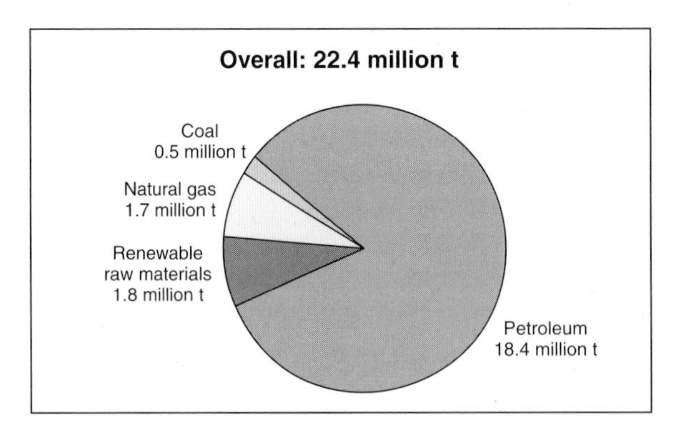

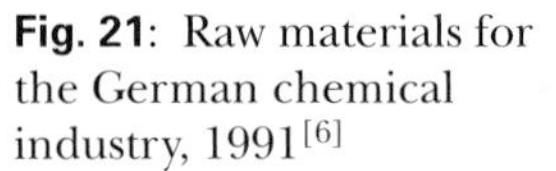
Fig. 21: Raw materials for the German chemical industry, 1991[6]

[6] data from: Verband der Chemischen Industrie (VCI).

Interestingly, only 6–7% of the annual world output of petroleum is directed toward enrichment. Over 93% is consigned to direct combustion in power plants, furnaces, and motors as a source of energy.

Renewable Raw Materials

Since the oil crises in the 1970s there has once again been increasing utilization of renewable raw material sources, reversing a steady downward trend that began in the opening years of the twentieth century. In the German chemical industry, renewable raw material consumption rose from about one million tons annually to 1.8 million tons in 1991, corresponding to ca. 8% of the total consumption of raw materials of all types. The **leading substances** in this category are oils and fats, most of which serve as starting points for surfactants and fabric softeners (Fig. 22).

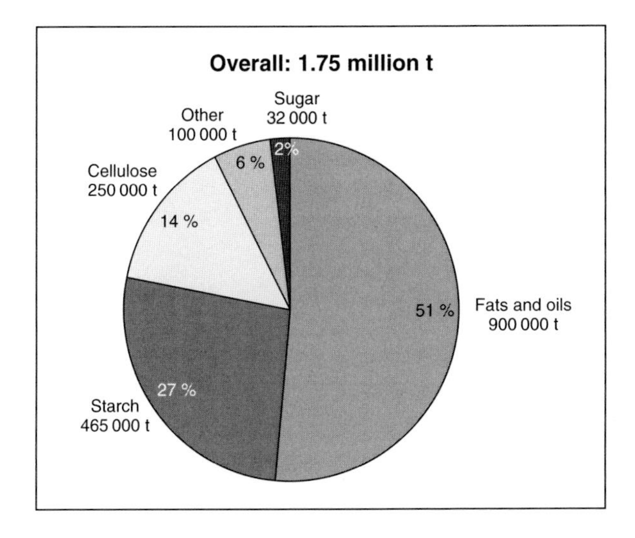

Fig. 22: Renewable raw materials in the German chemical industry, 1991[6]

The use of renewable raw materials has failed to increase markedly since 1991 mainly because the most attractive applications from a technical and economic standpoint have already largely been saturated. These raw materials are most suitable for situations in which nature's synthetic skills can be utilized efficiently – in other words, for transformations that retain natural structural elements and which are otherwise difficult to realize synthetically. It is from this perspective that renewable resources represent an interesting and forward-looking supplement to further petrochemical exploitation.

Alkylpolyglycosides – The Only Surfactant Category Prepared on an Industrial Scale from Sugars and Vegetable Oils

Worldwide consumption in 1995 of laundry detergents, cleansing agents, and personal hygiene products was estimated at 43 million metric tons. Surfactants represent an important constituent of all these products. The **surfactant with the longest history**, soap, faced its first serious competition early in the present century with the introduction of mechanical laundry equipment. This competition took the form of more effective substitutes or additives, such as fatty alcohol sulfates and fatty alcohol polyethyleneglycol ethers. A surfactant is characterized by a hydrophilic "head" and a hydrophobic hydrocarbon "tail" to bring together water and water-insoluble residues. By collecting at surfaces between water and air or water and oil, surfactants act to lower the surface tension. In the case of soaps and detergents, this promotes contact between the wash water and soiled materials (Fig. 23).

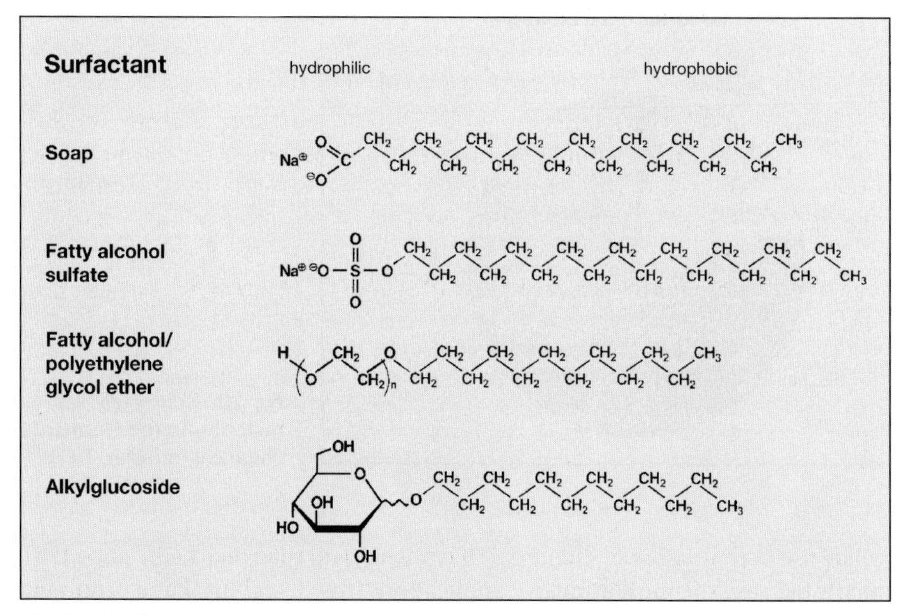

Fig. 23: Surfactants

Renewable raw materials like natural fats and oils can play a role comparable to that of petrochemicals as a basis for the hydrophobic hydrocarbon tail in a surfactant. The combination of fatty alcohols from coconut or palm kernel oil with glucose from cornstarch as a hydrophilic head leads to the class of compounds known as alkylpolyglycosides (APGs), which recently made history as the first surfactants prepared on an industrial scale exclusively from renewable raw materials.[7]

Emil Fischer (1852–1914, winner of the Nobel Prize for chemistry in 1902) first investigated the synthesis of alkylglycosides from short-chain alcohols and glucose nearly a hundred years ago. Together with Burckhardt Helferich (1887–1982) he was able to synthesize alkylglycosides with longer hydrocarbon chains and surfactant properties in a multistep process in 1911. No serious thought was given to large-scale production, however, because the process involved several intermediates and considerable effort in purification. This class of surfactants thus remained a rarity for a long time, with only limited applications.

Interest in the subject was rekindled a few years ago with a view to producing environmentally friendly laundry and cleansing agents as well as cosmetics. Application studies revealed that alkylpolyglycosides with hydrocarbon chains 8 to 18 carbon atoms in length would be capable of replacing certain surfactants then in common use. Moreover, in particular combinations with other surfactants the new materials displayed valuable synergistic properties. For example, the overall surfactant level in formulations for high-quality detergents could be reduced by their introduction without any penalty in terms of performance.

Toxicological and ecological laboratory tests also produced positive results. Alkylpolyglycosides are tolerated well by the eyes, the skin, and the mucous membranes, and they even reduce the irritation otherwise provoked by surfactant combinations. Moreover, these compounds are **completely biodegradable** both aerobically and anaerobically. For this reason, alkylpolyglycosides are entitled to a second round of superlatives: they represent the **first category of surfactants to be assigned to "Class 1" as the least damaging to water quality**.

[7] *Alkyl Polyglycosides – Technology, Properties, and Applications* (K. Hill, W. von Rybinski, G. Stoll, Eds.), VCH, Weinheim, **1997**; also W. von Rybinski, K. Hill. *Angew. Chem.* **1998**, *110*, 1394–1412; *Angew. Chem. Int. Ed. Engl.* **1998**, *37*, 1328–1345.

The Most Stable Carbanions

Anionic intermediates derived from carbon-containing compounds are extremely important in organic synthesis. Countless synthetic sequences depend on such alkyl anions as the strong, homogeneous base butyl lithium, but beyond that the enormous category of enolates with all their synthetic potential also lies within the realm of carbanion chemistry. The structural and chemical characteristics of carbanions are heavily dependent on what counterion is present (typically a metal ion) as well as on the solvent, which makes it much more difficult to ascribe clear-cut record features to carbanions themselves.

Carbon–metal bonds usually display a considerable amount of covalent-bonding character, so investigation of free carbanions is possible only under special conditions. Such conditions were achieved by Olmstead and Power[1] by complexing the cations from $Li^+[Ph_2CH]^-$ and $Li^+[Ph_3C]^-$ with a crown ether, thereby permitting the acquisition of X-ray crystal structures of the free carbanions **1** and **2** (Fig. 1). It turned out that, as a result of extensive delocalization with the adjacent π systems, the methide carbons in **1** and **2** are planar and exist in a trigonal-planar environment. The complexed compounds are very stable in the solid state at 0 °C in the absence of air and moisture, but decompose in THF solution after a few days even at –20 °C.

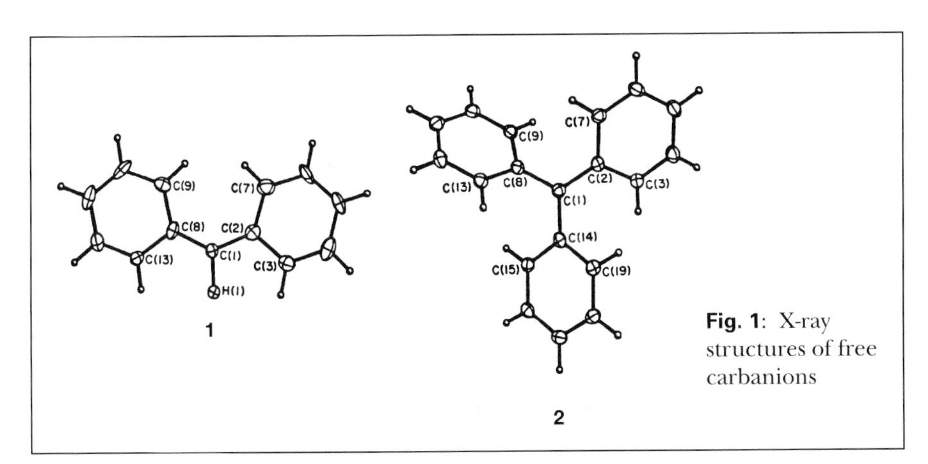

Fig. 1: X-ray structures of free carbanions

A Hydrocarbon Salt

One very noteworthy **chemical curiosity** is attributable to the research team of Okamoto,[2] which prepared the **especially stable carbanion 3** (Fig. 2) in the

[1] M. M. Olmstead, P. P. Power, *J. Am. Chem. Soc.* **1985**, *107*, 2174–2175.
[2] K. Okamoto, T. Kitagawa, K. Takeuchi, K. Komatsu, T. Kinoshita, S. Aonuma, M. Nagai, A. Miayabo, *J. Org. Chem.* **1990**, *55*, 996–1002.

presence of several stable carbocations, including the tricyclopropylcyclopropenyl ion **4** (→ Reactive Intermediates, Carbocations). These species can then combine to give **salts comprised entirely of carbon and hydrogen**. That the resulting compounds really are salts was confirmed by UV/Vis and IR spectroscopy as well as by conductivity measurements in DMSO. The deep green hydrocarbon solids are stable for more than a year in the dark at 10 °C. Salt [**3** · **4**] displays unusual behavior in solution: NMR spectroscopy indicates that after several hours in $CHCl_3$ at 20 °C its ions unite to form a covalent compound, conferring a brown color on the solution. If the chloroform is subsequently removed *in vacuo* or the solution is cooled to –78 °C the green hydrocarbon salt reappears quantitatively. This hydrocarbon is therefore the **only known covalent compound that exists only in chloroform solution**.

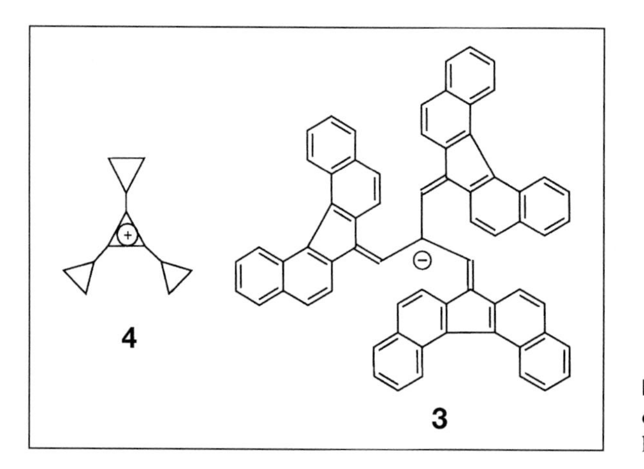

Fig. 2: The anion (and counterion) of a stable hydrocarbon salt

The Carbanion with the Highest Charge

The discovery of fullerenes added a new dimension to the chemistry of **carbon polyanions**. The **molecule with the highest electrical charge** so far observed is the hexaanion of fullerene C_{60}. Since C_{60} has unoccupied low-lying molecular orbitals with threefold degeneracy, it is very easy to reduce electrochemically to a hexaanion.[3] A synthesis of this anion has also been carried out in the solid state.[4] Reaction of elemental potassium with C_{60} leads to the compound K_6C_{60}, which has been characterized in various ways, including by solid-state NMR spectroscopy.

[3] Q. Xie, E. Pérez-Cordero, L. Echegoyen, *J. Am. Chem. Soc.* **1992**, *114*, 3978–3980.
[4] R. Tycko, G. Dabbagh, M. J. Rosseinsky, D. W. Murphy, R. M. Fleming, A. P. Ramirez, J. C. Tully, *Science* **1991**, *253*, 884–886.

The Most Stable Carbenes

It was long considered self-evident that carbon can only form stable compounds at room temperature in its tetravalent state. Divalent carbon compounds, so-called carbenes, were indeed recognized as very flexible species for preparative applications, but until a few years ago they were regarded as so highly reactive that they could be detected only as intermediates in the gas phase or at low temperature. This assumption was called into question in 1991 by the work of Arduengo, et al.[1] In one of the most exciting developments in recent years in organic structural chemistry, this research group succeeded in synthesizing the crystalline carbene **1**, which proved to be stable even at room temperature, and in examining its solid-state structure by X-ray crystallography (Fig. 1). Carbene **1** can be stored indefinitely in the absence of air and moisture. It can be recrystallized from toluene, has a melting point of 240–241 °C, and can be recovered unchanged from its melt!

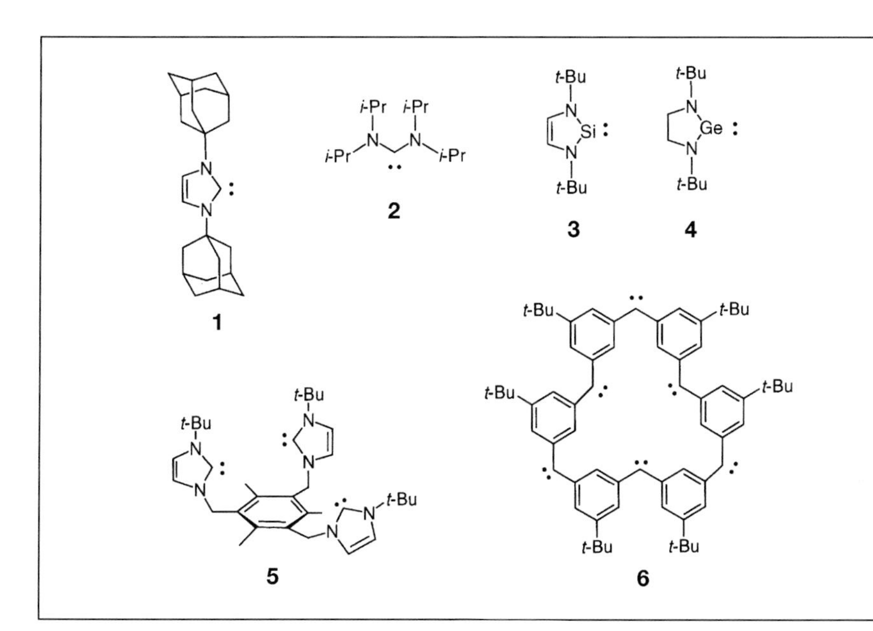

Fig. 1: Stable carbenes, silylenes, and germylenes

This discovery opened the door widely to further developments in the chemistry of stable carbenes. In the meantime, several derivatives of this type of carbene have been synthesized, and even the open-chain species **2** was found

[1] A. J. Arduengo III, R. L. Harlow, M. Kline, *J. Am. Chem. Soc.* **1991**, *113*, 361–363. A. J. Arduengo III, H. V. Rasika Dias, R. L. Harlow, M. Kline, *J. Am. Chem. Soc.* **1992**, *114*, 5530–5534.

to be stable.[2] The analogous silylene **3**[3] and germylene **4**[4] have also been made. The stability of carbenes of the imidazol-2-ylidene type raised the possibility as well that polycarbenes might be prepared through utilization of the right type of backbone. The **record holder in this category** so far is the tricarbene **5**, synthesized by Dias and Jin.[5] That this is far from the last word in this particular chain of developments is obvious from the hexacarbenes prepared by Iwamura, et al.,[6] for which **6** serves as an example. As a representative of the diarylmethylidene carbene family, **6** is not nearly as stable as **1**; its photolytic preparation from hexadiazo precursors and its investigation require working at 2 K (–271 °C) in solid 2-methyltetrahydrofuran. The benzene rings appear to behave as if they were isolated, so the six carbene centers do not react with each other in any way. Molecules of this type are of interest as potential organic ferromagnets.

[2] R. W. Alder, P. R. Allen, M. Murray, A. G. Orpen, *Angew. Chem.* **1996**, *108*, 1211–1213; *Angew. Chem. Int. Ed. Engl.* **1996**, *35*, 1121–1123.
[3] M. Denk, R. Lennon, R. Hayashi, R. West, A. V. Belyakov, H. P. Verne, A. Haaland, M. Wagner, N. Metzler, *J. Am. Chem. Soc.* **1994**, *116*, 2691–2692.
[4] W. A. Herrmann, M. Denk, J. Behm, W. Scherer, F.-R. Klingan, H. Bock, B. Solouki, M. Wagner, *Angew. Chem.* **1992**, *104*, 1489–1492; *Angew. Chem. Int. Ed. Engl.* **1992**, *31*, 1485.
[5] H. V. R. Dias, W. Jin, *Tetrahedron Lett.* **1994**, *35*, 1365–1366.
[6] K. Matsuda, N. Nakamura, K. Takahashi, K. Inoue, N. Koga, H. Iwamura, *J. Am. Chem. Soc.* **1995**, *117*, 5550–5560.

The First Carbocations

Carbocations are considered alongside carbanions, carbenes, and radicals among the most important reactive intermediates in organic chemistry. Their history dates back to the early days of this century when in 1901 both Norris[1] and Kehrmann[2] independently noticed the yellow color of solutions of triphenylmethanol (**1**) in sulfuric acid (Fig. 1). The following year, Adolf von Baeyer[3] (→ Appendix, Nobel Prize Winners: in Chemistry; Molecular Energy, Strain) recognized that the resulting compound **2** must be a salt, the cation of which we would today characterize as a carbocation. Finally, Hans Meerwein in 1922 postulated that carbocations are reactive intermediates in the transformation of 2-chloro-2,3,3-trimethylbicyclo[2.2.1]heptane (**3**, "camphene hydrochloride") into 2-chloro-1,7,7-trimethylbicyclo[2.2.1]heptane (**4**, "isobornyl chloride"), as shown in Fig. 2.[4] He thus established the basis for investigation of Wagner–Meerwein rearrangements, named in his honor. From these humble beginnings, carbocations have developed into a central motif in organic chemistry. Several standard works are dedicated to this theme,[5] and G. A. Olah, one of the founders of modern carbocation chemistry, was recognized with the Nobel Prize in 1994 (→ Appendix, Nobel Prize Winners: in chemistry) for his discovery that carbocations can be stable in solution and thus lend themselves to spectroscopic investigation.

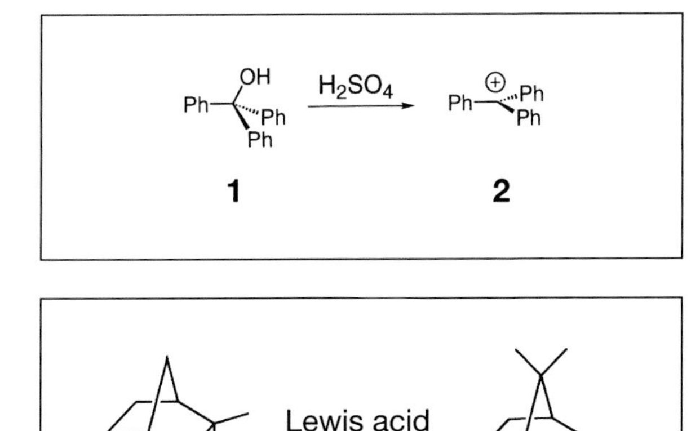

Fig. 1: The reaction of triphenylmethanol with sulfuric acid

Fig. 2: Transformation of camphene hydrochloride into isobornyl chloride

[1] J. F. Norris, *J. Am. Chem. Soc.* **1901**, *25*, 117.
[2] F. Kehrmann, F. Wentzel, *Ber. Dt. Chem. Ges.* **1901**, *34*, 3815–3819.
[3] A. Baeyer, V. Villiger, *Ber. Dt. Chem. Ges.* **1902**, *35*, 1189–1201.
[4] H. Meerwein, K. van Emster, *Chem. Ber.* **1922**, *55*, 2500–2528.
[5] Among others: G. A. Olah, P. v. R. Schleyer, *Carbonium Ions*, Vols. 1–5, Wiley, New York, **1968–1976**. P. Vogel, *Carbocation Chemistry*, Elsevier, Amsterdam, **1985**. M. Hanack (Ed.), *Houben-Weyl: Methoden der Organischen Chemie*, Vol. 19c, Thieme, Stuttgart, **1990**.

The Most Stable Carbocations

The pK_{R+} value[6] has established itself as the common measure of stability for long-lived carbocations, as defined by the position of equilibrium upon reaction of the carbocation with water:

$$R^+ + H_2O \rightleftharpoons ROH + H^+$$

$$pK_{R^+} = \log \frac{[R^+]}{[ROH]} + H_R$$

where H_R is the acidity function of the reaction medium (→ Atoms and Molecules, Acids and Bases). For example, in dilute aqueous solution the pK_{R+} value corresponds to the pH. It is apparent from this definition that carbocations are more stable the higher their pK_{R+} values. Table 1 provides a few representative values of pK_{R+}. The great stability of the triphenylmethyl cation, the structure of whose perchlorate salt has even been investigated in the solid state,[7] is a consequence of delocalization of the positive charge by the neighboring aromatic π system. It is therefore not surprising that introduction of electron-donating groups into the *para* positions in the phenyl rings increases the stability of the carbocation further. In this way it is possible to increase the pK_{R+} value from –6.63 in **2** (Fig. 1) to +0.82 in the trimethoxy derivative. An even more stable triarylmethyl cation is the triply substituted dimethylamino compound, with a pK_{R+} of +9.36.

Table 1: pK_{R+} values of selected carbocations.[a]

Carbocation	pK_{R+} value
triphenylmethyl (**2**)	–6.63
4-methoxytriphenylmethyl	–3.40
4,4′-dimethoxytriphenylmethyl	–1.24
4,4′,4″-trimethoxytriphenylmethyl	+0.82
4,4′,4″-tris(dimethylamino)triphenylmethyl	+9.36

[a] Taken from F. A. Carey, R. J. Sundberg, *Advanced Organic Chemistry*, 3rd ed., Plenum Press, New York, **1993**.

Still higher pK_{R+} values can be achieved with carbocations in which the cation is an integral part of an aromatic system in the sense of the Hückel rule. This is the case, for example, for the very smallest aromatic system, the cyclopropenyl cation with 2 π electrons, as well as for the cycloheptatrienyl cation with 6 π electrons. The **champions in stability** among carbocations (Fig. 3) are derivatives of these parent compounds, namely the tricyclopropylcyclopro-

[6] F. A. Carey, R. J. Sundberg, *Advanced Organic Chemistry*, 3rd ed., Plenum Press, New York, **1993** pp. 260–273.
[7] A. H. Gomes de Mesquita, C. H. MacGillavry, K. Eriks, *Acta Crystallogr.* **1965**, *18*, 437–443.

penyl cation **5**[8, 9] and a tropylium ion **6**[10] annelated with three bicyclic systems, with pK_{R+} values of 10.0 and 13.0, respectively. An illustration of the great stability of these carbocations is provided by the physical properties of **5**:[8] Its tetrafluoroborate is a white solid, stable to air and quite soluble in water. This salt has a melting point of 141–142 °C and can be recovered unchanged from its melt after an hour at 150 °C.

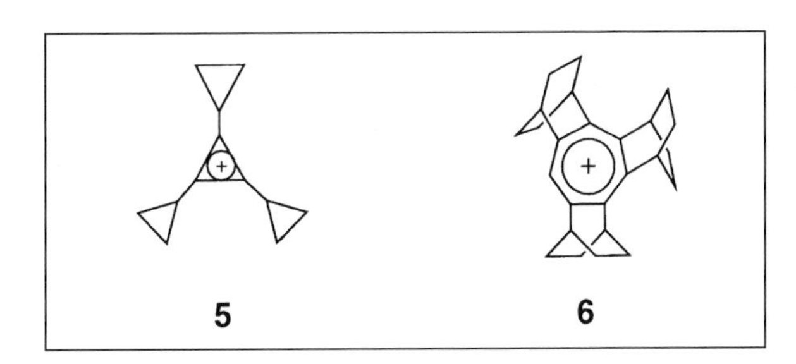

Fig. 3: Especially stable carbocations

Aliphatic carbocations are much less stable than the cations discussed above. The presence of the methyl cation, for example, has still not been demonstrated in solution.[11] One of George Olah's greatest accomplishments has been the generation of alkyl carbocations in the ionizing, non-nucleophilic media of the superacids (→ Atoms and Molecules, Acids and Bases) and then investigation of the solutions spectroscopically. A breakthrough came with NMR characterization of the *tert*-butyl cation **7** (Fig. 4).[12] Another milestone in this regard was the **first X-ray structural analysis of an aliphatic carbocation**, the 3,5,7-trimethyladamantan-1-yl ion **8** prepared by the methods of Olah.[13] Progress did not stop at this point, however; recently Olah et al.[14] reported the preparation of **9**, the first stable tetracation. Compound **9** is stable in the temperature range from –80 to –40 °C in a solution of SO_2ClF containing fluorosulfonic acid, decomposing only gradually at higher temperature.

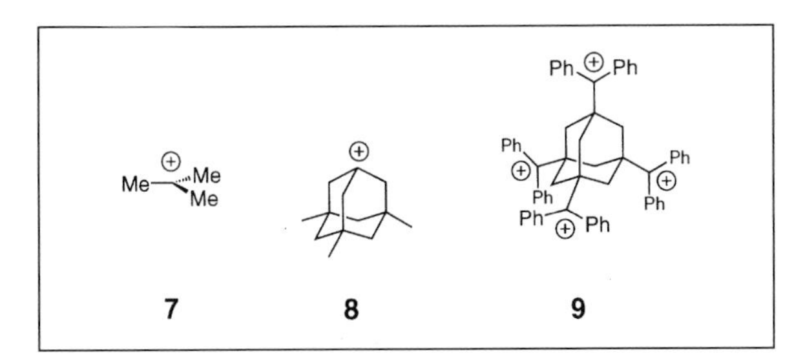

Fig. 4: Aliphatic carbocations

[8] K. Komatsu, I. Tomioka, K. Okamoto, *Tetrahedron Lett.* **1980**, *21*, 947–950.
[9] R. A. Moss, R. C. Munjal, *Tetrahedron Lett.* **1980**, *21*, 1221–1222.
[10] K. Komatsu, H. Akamatsu, Y. Jinbu, K. Okamoto, *J. Am. Chem. Soc.* **1988**, *110*, 633–634.
[11] See G. A. Olah, D. J. Donovan, H. C. Lin, *J. Am. Chem. Soc.* **1976**, *98*, 2661–2663.
[12] G. A. Olah, W. S. Tolgyesi, S. J. Kuhn, M. E. Moffatt, I. J. Bastien, E. B. Baker, *J. Am. Chem. Soc.* **1963**, *85*, 1328–1334.
[13] T. Laube, *Angew. Chem.* **1986**, *98*, 368–369; *Angew. Chem. Int. Ed. Engl.* **1986**, *25*, 349.
[14] N. J. Head, G. K. S. Prakash, A. Bashir-Hashemi, G. A. Olah, *J. Am. Chem. Soc.* **1995**, *117*, 12005–12006.

The long search for free cations of the higher homologues of carbon – silyl and germyl cations – also came recently to a successful end (Fig. 5). After more than ten years of experiments, Lambert and Zhao[15] managed to overcome the high electrophilicity that had prevented their earlier isolation of a free silyl cation. Thus, they were able to synthesize and characterize by NMR spectroscopy the trimesitylsilyl ion **10** with a tetrakis(pentafluorophenyl)borate counterion. Ion **10** is stable in solution at room temperature for several weeks. Almost at the same time, Sekiguchi et al.[16] obtained the first free germyl cation, **11**, by treatment of a cyclotrigermene with the triphenylmethyl cation (**2**). Analogous to a cyclopropenyl salt, the compound containing **11** is a crystalline solid sensitive to air and moisture whose structure has been determined by X-ray crystallography.

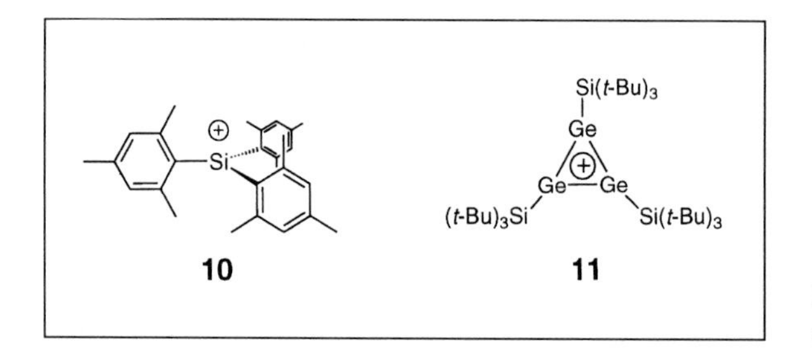

Fig. 5: Silyl and germyl cations

[15] J. B. Lambert, Y. Zhao, *Angew. Chem.* **1997**, *109*, 389–391; *Angew. Chem. Int. Ed. Engl.* **1997**, *36*, 400–401.
[16] A. Sekiguchi, M. Tsukamoto, M. Ichinohe, *Science* **1997**, *275*, 60–61.

Record-Threatening Radicals

Radicals in the chemical sense are atoms, molecules, or ions containing one or more unpaired electrons. The term is generally associated with high reactivity, and indeed many radicals are numbered among the unstable, reactive intermediates. But there also exist quite a few relatively stable (persistent) radicals. It would be difficult to overstate the importance of such radicals with respect to life itself. Without them the history of art and culture would be far less rich, sex would be much more difficult, and one can say without exaggeration that life as we know it would be absolutely impossible. To be more precise, without one particular **natural, stable radical**: a diradical, O_2. Without the oxygen in air we could not live. Another radical of this type is NO. The damaging effects of NO that accompany smog formation are well documented. For this reason attempts are made to remove it from automotive exhaust, the chief source of NO in the troposphere, with the aid of modern catalytic converters. It came as a complete surprise to learn that NO is also an extremely important biomolecule – a cyctotoxic agent in macrophages with neurotransmitter functions and the ability to regulate blood pressure.[1] NO is also very important as a physiological intermediary in male erection.[2] At present, NO is the **smallest known physiologically active neutral substance** (→ Pharmaceuticals, Hit List).[3] The remarkable fact that NO has reformed itself over the course of a few years from an "environmental poison" into a biological molecule with a fantastic diversity of functions induced the respected journal *Science* to declare this radical the "Molecule of the Year 1992."

The blue semiprecious stone *lapis lazuli*, incorporated for over five thousand years into jewelry, owes its color to the radical anion $S_3^{\cdot -}$.[4] The pigment ultramarine blue, famed for its stability to light and thus utilized by artists of every century despite its high cost, has been prepared from *lapis lazuli* since at least the seventh century (→ Pigments). The oldest German frescoes containing ultramarine are in a church built around 1130 by Bishop Sigwardus of Minden in Idensen in an area called the "Steinhuder Meer." Apart from these **stable inorganic radicals**, there is also known in the meantime a broad spectrum of stable organic radicals.[5] Some are even available commercially (Fig. 1), such as tetramethylpiperidin-*N*-oxide (TMPO) and 2,2-diphenyl-1-picrylhydrazyl (DPPH). Persistent organic radicals are actually present in living organisms! The active center of the R2 protein of the ribonucleotide reductase from the bacterium *Escherichia coli*, whose structure has been determined by X-ray crystallography,[6] contains a stable tyrosyl radical (Fig. 2).[7]

[1] See for example E. Culotta, D. E. Koshland, Jr., *Science* **1992**, *258*, 1862. H.-J. Galla, *Angew. Chem.* **1993**, *105*, 399; *Angew. Chem. Int. Ed. Engl.* **1993**, *32*, 378.
[2] A. L. Burnett, C. J. Lowenwstein, D. S. Bredt, T. S. K. Chang, S. H. Snyder, *Science* **1992**, *257*, 401.
[3] R. Henning, *Nachr. Chem. Tech. Lab.* **1993**, *41*, 412.
[4] F. Seel, G. Schäfer, H.-J. Güttler, G. Simon, *Chem. Unserer Zeit* **1974**, *8*, 65.
[5] See for example M. Ballester, J. Riera, J. Casta. S. K. Chang, S. H. Snyder, *J. Am. Chem. Soc.* **1971**, *93*, 2215.
[6] P. Nordlund, B.-M. Sjöberg, H. Eklund, *Nature (London)* **1990**, *345*, 593.
[7] See for example M. Fontecave, P. Nordlund, H. Eklund, P. Reichard, *Adv. Enzymol. Relat. Areas Mol. Biol.* **1992**, *65*, 147.

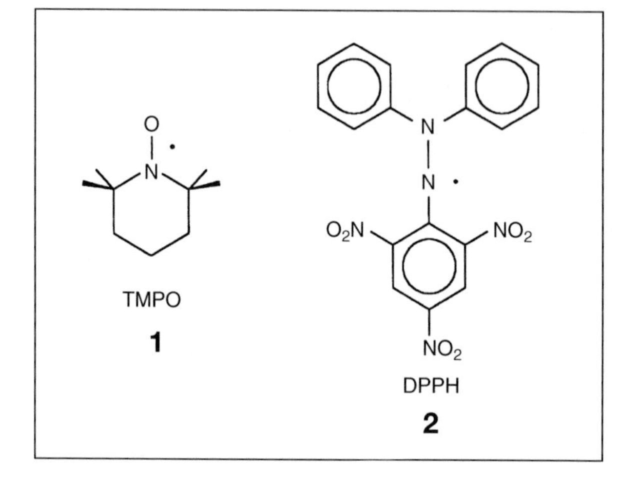

Fig. 1: Commercially available radicals

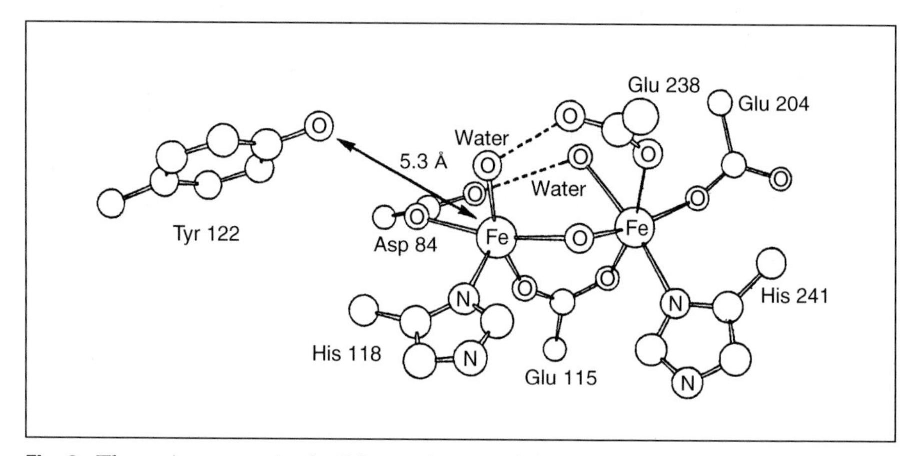

Fig. 2: The active center in the R2 protein, containing a stable tyrosyl radical (Tyr 122)

The Most Intense Aromas

"Chemistry is when it bangs and smells!" in the words of a popular definition of the discipline – though not one to be taken *too* seriously. It is certainly true, however, that the preparatively engaged chemist's sense of smell can be subject to a considerable amount of stress, at least in the case of a large laboratory ventilated only by a few rather sluggish fume hoods. But odors, and thus tastes as well, are a routine part of everyday life; chemists simply enjoy the special privilege of being in a position to provide precise identifications for the "stinkers" – i.e., those materials that stimulate our olfactory senses most disagreeably.

The chemistry of smells[a] has its own contributions to make under the heading of "records." For example, the definitive aromatic constituent of **grapefruit**, 1-*p*-menthen-8-thiol (**1**), classes as the compound with the **lowest known threshold value for taste**. And that value is a truly impressive one.[1] Thus, the (+)-*R* enantiomer of **1** is perceptible in water at a dilution of 2×10^{-5} ppb, which corresponds to a concentration of only 2 mg in 100 million liters of water (the volume of a 100 000-ton tanker), or 0.02 ng (i.e., two hundredths of a billionth of a gram) in a liter! It therefore comes as no surprise that the content of this substance in a grapefruit is measured in the sub-ppb range. It is also interesting that the human olfactory organs are able to distinguish the absolute configuration of the stereocenter in **1**, since the threshold value for the (–)-*S* enantiomer is roughly a factor of four higher. Qualitatively, both stereoisomers evoke the same typical grapefruit sensation.

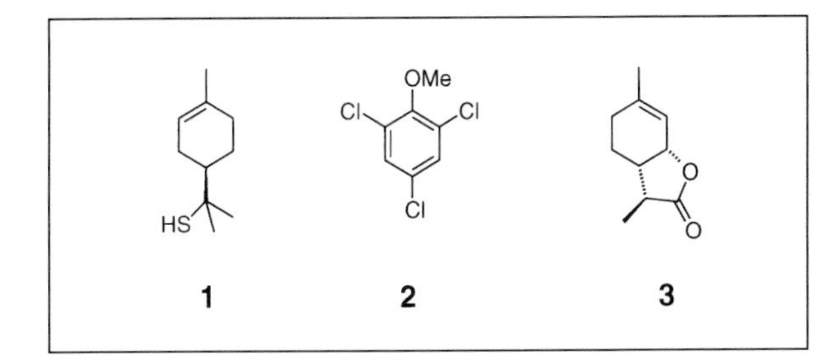

Fig. 1: Compounds with low threshold values for taste and smell

The Substance with the Lowest Aroma Threshold is found in ... Wine

Connoisseurs with sensitive noses and palates were long ago attracted to the charms of wine – and not without reason, since some of the **substances perceptible at the very lowest concentrations** are present in this fruit of the vine. One of them is better absent, however: 2,4,6-trichloroanisole (**2**), the

[1] E. Demole, P. Enggist, G. Ohloff, *Helv. Chim. Acta* **1982**, *65*, 1785–1794.

compound that confers on bad wines a typical "cork" flavor.[2] Worst of all, practiced wine tasters are able to detect **2** at concentrations as low as 10 ppt (or 10 ng per liter). The average wine imbiber would require the presence of about twice that much. It is estimated that the human nose is capable of perceiving as little as 10^{-12} g (one picogram). Compounds like **2** are presumably produced by microbial means directly within the cork itself,[3] or else they result from chorine-containing insecticides applied to the wood in wine cellars that somehow contaminate the wine via the corks.[4] On the other hand, another trace component is a very welcome guest in red and white wines. The so-called wine lactone **3** was recently[5] identified as the aromatic substance that confers a coconut-like, sweet, and slightly woody bouquet. The **low aroma threshold value** for this compound is again impressive. Wine lactone is detectable at a concentration of 0.01 picogram (10^{-14} g) in a liter of air. It thus exceeds the potency of the grapefruit chemical **1** by four orders of magnitude, a factor of ten thousand! Just as in the case of **1**, the human olfactory system displays great selectivity with respect to the wine lactone, again distinguishing one stereoisomer from the other. The enantiomer of **3** has an odor threshold level that exceeds one milligram per liter of air, a difference of eleven powers of ten. Nature was thus well-advised to take advantage of the more effective of the two substances, since only minute traces are required to bring about the desired effects. Indeed, it would be necessary to extract ten liters of a good gewürztraminer wine in order to isolate one microgram of **3**.[6]

The "Hottest" Compound

The "hottest" compound known is probably capsaicin (**4**), the spicy constituent of peppers (*Capsicum annuum*), cayenne pepper, chili peppers, and other capsicum species (Fig. 2). Compound **4** was isolated as early as 1876,[7] so its activity spectrum has been extensively investigated. Anyone with a clear, tearful recollection of the initial burning sensation followed by nerve deadening that accompanies deposition of a drop of Tabasco Sauce on the tongue will not be particularly surprised to learn of attempts that have been made to employ **4** as a local anaesthetic.[8] In this context it is worth mentioning the discovery of the plant poison resiniferatoxin (**5**),[9] which at the neuron level displays the same mechanism of action as capsaicin, albeit with 100–10 000

[2] H. R. Buser, C. Zanier, H. Tanner, *J. Agric. Food Chem.* **1982**, *30*, 359–362.
[3] W. R. Sponholz, H. Muno, *Wein-Wiss.* **1994**, *49*(1), 17–22.
[4] P. Chatonnet, G. Guimberteau, D. Dubourdieu, J. N. Boidron, *J. Int. Sci. Vigne Vin* **1994**, *28*(2), 131–151 [*Chem. Abstr.* **1994**, *122*, 30 064].
[5] H. Guth, *Helv. Chim. Acta* **1996**, *79*, 1559–1571.
[6] H. Guth, personal communication.
[7] C. S. J. Walpole, R. Wrigglesworth, S. Bevan, E. A. Campbell, A. Dray, I. F. James, M. N. Perkins, D. J. Reid, J. Winter, *J. Med. Chem.* **1993**, *36*, 2362–2372.
[8] M. Tresh, *Pharm. J. Trans.* **1876**, 7–15.
[9] A. Szallasi, P. M. Blumberg, *Neuroscience* **1989**, *30*, 515–520.

times the effectiveness. It is unknown whether this advantage would be reflected in the compound's potency as a spice.

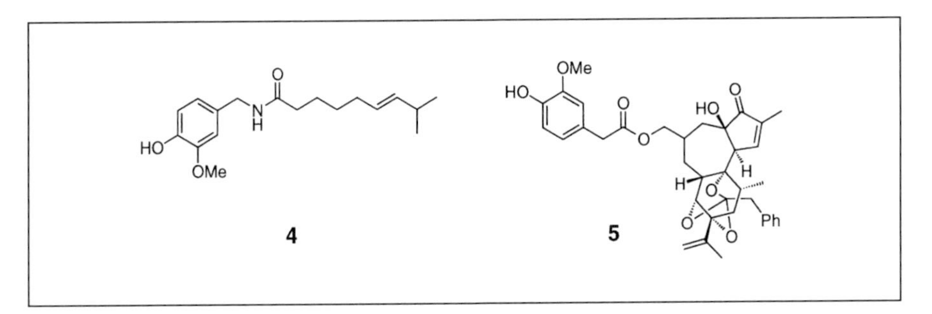

Fig. 2: "Hot" compounds

The Worst "Stinkers"

Whether a particular aroma is pleasant or foul is undoubtedly a matter shrouded in a great deal of subjectivity, and there are bound to be differences of opinion over whether the characteristic smell of garlic should be cited in the same context as that of the skunk. One thing the two have in common, however, irrespective of which is considered worst, is that both odors are derived from **sulfur-containing compounds** (Fig. 3), and in neither case has anyone yet shown a willingness to establish an odor threshold value. Garlic (*Allium sativum*) has the interesting attribute that while most people enjoy its taste, very few are impressed by reminders of it on someone else's breath. The aromatic agent in this case is the compound allicin (**6**). More and more people now seem willing to tolerate garlic's olfactory drawbacks in order to profit from the apparent health benefits conferred by this onion-like bulb. It is reported, for example, that garlic components are able to lower blood-cholesterol levels and reduce platelet aggregation, and both antimicrobial properties and a possible reduction in the threat from colon cancer have also been discussed. For reasons such as these, garlic products are now second only to echinacea extracts in the American market for plant-derived health products (→ Pharmaceuticals).[10] The potent secretions released by the skunk (*Mephitis mephitis*) in times of acute danger generate a far less ambiguous response. It has been known since 1975 that the culprits in this case are also sulfur-containing compounds, primarily the three substances **7–9**.[11] Anyone who has personally experienced the penetrating pungency of these compounds will welcome the knowledge that a patent (→ Patents) has been issued in defense of an anti-skunk shampoo.[12]

[10] R. Rawls, *Chem. Eng. News*, September 23, **1996**, pp. 53–60.
[11] K. K. Andersen, D. T. Bernstein, *J. Chem. Ecol.* **1975**, *1*, 493–499.
[12] C. J. Wiesner, US Patent 4834901 A 890530 [*Chem. Abstr.* **1989**, *111*, 102520].

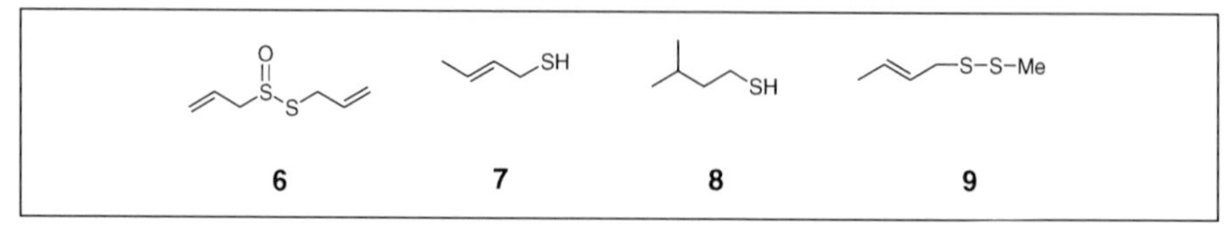

Fig. 3: Sulfur-containing compounds

The principle underlying this sensible invention is a 2% potassium iodate solution, which oxidizes compounds **7–9** to water-soluble, non-odoriferous sulfoxides, sulfates, or sulfones. A rather peculiar imagination must be the source of another patent,[13] however, one that covers a mobile synthetic chemistry kit for imitating "essence of skunk" as a way of covering up unpleasant human odors.

Secret Tip for Children and Other Snack Fans: The Sweetest Compounds

The trend in recent decades in the industrialized nations toward a diet rich in carbohydrates and fats, coupled with a simultaneous decrease in physical activity, has had obvious consequences with respect to corpulence.

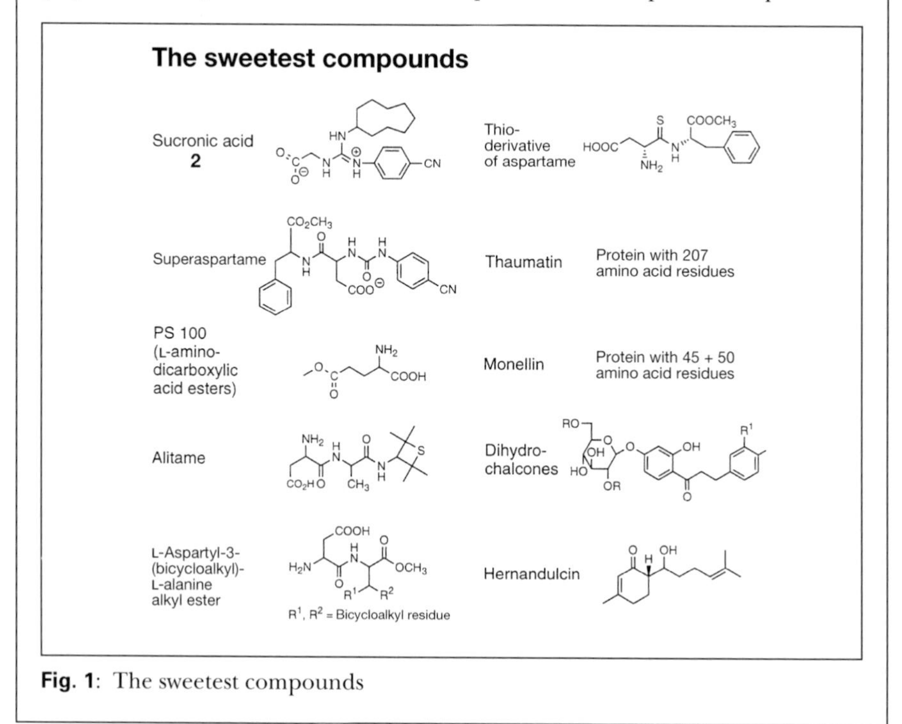

Fig. 1: The sweetest compounds

[13] A. F. Isbell, US Patent 4213875 800722 [*Chem. Abstr.* **1980**, *93*, 185748].

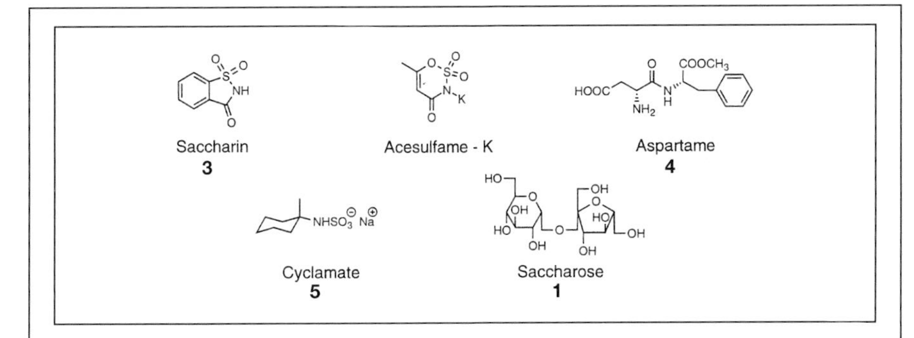

Table 1: Overview of sweeteners

Sweetener	Relative sweetness (sucrose = 1)
sucronic acid (the sweetest substance known)	200 000
thioureido derivatives of aspartame	50 000
superaspartame	8 000
thaumatin	2 000 – 3 000
PS 100	2 200
monellin	1 500 – 2 000
brazzein	500 – 2 000
alitame	2 000
dihydrochalcones	300 – 2 000
L-aspartyl-3-(bicycloalkyl)-L-alanine alkyl esters	1 900
saccharin	300
acesulfame-K	200
aspartame	180
cyclamate (cyclohexylsulfamic acid)	30
L-sugar (saccharose)	1

S. Marie, J. R. Piggott, in *Handbook of Sweeteners*, Blackie, London, **1991**.
T. H. Grenby in *Advances in Sweeteners*, Chapman & Hall, London, **1996**.

Partly for this reason, but also out of medical necessity associated with diabetics and others with metabolic problems, chemists have been actively searching for substances that possess sweetening powers exceeding those of the common sugar **1** derived from cane or beets and known chemically as sucrose. Relative sweetening power is established by quizzing participants in taste panels, who are provided with dilute samples of the compounds in question and then asked to rate subjectively the level of sweetness they perceive. By far the sweetest compound known is sucronic

[a] Since smell and taste are physiologically very closely coupled, both sensations have been treated together under odoriferous materials.

acid (**2**), which surpasses the sweetness of sugar by about 200 000-fold. Better-known synthetic sweeteners include saccharin (**3**; relative sweetness compared to sugar = 300), aspartame (**4**; 180), and cyclamate (**5**; 30). These are frequently found in products described as "light," "diet," or "low-calorie," but they offer nothing like the almost unimaginable effect on the tongue of sucronic acid.

The Smallest Chemical Signal

Plants and animals that are not equipped to communicate through sound utilize chemical compounds as a vehicle for transmitting signals into their environment. The **smallest known chemical signaling agent** is CO_2, a ubiquitous molecule that fulfills a long list of signaling functions. For example, it is an aggregation pheromone for fire ants of the genus *Solenopsis saevissima.*[1] Worker ants who find themselves outside their nests wander along a CO_2-concentration gradient to the point of highest CO_2 concentration, which is generally the nearest ant colony. Similarly, CO_2 serves as an attractant for larvae of the corn root worm *Diabrotica virgifera.*[2] Once they have hatched, the tiny larvae penetrate the earth for a distance of up to one meter in their search for nourishment, impelled by the attraction of CO_2 streaming out from corn roots. This apparently simple CO_2 molecule plays a remarkably complex role in the interaction between figs (*Ficus religiosa*) and fig wasps (*Blastophaga quadraticeps*) that live inside the ripening fruit. The ripening process itself causes a 10% increase in the CO_2 content of the figs relative to their surroundings, and this suffices to put female wasps into a slumbering state. The males, on the other hand, remain active, fertilize the females, and then leave the fruit. The hole they create allows the CO_2 level in the fruit to equilibrate with that outside, causing the females to awaken, at which point they leave the fruit as well, carrying pollen from the plant along with them.

The Lightest Signaling Agent

CO_2 may be the **simplest of all signaling agents**, since it consists of only three atoms, but the **lightest** is ethene (ethylene), C_2H_4 (→ Chemical Products, Hit List). Whereas ethene's function as a ripening hormone for plants has long been known, Bowles and coworkers[4] showed only recently that the same simple hydrocarbon also plays a central role in the healing process of plant tissue. For example, if leaves of the tomato plant are damaged by grazing insects, the plant begins to generate ethylene in a biochemical process. This ethylene then induces (together with jasmonic acid) the expression of proteinase inhibitor genes, which in turn direct the synthesis of proteins that inhibit insect feeding and also stimulate repairs.

---

[1] B. Hölldobler, E. O. Wilson, *The Ants*, Springer, Heidelberg, **1990**, p. 289.
[2] B. E. Hibbard, L. B. Bjostad, *J. Chem Ecol.* **1988**, *14*, 1523–1539.
[3] J. Galil, M. Zeroni, D. Bar Shalom, *New Phytol.* **1973**, *72*, 1113–1127.
[4] P. J. O'Donnell, C. Calvert, R. Atzkorn, C. Wasternack, H. M. O. Leyser, D. J. Bowles, *Science* **1996**, *274*, 1914–1917.

Another potential record-holder in the same category is NO, which fulfills a number of signaling functions in animal organs. Research in this area is still in full force, so the presentation of an award will need to be postponed for a later edition (→ Reactive Intermediates; Radicals)

Deadly CO_2 Gas Cloud from a Crater Lake

On August 21, 1986, a sudden eruption released a mighty cloud of CO_2 gas from Lake Nyos, a crater lake in Cameroon, that led to the death of 1700 people and over 3000 animals within a radius of 10 km around the lake. The gas had formed over the course of thousands of years in magma under the lake, and the lake water was completely saturated with it – probably even supersaturated. In this unstable situation some no longer precisely identifiable triggering event caused a rapid, eruptive gas release from the water which lasted only 15–20 seconds. It is estimated that a total of 1.2 cubic kilometers of CO_2 emerged, and because of its higher density relative to air this quickly distributed itself in topographically low-lying areas surrounding the lake and thereby caused the tragic deaths by suffocation.

G. W. Kling, M. A. Clark, H. R. Compton, J. D. Devine, W. C. Evans, A. M. Humphrey, E. J. Koenigsberg, J. P. Lockwood, M. L. Tuttle, G. N. Wagner, *Science* **1987**, *236*, 169–175.

Pheromones: Minute Quantities – Tremendous Effects

From the standpoint of simplicity, male hamsters turn out to be exceptionally easily satisfied.[5] Thus, a female hamster needs only to spray two nanograms of dimethyldisulfide (MeSSMe), a compound we humans find very offensive, in order to arouse sexual lust in a male hamster. By comparison, pheromone **10** (Fig. 1) of the ant genus *Atta texana* seems quite complex. The ants use this substance to mark their trails in the course of their wanderings.[6] As little as one milligram of it would suffice to plant such a track three times around the earth, so the ants easily make do with a supply of only ca. 3.3 ng as an accompaniment in their daily journeys.

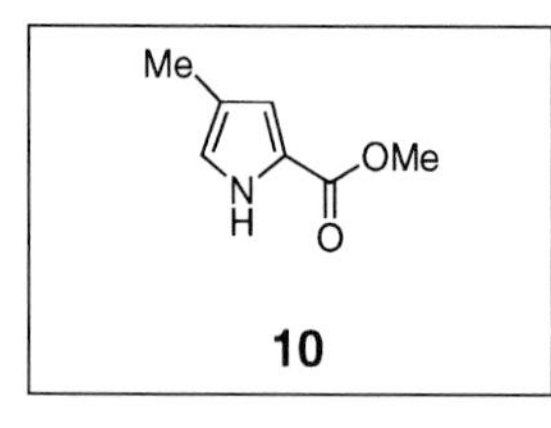

Fig. 1: An ant pheromone

---

[5] A. G. Singer, W. C. Agosta, R. J. O'Connell, C. Pfaffmann, D. V. Bowen, F. H. Field, H. Frank, *Science* **1976**, *191*, 948–950.
[6] Ref. [1], p. 246.

Records in Organic Synthesis

It would certainly not be difficult to list masterpieces in organic synthesis; indeed any attempt to provide a halfway complete account would quickly burst the bounds of this book. K. C. Nicolaou in his work *Classics in Total Synthesis*[1] has done such a didactically superb job of addressing excerpts from this theme that the present chapter is definitely not an attempt to follow his lead. We instead restrict ourselves to a few curiosities and records culled from the area of organic synthesis.

The Smallest Chiral Molecules

The tetrahedral arrangement of substituents about an sp^3-hybridized carbon atom produces the phenomenon of chirality ("handedness") with compounds that contain at least one carbon to which four mutually distinct groups are attached. Despite the differing spatial arrangements of the groups in the resulting enantiomers, the latter display identical physical and chemical properties in an isotropic medium such as a typical solvent. They differ dramatically in an anisotropic environment, however, as when they interact with polarized light or with a chiral binding site on an enzyme. The **smallest chiral molecules** (Fig. 1) have only one chiral center, as in the ethane **1** with two isotopic labels, or bromochlorofluoromethane (**2**). The chiral compound with the lowest molecular weight is **1**, possessing a molecular weight of only 33.08 g mol^{-1}. The molecule was prepared to facilitate a study of the stereochemical course of the oxidation of ethane to ethanol by the enzyme methanemonooxygenase.[2] A racemic mixture of substance **2** was synthesized as early as 1893,[3] but it was not separated gas chromatographically into its enantiomers until quite recently.[4] This compound is the smallest chiral species whose chiral center is not a function of multiple isotopes of a single element.

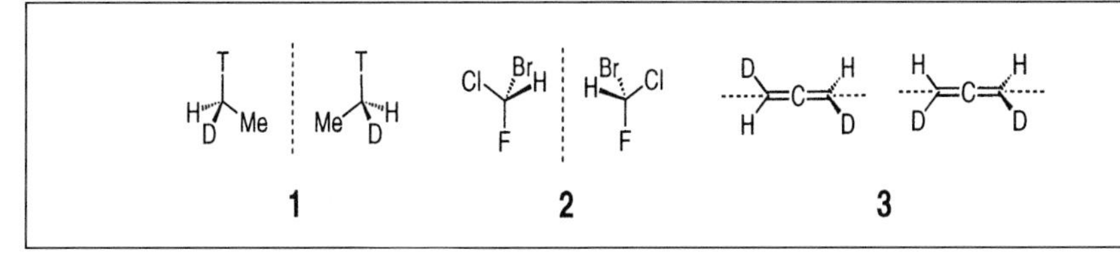

Fig. 1: The smallest chiral molecules

[1] K. C. Nicolaou, E. J. Sorensen, *Classics in Total Synthesis*, VCH, Weinheim, **1996**.

[2] A. M. Valentine, B. Wilkinson, K. E. Liu, S. Komar-Panicucci, N. D. Priestley, P. G. Williams, H. Morimoto, H. G. Floss, S. J. Lippard, *J. Am. Chem. Soc.* **1997**, *119*, 1818–1827.

[3] F. Swarts, *Ber. Dt. Chem. Ges.* (Referate) **1893**, *26*, 782.

[4] H. Grosenick, V. Schurig, J. Constante, A. Collet, *Tetrahedron Asymmetry* **1995**, *6*, 87–88.

The Simplest Axial-Chiral Compound

Chirality is not a function exclusively of stereogenic centers. It can also arise from an axis of chirality relative to which four substituents are arranged pairwise in an asymmetric way. The **simplest axial-chiral compound** is the doubly deuterated allene **3**,[5] a molecule that is at the same time the **lightest chiral compound** composed entirely of stable (i.e., nonradioactive) isotopes, with a molecular weight of 42.08 g mol^{-1}.

The Synthetic Molecule with the Largest Number of Stereocenters

Since most of the physiologically active compounds in nature are chiral but isolable only with great difficulty or only in minute quantities, considerable attention has been directed toward synthetic processes that permit the introduction of a stereogenic center with a particular desired absolute configuration. The **molecule with the largest number of stereocenters ever synthesized in the laboratory** may be the coral poison palytoxin (**4**; Fig. 2) (→ Poisons) with 64 stereocenters, prepared by Kishi et al.[6] It will be recalled that with n centers of chirality the number of possible stereoisomers is 2^n, so **4** should be capable of existing in 2^{64} different forms (without considering geometry about the double bonds). That means there might be ca. 1.8×10^{19} or 18 billion stereoisomers of **4**, only one of which is identical to the natural product. The stunning achievement embodied in the successful palytoxin synthesis underscores the remarkable level of stereochemical control exercised in construction of the complex molecular skeleton.

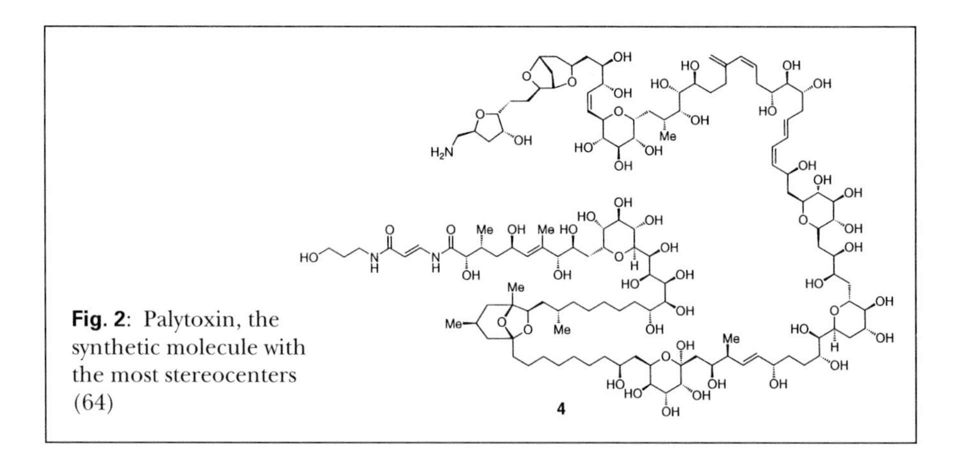

Fig. 2: Palytoxin, the synthetic molecule with the most stereocenters (64)

[5] G. M. Keserü, M. Nchurig, J. Constante, A. Collet, *Tetrahedron* **1997**, *53*, 2049–2054.
[6] E. M. Suh, Y. Kishi, *J. Am. Chem. Soc.* **1994**, *116*, 11 205–11 206 and references cited.

The Most Efficient Stereoselective Synthesis

Given the great flexibility of open-chain systems, controlling their stereochemistry can present a considerable challenge. What is the minimum number of **synthetic steps necessary for the selective construction of a stereocenter**? This is a question to which every synthetic chemist is likely to supply a different answer with respect to a particular target molecule. A possible record is embodied in the small number of steps Paterson required for the construction of one unique set of four stereocenters. In the course of synthesizing the antibiotic oleandomycin[7] his research group reacted the enolate of chiral ketone **5** (enantiomeric excess 97%) with 2-methylpropenal and subjected the product to catalytic hydrogenation to obtain in only two steps the C_7 building block **6** (diastereomeric excess 92%) containing four new stereocenters (Fig. 3). In other words, they reached their goal with only 0.5 reaction steps per stereocenter!

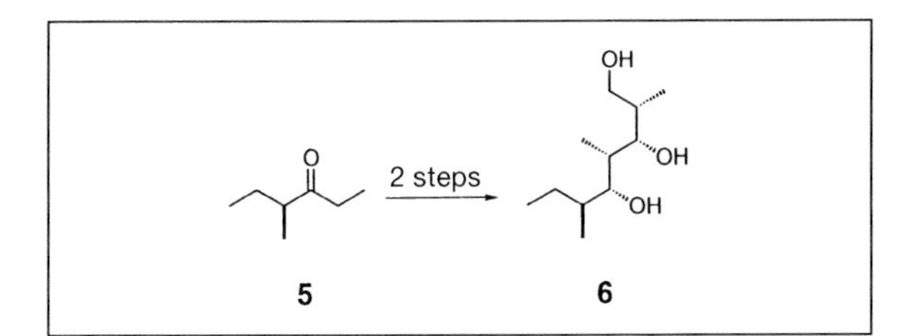

Fig. 3: Synthetic sequence containing the least steps per newly introduced stereocenter

Enzymes: Record-Holders for Efficiency and Selectivity

Enzymes manage to create complex molecules with even greater efficiency, and starting from relatively simple structures. An impressive example is vitamin B_{12} (**7**; → Chemical Products, Hit List: Vitamins), the classic total synthesis of which by Woodward's research group at Harvard and that of Eschenmoser at the ETH in Zurich kept nearly 130 graduate students and postdocs busy for ten years.[8] An enzymatic route (Fig. 4) to a direct precursor of **7**, hydrobyrinic acid **9**, was successfully pursued by Scott and coworkers.[9] Here, twelve enzymes were allowed to act upon 5-aminolevulinic acid (**8**) in a single reaction vessel, with the subsequent isolation of **9** in ca. 20% yield after 15 h, in the course of which 17 synthetic steps occurred.

[7] (a) I. Paterson, M. A. Lister, R. D. Norcross, *Tetrahedron Lett.* **1992**, *33*, 1767–1770. (b) I. Paterson, R. D. Norcross, R. A. Ward, P. Romea, M. A. Lister, *J. Am. Chem. Soc.* **1994**, *116*, 11 287–11 314.

[8] R. B. Woodward, *Pure Appl. Chem.* **1973**, *33*, 145; A. Eschenmoser, C. Wintner, *Science* **1977**, *196*, 1410.

[9] C. A. Roessner, J. B. Spencer, N. J. Stolowich, J. Wang, G. P. Nayar, P. J. Santander, C. Pichon, C. Min, M. T. Holderman, A. I. Scott, *Chem. Biol.* **1994**, *1*, 119–124.

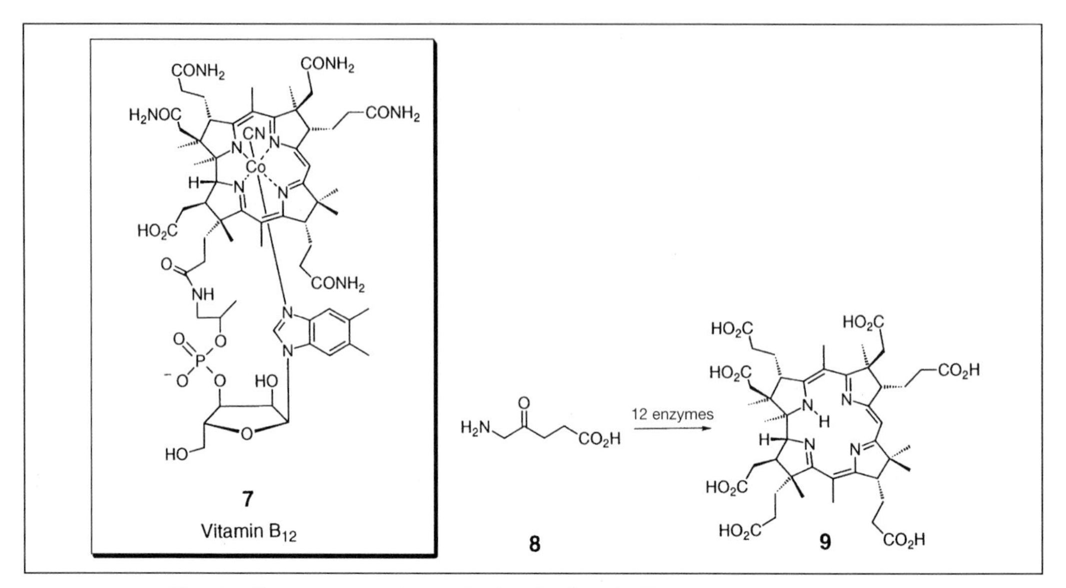

Fig. 4: Chemoenzymatic synthesis of a vitamin B_{12} precursor

The Reaction with the Most Components

In a similar way, Ugi introduces entire cocktails of chemical reagents into the **multicomponent reactions** named in his honor (Fig. 5). At the moment the award-winning representative takes advantage of a mixture of seven (!) components[10] in a single reaction vessel to generate as the sole major product along with sodium bromide and water the 1,3-thiazolidine **10** in 43% yield.

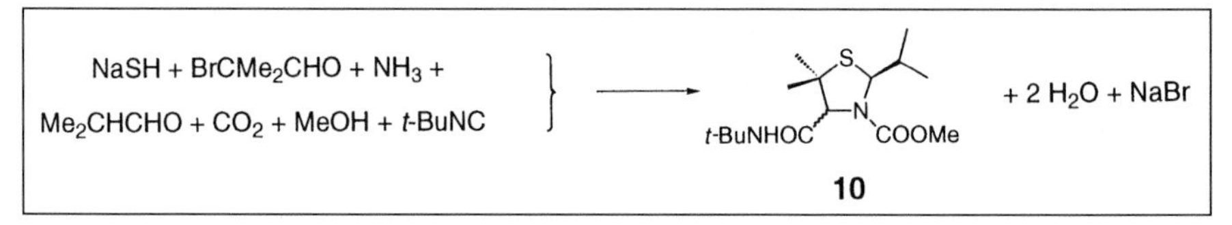

Fig. 5: Seven-component reaction developed by Ugi

The First "Computer Synthesis"

Quite a different approach to synthesis (more accurately: synthetic planning) has been pursued by Herges and his team.[11] With the aid of a computer and in the context of graph theory they conducted a search for possible pericyclic

[10] A. Dömling, I. Ugi, *Angew. Chem.* **1993**, *105*, 634–635; *Angew. Chem. Int. Ed. Engl.* **1993**, *32*, 960.
[11] R. Herges, C. Hoock, *Science* **1992**, *255*, 711–713.

reactions leading to conjugated dienes. Of the 72 reactions proposed by the computer, two were examples of reaction types at that time unknown. After further quantum chemical optimization, both processes were in fact realized in the laboratory. One resulted in 1,3-butadiene (**12**) and CS_2 in 95% yield from the trithiocarbonate **11** by heating with a phosphine. This butadiene synthesis thus qualifies as the **first experimentally verified new chemical reaction devised entirely by a computer** (Fig. 6).

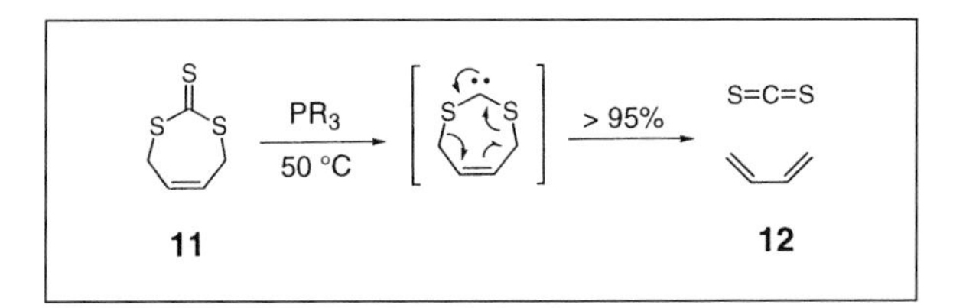

Fig. 6: A successful computer-developed synthetic strategy

The Reaction with the Longest Name

They started for generations of chemistry students as unpleasant, brain-tiring exercises in memorization, but were gradually transferred after repeated utilization into the realm of augury. What we're alluding to here are "name reactions," which have a long tradition in chemistry, especially in organic chemistry. It would be difficult to say which name reactions over the years have proven the most difficult to memorize, but certainly the **length of the name itself** would not contribute in a positive way to the process (let alone to an understanding of the reaction). Among the longest names still in current use for organic reactions is the **Meerwein–Ponndorf–Verley reduction**[12] (34 letters), which represents the reduction of a ketone by propan-2-ol and aluminum triisopropanolate (Fig. 7).

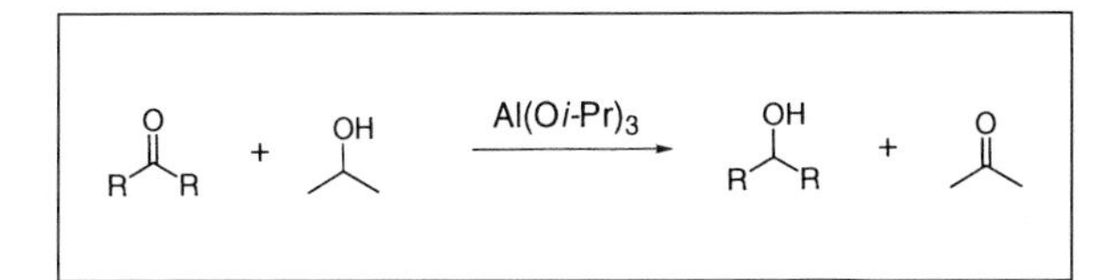

Fig. 7: The Meerwein–Ponndorf–Verley reduction

Somewhat longer still is the name applied to a rearrangement from carbohydrate chemistry: the **Lobry-de-Bruyn–von-Ekenstein reaction**[13] (37 letters), which describes the equilibrium between glucose, fructose, and mannose (Fig. 8).

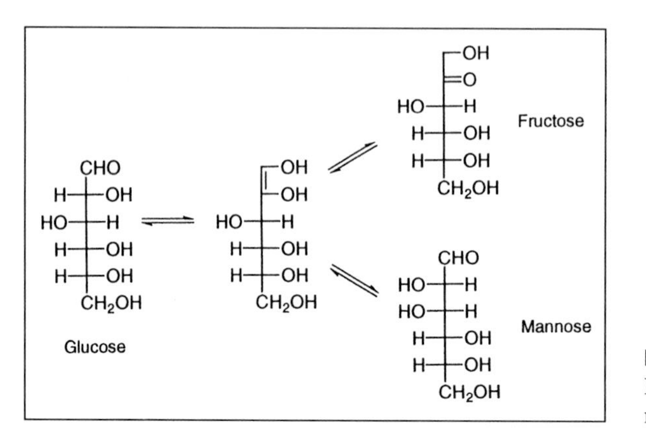

Fig. 8: The Lobry-de-Bruyn–von-Ekenstein reaction

The probable champion in this category is somewhat less generally familiar. It is the **Buchner–Curtius–Schlotterbeck reaction** (38 letters), with the aid of which unsymmetrical ketones can be prepared from diazo compounds and aldehydes.[14]

$$RCH_2N_2 + R'CHO \longrightarrow RCH_2COR' + N_2$$

Fig. 9: The Buchner–Curtius–Schlotterbeck reaction

Clearly the length of a reaction's name has nothing whatsoever to do with its utility. How might one in fact establish which is the **synthetically most important of the host of name reactions**? Some indication can be obtained from a scrutiny of the *Chemical Abstracts Database* (→ Literature). There one finds that the **Grignard reaction**, with 19 241 citations, commands a decisive lead over both the **Diels–Alder** (12 992 citations) and the **Wittig** (11 536 citations) reactions. It is no coincidence that all four of the chemists singled out in the names of these reactions have been awarded Nobel Prizes in chemistry (→ Nobel Prize, Records).

[12] T. Laue, A. Plagens, *Namen and Schlagwort-Reaktionen der Organischen Chemie*, Teubner, Stuttgart, **1994**, p. 221.
[13] P. Collins, R. Ferreir, *Monosaccharides*, Wiley, Chichester, **1995**, p. 139.
[14] J. B. Bastus, *Tetrahedron Lett.* **1963**, 955–958.

Appendices

1998	Walter Kohn (USA)	Development of the density-functional theory
	John A. Pople (USA)	Development of computational methods in quantum chemistry
1997	Paul D. Boyer (USA) John E. Walker (United Kingdom) Jens C. Skou (Denmark)	First discovery of an ion-transporting enzyme, Na^+/K^+-ATPase
1996	Robert F. Curl, Jr. (USA) Sir Harold W. Kroto (United Kingdom) Richard E. Smalley (USA)	Discovery of fullerenes
1995	Paul J. Crutzen (Netherlands) Mario J. Molina (USA) F. Sherwood Rowland (USA)	Chemistry of the atmosphere, especially with respect to the formation and decomposition of ozone
1994	George A. Olah (Hungary)	Contributions to the chemistry of carbocations
1993	Kary B. Mullis (USA)	Development of the polymerase chain reaction for replication of DNA
	Michael Smith (United Kingdom)	Invention of cite-specific mutagenases based on oligonucleotides and their application in the study of proteins
1992	Rudolph A. Marcus (Canada)	Contributions to the theory of electron transfer processes in chemical systems
1991	Richard R. Ernst (Switzerland)	Contributions to development of methods in high-resolution nuclear magnetic resonance (NMR) spectroscopy
1990	Elias James Corey (USA)	Development of the theory and methodology of organic synthesis
1989	Thomas R. Cech (USA) Sidney Altman (USA)	Discovery of the catalytic activity of RNA molecules (ribozymes)

1988	Johann Deisenhofer (Germany) Robert Huber (Germany) Hartmut Michel (Germany)	Crystallization and X-ray structural analysis of the photosynthetic reaction center of the bacterium *Rhodopseudomonas viridis*
1987	Donald J. Cram (USA) Jean-Marie Lehn (France) Charles J. Pedersen (Norway)	Development and utilization of molecules with highly selective structure-specific interactions
1986	Dudley R. Herschbach (USA) Yuan T. Lee (Taiwan) John C. Polanyi (Canada)	Contributions to the dynamics of elementary chemical processes
1985	Herbert A. Hauptman (USA) Jerome Karle (USA)	Development of "direct methods" for the determination of crystal structures
1984	Robert Bruce Merrifield (USA)	Development of a method for chemical synthesis in a solid matrix
1983	Henry Taube (Canada)	Mechanisms of electron transfer, especially in metal complexes
1982	Sir Aaron Klug (Lithuania)	Development of crystallographic electron microscopy and the structural elucidation of biologically important nucleic acid–protein complexes
1981	Kenichi Fukui (Japan) Roald Hoffmann (USA)	Development of theories regarding the course of chemical reactions
1980	Paul Berg (USA)	Biochemistry of nucleic acids, especially with respect to recombinant DNA
	Walter Gilbert (USA) Frederick Sanger (United Kingdom)	Determination of the base sequences in nucleic acids
1979	Herbert C. Brown (United Kingdom) Georg Wittig (Germany)	Development of boron and phosphorus compounds into important reagents in organic synthesis

1978	Peter D. Mitchell (United Kingdom)	Contributions to the understanding of biological energy-transfer processes through the chemiosmotic theory
1977	Ilya Prigogine (USSR)	Nonequilibrium thermodynamics, especially the theory of dissipative structures
1976	William N. Lipscomb (USA)	Investigation of the structure of boranes in order to clarify problems related to chemical bonding
1975	Sir John Warcup Cornforth (United Kingdom)	Stereochemistry of enzyme-catalyzed reactions
	Vladimir Prelog (Bosnia)	Stereochemistry of organic molecules and reactions
1974	Paul J. Flory (USA)	Fundamental accomplishments both theoretical and experimental in the physical chemistry of macromolecules
1973	Ernst Otto Fischer (Germany) Sir Geoffrey Wilkinson (United Kingdom)	Chemistry of organometallic sandwich compounds
1972	Christian B. Anfinsen (USA)	Studies of ribonuclease, especially regarding the relationship between amino acid sequence and biologically active conformations
	Stanford Moore (USA) William H. Stein (USA)	Contributions to an understanding of the relationship between chemical structure and catalytic activity in the active centers of ribonucleases
1971	Gerhard Herzberg (Germany)	Contributions to an understanding of the electronic structures and geometries of molecules, especially radicals
1970	Luis F. Leloir (Argentina)	Discovery of sugar nucleotides and their role in the biosynthesis of carbohydrates

1969	Sir Derek H. R. Barton (United Kingdom) Odd Hassel (Norway)	Development of the concept of conformation and its applications in chemistry
1968	Lars Onsager (Norway)	Discovery of the reciprocity relationship named for him, one of the fundamental laws of the thermodynamics of irreversible processes
1967	Manfred Eigen (Germany) Ronald George Wreyford Norrish (United Kingdom) Lord George Porter (United Kingdom)	Investigation of extremely fast reactions by the disruption of equilibrium through pulsed energy input
1966	Robert S. Mulliken (USA)	Fundamental work related to chemical bonding and the electronic structure of molecules through the molecular orbital (MO) theory
1965	Robert Burns Woodward (USA)	Outstanding achievements in the art of organic synthesis
1964	Dorothy Crowfoot Hodgkin (United Kingdom)	X-ray structural analysis of biochemically important molecules
1963	Karl Ziegler (Germany) Giulio Natta (Italy)	Discoveries in the chemistry and technology of high-polymeric compounds
1962	Sir John Cowdery Kendrew (United Kingdom) Max Ferdinand Perutz (Austria)	Investigations into the structure of globular proteins
1961	Melvin Calvin (USA)	Studies related to CO_2 assimilation in plants
1960	Willard Frank Libby (USA)	The use of ^{14}C for age determinations in archeology, geology, geophysics, and other disciplines
1959	Jaroslav Heyrovsky (Czechoslovakia)	Discovery and development of polarography as a method in analytical chemistry

1958	Frederick Sanger (United Kingdom)	Structure of proteins, especially insulin
1957	Lord Alexander R. Todd (United Kingdom)	Nucleotides and nucleotide coenzymes
1956	Sir Cyril Norman Hinshelwood (United Kingdom) Nikolay Nikolaevich Semenov (USSR)	Mechanisms of chemical reactions
1955	Vincent du Vigneaud (USA)	Synthesis of biologically important sulfur compounds, especially the synthesis of the first polypeptide hormone
1954	Linus Carl Pauling (USA)	Studies of the nature of the chemical bond and their application to the structural elucidation of complex compounds
1953	Hermann Staudinger (Germany)	Discoveries in the field of macromolecular chemistry
1952	Archer John Porter Martin (United Kingdom) Richard Laurence Millington Synge (United Kingdom)	Development of distribution chromatography
1951	Edwin Mattison McMillan (USA) Glenn Theodore Seaborg (USA)	Discovery of the transuranium elements
1950	Otto Paul Hermann Diels (Germany) Kurt Alder (Germany)	Discovery and development of the diene synthesis
1949	William Francis Giauque	Contributions in the field of chemical thermodynamics, especially with regard to the behavior of compounds at extremely low temperatures
1948	Arne Wilhelm Kaurin Tiselius (Sweden)	Studies of electrophoresis and adsorption analysis, and especially the discovery of the complex structure of serum proteins
1947	Sir Robert Robinson (United Kingdom)	Investigation of plant constituents, especially alkaloids, with respect to their biological significance

1946	James Batcheller Sumner (USA)	Discovery that enzymes can be crystallized
	John Howard Northrop (USA) Wendell Meredith Stanley (USA)	Isolation of enzymes and virus proteins in pure form
1945	Artturi Ilmari Virtanen (Finland)	Studies in the field of agricultural and food chemistry, and especially methods for conserving fodder
1944	Otto Hahn (Germany)	Discovery of nuclear fission
1943	George de Hevesy (Hungary)	Use of isotopes as markers in the study of chemical reactions
1942	No prizes awarded	
1941	No prizes awarded	
1940	No prizes awarded	
1939	Adolf Friedrich Johann Butenandt (Germany)	Studies of sexual hormones
	Leopold Ruzicka (Austria–Hungary)	Studies of polymethylene compounds and higher terpenes
1938	Richard Kuhn (Austria)	Studies of carotenoids and flavins
1937	Sir Walter Norman Haworth (United Kingdom)	Investigations in the field of carbohydrates and vitamin C
	Paul Karrer (Switzerland)	Investigations in the field of carotenoids, flavin, and vitamins A and B_2
1936	Petrus Josephus Wilhelmus Debye (Netherlands)	Contributions to the understanding of molecular structure through the investigation of dipole moments, and of the diffraction of X-rays and electrons in gases
1935	Frédéric Joliot (France) Irène Joliot-Curie (France)	Synthesis of new radioactive elements
1934	Harold Clayton Urey (USA)	Discovery of heavy hydrogen
1933	No prizes awarded	

Year	Laureate	Achievement
1932	Irving Langmuir (USA)	Discoveries and investigations in the field of surface chemistry
1931	Carl Bosch (Germany) Friedrich Bergius (Germany)	Discovery and development of methods in high-pressure chemistry
1930	Hans Fischer (Germany)	Constitution of hematin and chlorophyll, and especially the synthesis of hematin
1929	Sir Arthur Harden (United Kingdom) Hans Karl August Simon von Euler-Chelpin (Germany)	Investigations of sugar fermentation and fermentation enzymes
1928	Adolf Otto Reinhold Windaus (Germany)	Constitution of sterols and their relationship to vitamins
1927	Heinrich Otto Wieland (Germany)	Constitution of the bile acids and related compounds
1926	Theodor Svedberg (Sweden)	Studies of disperse systems
1925	Richard Adolf Zsigmondy (Austria)	Proof of the heterogeneous nature of colloidal solutions by methods that have since become the basis of modern colloid chemistry
1924	No prizes awarded	
1923	Fritz Pregl (Austria)	Discovery of the microanalysis of organic compounds
1922	Francis William Aston (United Kingdom)	Mass spectrometric discovery of isotopes of a large number of nonradioactive elements, and formulation of the integer rule
1921	Frederick Soddy (United Kingdom)	Contributions to an understanding of the chemistry of radioactive compounds and investigations into the origin and properties of isotopes
1920	Walter Hermann Nernst (Germany)	Studies in thermochemistry

1919	No prizes awarded	
1918	Fritz Haber (Germany)	Synthesis of ammonia from its elements
1917	No prizes awarded	
1916	No prizes awarded	
1915	Richard Martin Willstätter (Germany)	Studies of plant pigments, especially chlorophyll
1914	Theodore William Richards (USA)	Exact determination of the atomic weights of a large number of chemical elements
1913	Alfred Werner (Germany)	Studies of the bonds between atoms in molecules, shedding new light on previous investigations and opening new research areas especially in inorganic chemistry
1912	Victor Grignard (France)	Discovery of the Grignard reaction
	Paul Sabatier (France)	Method for the hydrogenation of organic compounds through the presence of finely divided metals
1911	Marie Curie neé Sklodowska (Poland)	Discovery of the elements radium and polonium, isolation of radium, and investigations of the properties and compounds of this element
1910	Otto Wallach (Germany)	Fundamental studies in the area of alicyclic compounds
1909	Wilhelm Ostwald (Latvia)	Studies of catalysis and investigations into the fundamental laws of chemical equilibrium and rates of chemical reactions
1908	Lord Ernest Rutherford (New Zealand)	Investigations into the fission of elements and the chemistry of radioactive compounds
1907	Eduard Buchner (Germany)	Studies in the field of biochemistry and cell-free fermentation

1906	Henri Moissan (France)	Isolation of fluorine and development of an electric furnace
1905	Johann Friedrich Wilhelm Adolf von Baeyer (Germany)	Advancements in organic chemistry and the chemical industry through studies of organic dyes and hydroaromatic compounds
1904	Sir William Ramsay (United Kingdom)	Discovery of the noble gases as constituents of atmospheric air and of their position in the periodic table of the elements
1903	Svante August Arrhenius (Sweden)	Theory of electrolytic dissociation
1902	Hermann Emil Fischer (Germany)	Synthesis of sugar and purine compounds
1901	Jacobus Henricus van't Hoff (Netherlands)	Discovery of the laws of chemical dynamics as well as of osmotic pressure in solutions

1998	Robert F. Furchgott (USA) Louis J. Ignarro (USA) Ferid Murad (USA)	Discoveries concerning nitric oxide as a signaling molecule in the cardiovascular system
1997	Stanley B. Prusiner (USA)	Discovery of prions, a new biological principle of infection
1996	Peter C. Doherty (Austria) Rolf M. Zinkernagel (Switzerland)	Discovery of the specificity of cell-mediated immune response
1995	Edward B. Lewis (USA) Christian Nüsslein-Volhard (Germany) Eric F. Wieschaus (USA)	Discoveries in the field of genetic control in the early stages of embryo development
1994	Alfred G. Gilman (USA) Martin Rodbell (USA)	Discovery of the G-proteins and their role in signal transduction in cells
1993	Richard J. Roberts (United Kingdom) Philip A. Sharp (USA)	Independent discovery of mosaic genes
1992	Edmond H. Fischer (China) Edwin G. Krebs (USA)	Discovery of reversible protein phosphorylation as a biological regulating mechanism
1991	Erwin Neher (Germany) Bert Sakman (Germany)	Discovery of the function of single-ion channels in cells
1990	Joseph E. Murray (USA) E. Donnall Thomas (USA)	Basic studies related to organ and cell transplants
1989	J. Michael Bishop (USA) Harold E. Varmus (USA)	Discovery of the cellular origin of retroviral oncogenes
1988	Sir James W. Black (United Kingdom) Gertrude B. Elion (USA) George H. Hitchings (USA)	Discovery of important principles in the treatment of disease with medication
1987	Susumu Tonegawa (Japan)	Discovery of genetic principles underlying the diversity of antibodies
1986	Stanley Cohen (USA) Rita Levi-Montalcini (Italy)	Discovery of growth factors
1985	Michael S. Brown (USA) Joseph L. Goldstein (USA)	Discoveries related to the regulation of cholesterol metabolism

Year	Laureates	Discovery
1984	Niels K. Jerne (Denmark) Georges F. Köhler (Germany) César Milstein (Argentina)	Theories regarding specificity in the development and control of the immune system; preparation of monoclonal antibodies
1983	Barbara McClintock (USA)	Discovery of mobile genetic elements
1982	Sune K. Bergström (Sweden) Bengt I. Samuelsson (Sweden) Sir John R. Vane (United Kingdom)	Discoveries in the field of prostaglandins and related biologically active substances
1981	Roger W. Sperry (USA)	Discovery of functional specialization in the hemispheres of the brain
	David H. Hubel (Canada) Torsten N. Wiesel (Sweden)	Discoveries related to information processing in visual systems
1980	Baruj Benacerraf (Venezuela) Jean Dausset (France) George D. Snell (USA)	Discoveries regarding genetically determined structures on cell surfaces that control immunological reactions
1979	Alan M. Cormack (South Africa) Sir Godfrey N. Hounsfield (United Kingdom)	Development of computer-assisted tomography
1978	Werner Arber (Switzerland) Daniel Nathans (USA) Hamilton O. Smith (USA)	Discovery of restriction enzymes and their application in molecular genetics
1977	Roger Guillemin (France) Andrew V. Schally (Poland)	Discovery of peptide hormone production in the brain
	Rosalyn Yalow (USA)	Development of radioimmunoassays for peptide hormones
1976	Baruch S. Blumberg (USA) D. Carleton Gajdusek (USA)	New mechanisms for the origin and spread of infectious diseases
1975	David Baltimore (USA) Renato Dulbecco (Italy) Howard Martin Temin (USA)	Interaction between tumor viruses and cellular genetic material
1974	Albert Claude (Belgium) Christian de Duve (Belgium) George E. Palade (Rumania)	Discoveries related to the structural and functional organization of the cell

Year	Laureates	Discovery
1973	Karl von Frisch (Austria) Konrad Lorenz (Austria) Nikolaas Tinbergen (Netherlands)	Discoveries related to the organization and causes of individual and social behavioral patterns
1972	Gerald M. Edelman (USA) Rodney R. Porter (United Kingdom)	Discovery of the chemical structure of antibodies
1971	Earl W. Sutherland, Jr. (USA)	Discovery of the mechanism of action of hormones
1970	Sir Bernard Katz (United Kingdom) Ulf von Euler (Sweden) Julius Axelrod (USA)	Discovery of humoral transmitters at nerve endings and of mechanisms for their storage, release, and inactivation
1969	Max Delbrück (Germany) Alfred D. Hershey (USA) Salvadore E. Luria (Italy)	Discovery of replication mechanisms and the genetic structure of viruses
1968	Robert W. Holley (USA) Har Gobind Khorana (India) Marshall W. Nirenberg (USA)	Interpretation of the genetic code and its function in protein biosynthesis
1967	Ragnar Granit (Finland) Haldan Keffer Hartline (USA) George Wald (USA)	Discovery of the primary physiological and chemical events in visual processes of the eye
1966	Peyton Rous (USA)	Discovery of tumor viruses
	Charles Brenton Huggins (USA)	Development of a hormone treatment for prostate cancer
1965	François Jacob (France) André Lwoff (France) Jacques Monod (France)	Discovery of the genetic control of enzymes and of virus synthesis
1964	Konrad Bloch (Germany) Feodor Lynen (Germany)	Discovery of the function and regulation of cholesterol and fatty acid metabolism
1963	Sir John Carew Eccles (Austria) Sir Alan Lloyd Hodgkin (United Kingdom) Sir Andrew Fielding Huxley (United Kingdom)	Discovery of ionic mechanisms involved in the excitation and inhibition of peripheral and central regions of the nerve cell membrane

1962	Francis Harry Compton Crick (United Kingdom) James Dewey Watson (USA) Maurice Hugh Frederick Wilkins (United Kingdom)	Discovery of the molecular structure of nucleic acids and its significance in information transfer in living systems
1961	Georg von Békésky (Hungary)	Discovery of the physical mechanism of the impulse conduction system in the inner ear
1960	Sir Frank MacFarlane Burnet (Australia) Sir Peter Brian Medawar (United Kingdom)	Discovery of inherited immunotolerance
1959	Severo Ochoa (Spain) Arthur Kronberg (USA)	Discovery of the biological synthetic pathway for ribo- and deoxyribonucleic acids
1958	George Wells Beadle (USA) Edward Lawrie Tatum (USA)	Discovery that the effectiveness of genes depends on control of specific chemical events
	Joshua Lederberg (USA)	Discoveries regarding genetic recombination and the organization of genetic material in bacteria
1957	Daniel Bovet (Switzerland)	Discovery that the function of specific physiological substances, especially within the blood vessels and skeletal muscles, can be inhibited by synthetic compounds
1956	André Frédéric Cournand (France) Werner Forssmann (Germany) Dickinson W. Richards (USA)	Discovery of the heart catheter and pathological changes in the circulatory system
1955	Axel Hugo Theodor Theorell (Sweden)	Discovery of the nature and method of action of oxidative enzymes
1954	John Franklin Enders (USA) Thomas Huckle Weller (USA) Frederick Chapman Robbins (USA)	Discovery that poliomyelitis viruses can be grown in cultures of various tissues
1953	Sir Hans Adolf Krebs (Germany)	Discovery of the citric acid cycle

1953	Fritz Albert Lipmann (Germany)	Discovery of coenzyme A and its significance in intermediary metabolism
1952	Selman Abraham Waksman (USSR)	Discovery of streptomycin, the first antibiotic effective against tuberculosis
1951	Max Theiler (South Africa)	Discovery of yellow fever and means of combating it
1950	Edward Calvin Kendall (USA) Tadeus Reichstein (Poland) Philip Showalter Hench (USA)	Discovery of the adrenal cortex hormones, their structures, and biological activity
1949	Walter Rudolf Hess (Switzerland)	Discovery of the functional organization of the interbrain as coordinator of the functions of internal organs
	Antonio Caetano de Abreu Friere Egas Monitz (Portugal)	Discovery of the therapeutic value of leucotomy for certain psychoses
1948	Paul Hermann Müller (Switzerland)	Discovery of the great effectiveness of DDT as a contact poison targeting various arthropods
1947	Carl Ferdinand Cori (Austria) Gerty Theresa Cori (Austria)	Discoveries related to glycogen metabolism
	Bernardo Alberto Houssay (Argentina)	Discovery of the role of pituitary hormones in carbohydrate metabolism
1946	Hermann Joseph Muller (USA)	Discovery that X-rays can cause mutations
1945	Sir Alexander Fleming (United Kingdom) Sir Ernst Boris Chain (Germany) Lord Howard Walter Florey (Australia)	Discovery of penicillin and its curative powers with respect to various infectious diseases
1944	Joseph Erlanger (USA) Herbert Spencer Gasser (USA)	Discovery of the highly differentiated functions of individual nerve fibers
1943	Henrik Carl Peter Dam (Denmark)	Discovery of vitamin K
	Edward Adelbert Doisy (USA)	Discovery of the chemical structure of vitamin K
1942	No prizes awarded	

1941	No prizes awarded	
1940	No prizes awarded	
1939	Gerhard Domagk (Germany)	Discovery of the antibacterial activity of prontosil
1938	Corneille Jean François Heymans (Belgium)	Discovery of the role of the arterial sinus and the aorta mechanism in the regulation of respiration
1937	Albert Szent-Györgyi (Hungary)	Discoveries in the context of biological combustion processes, especially with respect to vitamin C and fumaric acid
1936	Sir Henry Hallett Dale (United Kingdom) Otto Loewi (Germany)	Discoveries regarding the chemical transmission of nerve impulses
1935	Hans Spemann (Germany)	Discovery of the organizer effect in embryonic development
1934	George Hoyt Whipple (USA) George Richards Minot (USA) William Parry Murphy (USA)	Discovery of liver therapy for treatment of anemia
1933	Thomas Hunt Morgan (USA)	Discoveries related to the function of the chromosomes in inheritance
1932	Sir Charles Scott Sherrington (United Kingdom) Lord Edgar Douglas Adrian (United Kingdom)	Discoveries related to the function of neurons
1931	Otto Heinrich Warburg (Germany)	Discoveries regarding the nature and function of enzymes in the respiratory chain
1930	Karl Landsteiner (Austria)	Discovery of human blood types
1929	Christiaan Eijkman (Netherlands)	Discovery of the antineuritic vitamin
	Sir Frederick Gowland Hopkins (United Kingdom)	Discovery of the growth-stimulating vitamins
1928	Charles Jules Henri Nicolle (France)	Studies of typhus

Year	Laureate(s)	Achievement
1927	Julius Wagner-Jauregg (Austria)	Discovery of the therapeutic value of malarial infections in the treatment of tertiary syphilis (*dementia paralytica*)
1926	Johannes Andreas Grib Fibiger (Denmark)	First experimental induction of cancer in an animal model
1925	No prizes awarded	
1924	Wilhelm Einthoven (Netherlands)	Discovery of the electrocardiogram
1923	Sir Frederick Grant Banting (Canada) John James Richard Macleod (United Kingdom)	Discovery of insulin
1922	Sir Archibald Vivian Hill (United Kingdom)	Discovery of the evolution of heat in muscle
	Otto Fritz Meyerhoff (Germany)	Discovery of the relationship between oxygen uptake and lactic acid metabolism in muscle
1921	No prizes awarded	
1920	Schack August Steenberger Krogh (Denmark)	Studies in respiratory metabolism
1919	Jules Bordet (Belgium)	Discoveries in the field of immunity
1918	No prizes awarded	
1917	No prizes awarded	
1916	No prizes awarded	
1915	No prizes awarded	
1914	Robert Bárány (Austria)	Studies on the physiology and pathology of the vestibular apparatus in the ear
1913	Charles Robert Richet (France)	In recognition of his studies on anaphylaxis
1912	Alexis Carrel (France)	In recognition of his studies on the transplantation of blood vessels and organs
1911	Allvar Gullstrand (Sweden)	Studies of the optical properties of the eye

1910	Albrecht Kossel (Germany)	In recognition of his contributions to an understanding of the chemistry of the cell through his studies of proteins and the compounds present in cell nuclei
1909	Emil Theodor Kocher (Switzerland)	Studies of the physiology, pathology, and surgical treatment of the thyroid gland
1908	Ilya Ilyich Mechnikov (Russia) Paul Ehrlich (Germany)	In recognition of their work in the field of immunity
1907	Charles Louis Alphonse Laveran (France)	In recognition of his work on the role of protozoa as sources of disease
1906	Camillo Golgi (Italy) Santiago Ramon y Cajal (Spain)	In recognition of their work on the structure of the nervous system
1905	Robert Koch (Germany)	Studies and discoveries in the field of tuberculosis
1904	Ivan Petrovich Pavlov (Russia)	In recognition of his work on the physiology of digestion
1903	Niels Ryberg Finsen (Denmark)	In recognition of his contributions to treatment of disease, especially the use of UV irradiation with skin tuberculosis
1902	Sir Ronald Ross (United Kingdom)	Studies related to malaria
1901	Emil Adolf von Behring (Germany)	Studies in serum therapy, especially its application with respect to diphtheria

Year	Laureates	Work
1998	Robert B. Laughlin (USA) Horst L. Störmer (USA) Daniel C. Tsui (USA)	Discovery of a new form of quantum fluid with fractionally charged excitations
1997	Steven Chou (USA) Claude Cohen-Tannoudji (France) William D. Phillips (USA)	Development of methods to cool and trap atoms with laser light
1996	David M. Lee (USA) Douglas D. Osheroff (USA) Robert C. Richardson (USA)	Discovery of the superfluidity of helium-3
1995	Martin L. Perl (USA)	Discovery of the tau lepton
	Frederick Reines (USA)	Experimental verification of the existence of the neutrino
1994	Bertram N. Brockhouse (Canada) Clifford G. Shull (USA)	Development of fundamental procedures for elastic and inelastic neutron scattering in the structural characterization of solids
1993	Russell A. Hulse (USA) Joseph H. Taylor, Jr. (USA)	Discovery of a new class of pulsars, which opened new possibilities for the investigation of gravitation
1992	Georges Charpak (Poland)	Discovery and development of particle detectors, especially the multiwire proportional chamber
1991	Pierre-Gilles de Gennes (France)	Application of methods developed for the investigation of orderliness phenomena in simple systems to such complex systems as liquid crystals or polymers
1990	Jerome I. Friedman (USA) Henry W. Kendall (USA) Richard E. Taylor (Canada)	Pioneer work in the field of inelastic scattering of high-energy electrons by protons and bound neutrons for the determination of their internal structure, which was of tremendous significance for the development of the quark model in particle physics

1989	Norman F. Ramsay (USA)	Invention of a precision method for electromagnetic excitation of atoms in separate microwave fields with fixed phase relationships and its application in hydrogen masers and other atomic clocks
	Hans G. Dehmelt (Germany) Wolfgang Paul (Germany)	Development of ion traps
1988	Leon M. Lederman (USA) Melvin Schwartz (USA) Jack Steinberger (Germany)	Development of neutrino beam methods and proof of the doublet structure of the lepton through discovery of the muon neutrino
1987	J. Georg Bednorz (Germany) K. Alexander Müller (Switzerland)	Important breakthrough in the discovery of high-temperature superconducting ceramic materials
1986	Ernst Ruska (Germany)	Fundamental contributions to electron optics and development of the first electron microscope
	Gerd Binnig (Germany) Heinrich Rohrer (Switzerland)	Development of the scanning tunneling microscope
1985	Klaus von Klitzing (Germany)	Discovery of the quantum Hall effect
1984	Carlo Rubbia (Italy) Simon van der Meer (Netherlands)	Decisive contributions to the overall project that led to the discovery of W- and Z-field particles, the intermediaries in weak interactions
1983	Subramanyan Chandrasekhar (India)	Theoretical studies of physical processes involved in the origin and evolution of stars
	William A. Fowler (USA)	Theoretical and experimental studies of nuclear reactions important in the formation of chemical elements in the universe
1982	Kenneth G. Wilson (USA)	Theory of the phenomenon of critical properties in phase transitions

Year	Laureates	Contribution
1981	Nicolaas Bloembergen (Netherlands) Arthur L. Schawlow (USA)	Contributions to the development of laser spectroscopy
	Kai M. Siegbahn (Sweden)	Contributions to the development of high-resolution electron microscopy
1980	James W. Cronin (USA) Val L. Fitch (USA)	Discovery of the violation of PC invariance in the decay of neutral K mesons
1979	Sheldon L. Glashow (USA) Abdus Salam (Pakistan) Steven Weinberg (USA)	Contributions to the mutual theory of weak and electromagnetic interactions between elementary particles, including among others prediction of weak neutral streams
1978	Pyotr Leonidovich Kapitsa (USSR)	Fundamental discoveries in the field of low-temperature physics
	Arno A. Penzias (Germany) Robert W. Wilson (USA)	Discovery of cosmic microwave background radiation
1977	Philip W. Anderson (USA) Sir Nevill F. Mott (United Kingdom) John H. van Vleck (USA)	Fundamental theoretical investigations into the electronic structure of magnetic disorderly systems
1976	Burton Richter (USA) Samuel C. C. Ting (USA)	Discovery of a new class of heavy elementary particles, the so-called psi particles
1975	Aage Bohr (Denmark) Ben Mottelson (USA) James Rainwater (USA)	Discovery of a relationship between collective and particle motions in atomic nuclei, as well as development of a corresponding theory of nuclear structure
1974	Sir Martin Ryle (United Kingdom) Antony Hewish (United Kingdom)	Fundamental research in the area of radioastrophysics
1973	Leo Esaki (Japan) Ivar Giaever (Norway)	Experimental observations of tunneling phenomena in semiconductors and superconductors

1973	Brian D. Josephson (United Kingdom)	Prediction of the properties of a supercurrent through tunneling barriers, especially the phenomenon known as the Josephson effect
1972	John Bardeen (USA) Leon N. Cooper (USA) J. Robert Schrieffer (USA)	Theory of superconductivity (the BCS theory)
1971	Dennis Gabor (Hungary)	Discovery and development of holographic methods
1970	Hannes Alfvén (Sweden)	Discoveries in the field of magnetohydrodynamics with wide-ranging applications in plasma physics
	Louis Néel (France)	Fundamental studies of antiferromagnetism and ferrimagnetism, which have led to applications in solid-state physics
1969	Murray Gell-Mann (USA)	Contributions to the classification of elementary particles and their interactions
1968	Luis W. Alvarez (USA)	Decisive contributions to elementary particle physics, especially discovery of a multitude of resonance states
1967	Hans Albrecht Bethe (Germany)	Contributions to the theory of nuclear reactions, especially discoveries related to energy production in stars
1966	Alfred Kastler (France)	Discovery and development of optical methods for the investigation of Hertzian resonance in atoms ("optical pumps")
1965	Sin-Itiro Tomonaga (Japan) Julian Schwinger (USA) Richard P. Feynman (USA)	Fundamental studies in quantum electrodynamics with profound consequences in elementary particle physics

1964	Charles H. Townes (USA) Nicolay Gennadiyevich Basov (USSR) Aleksandr Mikhailovich Prokhorov (USSR)	Fundamental studies in the field of quantum electronics, which led to the construction of oscillators and amplifiers based on the maser–laser principle
1963	Eugene P. Wigner (Hungary)	Contributions to the theory of the atomic nucleus and elementary particles, especially discovery and application of fundamental symmetry laws in quantum theory
	Maria Goeppert-Mayer (Germany) J. Hans D. Jensen (Germany)	Studies of the shell structure of the atomic nucleus
1962	Lev Davidovich Landau (USSR)	Pioneering theories of condensed material, especially liquid helium
1961	Robert Hofstadter (USA)	Investigations of electron scattering at atomic nuclei and the related discovery of the internal structure of nucleons
	Rudolf Ludwig Mössbauer (Germany)	Studies of resonance absorption of gamma rays and in this context discovery of the Mössbauer effect
1960	Donald A. Glaser (USA)	Invention of the bubble chamber
1959	Emilio Gino Segrè (Italy) Owen Chamberlain (USA)	Discovery of the antiproton
1958	Pavel Alekseyevich Cherenkov (USSR) Ilya Mikhailovich Frank (USSR) Igor Yevgenyevich Tamm (USSR)	Discovery and explanation of the Cherenkov effect
1957	Chen Ning Yang (China) Tsung-Dao Lee (China)	Investigations of the principle of maintenance of parity, which led to important discoveries in the field of elementary particle physics

1956	William Shockley (United Kingdom) John Bardeen (USA) Walter Houser Brattain (USA)	Studies of semiconductors and discovery of the transistor effect
1955	Willis Eugene Lamb (USA)	Discovery of the fine structure of the hydrogen spectrum
	Polykarp Kusch (Germany)	Precise determination of the magnetic moment of the electron
1954	Max Born (Germany)	Fundamental studies in quantum mechanics, especially statistical treatment of the wave function
	Walter Bothe (Germany)	Development of the coincidence method and resulting discoveries in the field of cosmic radiation and nuclear transformation
1953	Frits (Frederik) Zernike (Netherlands)	Development of phase-contrast methods, especially invention of the phase-contrast microscope
1952	Felix Bloch (Switzerland) Edward Mills Purcell (USA)	New methods for determining nuclear magnetic precession, as well as inventions introduced in this context
1951	Sir John Douglas Cockcroft (United Kingdom) Ernest Thomas Sinton Walton (Ireland)	Fundamental studies of the transmutation of atomic nuclei through artificially accelerated elementary particles
1950	Cecil Frank Powell (United Kingdom)	Development of photographic methods for investigating nuclear processes, and discovery with these methods of the meson
1949	Hideki Yukawa (Japan)	Prediction of the existence of mesons on the basis of theoretical studies of nuclear forces

Year	Laureate	Achievement
1948	Lord Patrick Maynard Stuart Blackett (United Kingdom)	Further development of the Wilson cloud-chamber method and resulting discoveries in the field of nuclear physics and cosmic radiation
1947	Sir Edward Victor Appleton (United Kingdom)	Investigations into the physics of the upper atmosphere, especially discovery of the so-called Appleton layer
1946	Percy Williams Bridgman (USA)	Invention of an apparatus for achieving ultrahigh pressures, and discoveries in this context in the field of high-pressure physics
1945	Wolfgang Pauli (Austria)	Discovery of the Pauli exclusion principle
1944	Isidor Isaac Rabi (Austria–Hungary)	Development of the resonance method for determining the magnetic properties of atomic nuclei
1943	Otto Stern (Germany)	Contributions to the development of the electron-beam method and discovery of the magnetic moment of the proton
1942	No prizes awarded	
1941	No prizes awarded	
1940	No prizes awarded	
1939	Ernest Orlando Lawrence (USA)	Discovery and development of the cyclotron as well results achieved with it, especially with respect to artificial radioactive elements
1938	Enrico Fermi (Italy)	Determination of new radioactive elements prepared by neutron bombardment, as well as for the discovery of nuclear reactions initiated by slow neutrons
1937	Clinton Joseph Davisson (USA) Sir George Paget Thomson (United Kingdom)	Discovery of electron diffraction by crystals

1936	Victor Franz Hess (Austria)	Discovery of cosmic rays
	Carl David Anderson (USA)	Discovery of the positron
1935	Sir James Chadwick (United Kingdom)	Discovery of the neutron
1934	No prizes awarded	
1933	Erwin Schrödinger (Austria) Paul Adrien Maurice Dirac (United Kingdom)	Discovery of new, fruitful forms of atomic theory
1932	Werner Heisenberg (Germany)	Founding of quantum mechanics, the application of which has led among other things to the discovery of allotropic forms of hydrogen
1931	No prizes awarded	
1930	Sir Chandrasekhara Venkata Raman (India)	Studies related to the scattering of light and discovery of the Raman effect
1929	Prince Louis-Victor de Broglie (France)	Discovery of the wave nature of the electron
1928	Sir Owen Willans Richardson (United Kingdom)	Studies of the thermoionic effect, especially for discovery of the law named in his honor
1927	Arthur Holly Compton (USA)	Discovery of the effect named in his honor
	Charles Thomson Rees Wilson (United Kingdom)	Methods for visualizing the paths of electrically charged particles through vapor condensation
1926	Jean Baptiste Perrin (France)	Studies of the discontinuous structure of matter, especially the discovery of the sedimentation equilibrium
1925	James Franck (Germany) Gustav Hertz (Germany)	Discovery of the laws describing the collision of an electron with an atom
1924	Karl Manne Georg Siegbahn (Sweden)	Discoveries in the field of X-ray spectroscopy

1923	Robert Andrews Millikan (USA)	Studies regarding the fundamental electrical charge and the photoelectric effect
1922	Niels Bohr (Denmark)	Studies of the structure of atoms and the radiation they emit
1921	Albert Einstein (Germany)	Contributions to theoretical physics, especially discovery of the laws describing the photoelectric effect
1920	Charles Edouard Guillaume (Switzerland)	In recognition of his achievements in the field of precise measurements in physics through discovery of anomalies in nickel-steel alloys
1919	Johannes Stark (Germany)	Discovery of the Doppler effect and the splitting of spectral lines by electric fields
1918	Max Karl Ernst Ludwig Planck (Germany)	In recognition of his contributions toward the development of physics as a consequence of his quantum theory
1917	Charles Glover Barkla (United Kingdom)	Discovery of the characteristic X-radiation of the elements
1916	No prizes awarded	
1915	Sir William Henry Bragg (United Kingdom) Sir William Lawrence Bragg (United Kingdom)	Studies leading to the analysis of crystal structures through the use of X-rays
1914	Max von Laue (Germany)	Discovery of the diffraction of X-rays by crystals
1913	Heike Kamerlingh-Onnes (Netherlands)	Investigations into the properties of materials at low temperature, which have led among other things to the preparation of liquid helium

Year	Laureate	Achievement
1912	Nils Gustaf Dalén (Sweden)	Invention of automatic regulators, which can be used together with gas accumulators in the illumination of lighthouses and buoys
1911	Wilhelm Wien (Germany)	Discovery of the laws describing thermal radiation
1910	Johannes Diderik van der Waals (Netherlands)	Studies of the equations of state of gases and liquids
1909	Guglielmo Marconi (Italy) Carl Ferdinand Braun (Germany)	In recognition of their contributions to wireless telegraphy
1908	Gabriel Lippmann (Luxembourg)	Methods for the photographic reproduction of colors based on the phenomenon of interference
1907	Albert Abraham Michelson (Germany)	Development of optically precise instrumentation, and the spectroscopic and metrological investigations their existence permitted
1906	Sir Joseph John Thomson (United Kingdom)	In recognition of the importance of his theoretical and experimental work on the electrical conductivity of gases
1905	Philip Eduard Anton Lenard (Hungary)	Studies of cathode rays
1904	Lord John William Strutt Rayleigh (United Kingdom)	Investigations of the densities of the most important gases, and the discovery in this context of argon
1903	Antoine Henri Becquerel (France)	Discovery of spontaneous radioactivity
	Pierre Curie (France) Marie Curie (Poland)	Studies related to the radiation phenomena discovered by Becquerel
1902	Hendrik Antoon Lorentz (Netherlands) Pieter Zeeman (Netherlands)	Influence of magnetism on radiation phenomena
1901	Wilhelm Conrad Röntgen (Germany)	Discovery of X-rays (also known as Röntgen rays)

January

1/1 **1852** Birth of Eugène Anatole Demarcay (†1903), the discoverer of europium.

1/1 **1872** The metric system is officially introduced throughout the German Empire.

1/2 **1765** Birth of Charles Hatchett, the discoverer of niobium.

1/2 **1863** The Meister Lucius & Co. dyeworks, later Hoechst AG, began operation in Hoechst.

1/4 **1643** Birth of Isaac Newton (†1727). He discovered the law of gravity and introduced the four *Newtonian axioms* of mechanics; relevant to the history of chemistry are his speculations regarding a systematic alchemy.

1/5 **1981** Death of Harold Clayton Urey (*1893), the discoverer of deuterium (Nobel Prize, 1934).

1/6 **1914** Birth of Kenneth Sanborn Pitzer. He introduced the notion of *Pitzer strain,* which occurs in molecules as a result of van der Waals forces between neighboring substituents that are not sufficiently staggered.

1/6 **1939** An article by Otto Hahn and Fritz Strassman appears in the journal *Naturwissenschaften* describing their observations upon bombarding ^{235}U with slow neutrons. Lise Meitner and Otto Frisch explained the results theoretically in terms of nuclear fission.

1/7 **1794** Eilhard Mitscherlich (†1863) is born. This chemist and mineralogist discovered isomorphism in crystals in 1819. His rule of isomorphology was important in atomic weight determinations and the early systematization of the elements.

1/9 **1868** Birth of Sören Peter Lauritz Sörenson (†1939), who introduced the concept of the pH value.

1/9 **1922** Birth of Har Gobind Khorana. He received the Nobel Prize for medicine or physiology in 1968 for his first total synthesis of a gene.

1/10 **1916** Birth of Sune Karl Bergström, winner of the Nobel Prize for medicine in 1982. He succeeded in 1957 in isolating several prostaglandins in crystalline form and explaining their structure and origin.

1/11 **1869** Carl Graebe and Carl Liebermann present the Berlin Chemical Society with the first samples of their synthetically prepared alizarin. The Badische Anilin- & Soda-Fabrik acquired the patent and transformed this synthesis into an industrial process based on a variant optimized by Heinrich Caro.

1/11 **1911** The "Kaiser Wilhelm Society for the Advancement of Science" is founded in Berlin, the forerunner of today's "Max Planck Society."

1/12 **1716** Birth of Antonio de Ulloa, the discoverer of platinum (†1795).

1/13 **1813** Birth of Henry Bessemer (†1898). He discovered the process for blast refining named in his honor, with which crude iron is transformed into steel by the introduction of a blast of air.

1/14 **1851** Ludwig Claisen is born (†1930). In 1881 he discovered the *Claisen condensation*, which bears his name. This is the transformation of an aromatic aldehyde with an aliphatic aldehyde or ketone into an unsaturated aromatic carbonyl compound. He was also the originator of the *Claisen head* for fractional distillation under reduced pressure, and he introduced the *Claisen rearrangement* of allyl phenyl and allyl vinyl ethers.

1/15 **1568** Birth of Johannes Hartmann (†1631), who established the first university chemistry laboratory in Germany in which students could carry out practical experiments.

1/15 **1895** Birth of Artturi I. Virtanen (†1973), who was awarded a Nobel Prize in 1945 for his work in the area of food and agricultural chemistry, especially for methods for increasing the storage potential of fodder.

1/16 **1806** Nicolas Leblanc (*1742), the discoverer of the process named in his honor for the production of soda, takes his own life in the poorhouse.

1/16 **1875** Leonor Michaelis is born (†1949). Together with Menten he developed the *Michaelis–Menten equation* describing the kinetics of enzyme-catalyzed reactions.

1/17 **1941** Death of Eduard Zintl (*1898), known for his work in the area of intermetallic compounds – *Zintl phases.*

1/18 **1861** Birth of Hans Goldschmidt (†1923). He developed in 1898 the thermite process, with which high-melting metals could be prepared free of carbon.

1/19 **1927** Death of Carl Graebe (*1841). Together with Carl Liebermann he synthesized for the first time a natural dye, alizirin.

1/20 **1834** Birth of Adolf Frank (†1916), who provided the impulse for opening of the potassium salt deposits at Stassfurt and the recovery of bromine from salt wastes. As director of a glass factory he introduced brown coloration into beer bottles as a way of protecting the contents against the effects of light. Together with Heinrich Caro he developed the *Frank–Caro method* for the preparation of potassium cyanamide.

1/21 **1912** Birth of Konrad Emil Bloch. This Nobel Laureate in medicine (1964) made substantial contributions to clarification of the biosynthesis of cholesterol and other steroids.

January

1/22 **1775** Birth of André Marie Ampère (†1836). He developed the electrodynamic theory and recognized the composition of hydrofluoric acid. The unit of electrical current is named in his honor.

1/23 **1796** Birth of Carl Claus (†1864), the discoverer of ruthenium.

1/23 **1876** Birth of Otto Diels, who together with Kurt Alder discovered the addition of activated unsaturated compounds (dienophiles) to dienes. For this *Diels–Alder reaction*, which also led to commercially important syntheses of cyclic compounds, Diels and Alder received the Nobel Prize in 1950.

1/23 **1929** Birth of John C. Polanyi. He was awarded the Nobel prize in 1986 for his investigations of elementary chemical reactions and transition states.

1/25 **1917** Ilya Prigogine is born. The Nobel Laureate for 1977 studied the thermodynamics of systems far from equilibrium, especially dissipative structures.

1/27 **1865** August Kekulé presents his structure for benzene to the Société Chimique in Paris.

1/27 **1881** Birth of Rudolf Fischer (†1957). In 1911 he applied for the basic patents in color photography, although his procedure was not realized in practice until 1935/1936.

1/28 **1873** The gold and silver separation facility opened in 1843 under his own name by Friedrich Roessler is transformed into the "Aktiengesellschaft Deutsche Gold- und Silber-Scheideanstalt." With the establishment of the German Empire in 1871 this company, known today by its acronym DEGUSSA, acquired an ample supply of raw material in the form of now invalid coins originally issued by the former German principalities.

1/29 **1938** The chemist Paul Schlack, working for I. G. Farben, succeeds in preparing polyamide-6 ("Perlon") from ε-caprolactam, a compound with properties very similar to those of polyamide-66 ("Nylon").

1/30 **1952** BASF is reorganized, and in March of 1953 it is entered into the Ludwigshafen commercial registry as the Badische Anilin- & Soda-Fabrik AG.

1/31 **1847** The chemist Ascanio Sobrero presents the Academy of Sciences in Turin with his discovery of nitroglycerine.

1/31 **1881** Birth of Irving Langmuir (†1957). In 1913 he introduced the idea of an inert atmosphere and a metal-spiral filament for light bulbs. His efforts in the area of surface chemistry earned him the Nobel Prize in 1932.

February

2/1 **1899** "Aspirin" is registered at the German Imperial Patent Office.

February

2/3 **1890** Birth of Paul Hermann Scherrer (†1969). Together with Debye he developed a method of crystal-structure analysis in which crystalline powder was used for lattice determinations with the aid of X-ray interference: the *Debye–Scherrer powder method.*

2/4 **1682** Birth of Johann Friedrich Böttger (†1719), an alchemist who, while searching for a means of producing gold, prepared the first porcelain in Europe.

2/4 **1896** Birth of Friedrich Hund, a cofounder of the orbital theory who introduced *Hund's rule.*

2/6 **1861** Birth of Nikolai Zelinski (†1953). His name is associated, for example, with the *Hell–Volhard–Zelinski reaction,* the α-halogenation of a carboxylic acid.

2/7 **1864** John A. R. Newlands publishes his first article on the "Law of Octaves," in which he summarizes the relationships between various groups of eight elements.

2/8 **1777** Bernhard Courtois, the discoverer of iodine, is born (†1838).

2/8 **1794** Friedlieb Ferdinand Runge (†1867) is born. He investigated and isolated natural products, including quinine. His work led to the discovery of aniline, for which reason he is considered among the pioneers in coal-tar dye chemistry.

2/10 **1847** Thomas Alva Edison is born (†1931). The versatile inventor registered over 2000 patents and developed among other things the carbon-fiber light bulb.

2/12 **1947** Death of Moses Gomberg (*1866). In 1900 he discovered the first free radical, the triphenylmethyl radical.

2/13 **1834** Birth of Heinrich Caro (†1919). As technical director of the Badische Anilin- & Soda-Fabrik he helped achieve the manufacture and commercialization of numerous compounds, including alizarin and indigo.

2/13 **1929** Alexander Fleming lectures on his research under the title "Cultures of a Penicillium" before the Medical Research Club.

2/14 **1917** Birth of Herbert Aaron Hauptman. He received a Nobel Prize in 1985 for developing direct methods for X-ray crystal-structure determination.

2/15 **1873** Birth of Hans von Euler-Chelpin (†1964), who received a Nobel Prize in 1929 for investigations into the enzymes responsible for sugar fermentation.

2/16 **1955** F. P. Bundy, H. T. Hall, H. M. Strong, and R. H. Wentoff of GE Research Laboratories make public the successful synthesis of diamonds.

February

2/17 **1869** Dmitri Mendeleev submits a publication on the periodic system of the elements, including predictions regarding elements not yet discovered.

2/18 **1745** Birth of Alessandro Volta (†1827). Around 1800 he discovered the voltaic pile, which constituted a convenient source of current for electrochemical experiments.

2/19 **1859** Birth of Svante August Arrhenius (†1927), who developed the theory of electrolytic dissocation (Nobel Prize, 1903).

2/20 **1901** Birth of Henry Eyring (†1981). He dedicated himself to the theory of reaction rates and applied quantum mechanics and statistical mechanics to kinetic problems.

2/20 **1937** Birth of Robert Huber, who received a Nobel Prize in 1988 for his work on the three-dimensional structures of proteins involved in photosynthesis.

2/21 **1791** Birth of John Mercer (†1866), who discovered the process of treating cotton with sodium hydroxide solution known as *mercerization.*

2/21 **1926** Death of Heike Kamerlingh-Onnes (*1853), who accomplished the first liquification of helium. In the course of investigations into the behavior of metals at extremely low temperature he discovered the phenomenon of superconductivity (Nobel Prize in physics, 1913).

2/22 **1828** Friedrich Wöhler reports in a letter to Jöns Jacob Berzelius that he has succeeded in the artificial preparation of urea. Thus, an animal metabolic product had for the first time been prepared in the laboratory. Wöhler's accidental discovery is regarded as the birth of scientific organic chemistry.

2/22 **1879** Birth of Johannes Nicolaus Brønsted, who developed an expanded definition of acids and bases as proton-donating and proton-accepting substances.

2/23 **1944** Death of Leo Hendrick Baekeland (*1863), the inventor of Bakelite.

2/24 **1950** The German Arbeitsgemeinschaft Chemische Industrie creates the "Fonds der Chemie" as a way of supporting research, science, and education in their state of economic need.

2/25 **1896** Birth of Ida Noddack (née Tacke; †1978) Together with her husband Walter she investigated element 75, rhenium, which the two of them detected in 1925 through X-ray spectra.

February

2/26 **1799** Benoît Pierre Emile Clapeyron is born (†1864). In 1834 he published his studies on the theory of heat, the further development of which by Rudolf Clausius resulted in 1850 in the *Clausius–Clapeyron equation* dealing with the thermal equilibria accompanying phase transformations.

2/26 **1948** The former Kaiser Wilhelm Society is reconstituted under the name "Max Planck Society."

2/27 **1856** Birth of Alfred Einhorn (†1917). He introduced novocaine as an anaesthetic and accomplished important work in the field of preparative organic chemistry. He is known, for example, for a reaction named in his honor by which sensitive alcohols are acylated with acid chlorides in pyridine.

2/28 **1901** Birth of Linus Carl Pauling (†1994). His efforts in the application of quantum mechanics to molecular structure and chemical bonding led to the valence-bond theory, which he utilized in elucidating the structure of complex molecules (Nobel Prize, 1954). After the first American atomic bomb was dropped on Hiroshima he engaged in a quest for disarmament and peace, for which he received the Nobel Peace Prize in 1962.

2/28 **1935** The chemist Wallace Hume Carothers of the DuPont Corporation synthesizes polyamide 66 from hexamethylenediamine and adipic acid. This compound, familiar by the name "nylon," can be drawn into threads and is distinguished for its high strength. Commercial production commenced in 1938.

March

3/1 **1910** Archer John Porter Martin is born. Together with Richard Synge he developed partition chromatography, for which both received the Nobel Prize in 1952.

3/2 **1848** Birth of Philippe Antoine François Barbier (†1922). He developed syntheses with organozinc and organomagnesium compounds that were further pursued by his student Victor Grignard.

3/2 **1896** Henri Becquerel recognizes that radiation emitted by uranium salts is capable of blackening photographic plates even through barriers. Marie Curie applied the term "radioactivity" to this phenomenon.

3/3 **1918** Birth of Arthur Kornberg, who carried out fundamental research in the field of DNA biosynthesis and in 1959 received the Nobel Prize in Medicine.

3/4 **1887** Gottlieb Wilhelm Daimler drives a motorized four-wheeled vehicle of his own design from Cannstatt to Stuttgart at an average speed of 11 mph. The subsequent construction of more efficient engines is closely related to the development of chemical additives for improving the antiknock characteristics of the fuel.

3/4 **1927** Death of Ira Remsen (*1846). Together with Fahlberg he discovered saccharine, and he was one of the founders of the "American Chemical Journal."

3/5 **1808** Birth of Petrus Jacobus Kipp (†1864), developer of the *Kipp apparatus* that bears his name.

3/5 **1877** Carl Mannich is born (†1947). He is familiar from an aminomethylation reaction with formaldehyde as the carbonyl component, which bears his name.

3/6 **1787** Birth of Joseph von Fraunhofer (†1826), after whom the *Fraunhofer lines* in the solar spectrum (which he discovered) are named. This inventor of the diffraction grating also worked on the improvement of optical devices.

3/6 **1862** Birth of René Bohn (†1922). In 1901 he discovered indanthrene, which developed into one of the most important categories of dyes of that period.

3/7 **1857** Birth of Arthur Rudolf Hantzsch (†1935), the originator of the *Hantzsch pyridine and pyrrole syntheses.*

3/8 **1798** Birth of Heinrich Wilhelm Wackenroder (†1854), the discoverer of *Wackenroder's solution*, an aqueous solution of polythionic acids.

3/8 **1839** Birth of James Mason Crafts (†1917), who together with Charles Friedel discovered the catalytic activity of aluminum chloride in the reaction of aromatic compounds with alkyl and acyl halides – the *Friedel–Crafts reaction.*

3/9 **1960** Death of Hermann Otto Laurenz Fischer (*1888). The son of Emil Fisher, he succeeded in 1932 in establishing the structure of quinic acid. He also developed procedures for lengthening and shortening the chains of aldoses.

3/10 **1852** Richard Anschütz is born. He developed vacuum distillation into a generally applicable procedure in chemistry.

3/11 **1870** Birth of Rudolf Schenk (†1965), the discoverer of the bright red (*Schenck*) modification of phosphorus.

3/11 **1954** An entry in Giulio Natta's notebook documents the successful preparation of polypropylene using the low-pressure process and catalysts discovered by Karl Ziegler.

3/12 **1824** Birth of Gustav Robert Kirchhoff. Together with Bunsen he introduced the concept of spectral analysis and discovered the elements rubidium and cesium. He also identified the Fraunhofer lines as constituting absorption spectra. He engaged himself with the theory of radiation and developed the *Kirchhoff formula for radiation* and, in part with the help of others, *Kirchhoff's laws.*

March

3/12 **1838** Birth of William Henry Perkin, Sen. (†1907), the discoverer of mauveine, the first coal-tar dye. He also founded a company for the manufacture and marketing of this material. Several reactions are named after him and his son, including the *Perkin cinnamic acid synthesis.*

3/13 **1845** John Frederic Daniell dies. In 1836 he invented the zinc–copper cell – the *Daniell cell* – the output of which remains constant for a relatively long period of time.

3/14 **1854** Birth of Paul Ehrlich (†1915). This chemist and bacteriologist introduced new histological dyes and thus simplified the diagnosis of hematological diseases. He provided a method for identifying tuberculosis bacteria and laid the groundwork for chemotherapy with the use of parasitotropic substances (Nobel Prize in medicine, 1908).

3/15 **1821** Birth of Joseph Loschmidt (†1895). In 1865 he presented a theoretical derivation for the number of molecules in a cubic centimeter of gas during the course of a lecture at the Academy of Sciences in Vienna. Today the *Loschmidt number* refers to the number of molecules in a mole.

3/15 **1854** Emil von Behring is born (†1917). This bacteriologist introduced serum therapy and the study of immunity, and was the first to prepare sera to counter diphtheria and tetanus. He was awarded the first Nobel Prize in medicine in 1901.

3/16 **1670** Johann Rudolph Glauber dies (*1604). In his small factory in Amsterdam he worked out procedures for the preparation of mineral acids and salts, and *Glauber's salt* (sodium sulfate) still bears his name. He is said to be the first chemist who was able to live off the sale of products prepared in his own laboratory.

3/17 **1871** Birth date of Alexei Tschitschibabin. He investigated and synthesized various alkaloids, developed in 1906 a general pyridine synthesis involving the treatment of aldehydes or ketones with ammonia, and discovered the nucleophilic pyridine amination with sodium amide that bears his name.

3/19 **1883** Birth of Walter Norman Haworth (†1950). He was the originator of the *Haworth projection,* especially useful for representing three-dimensional configurations of carbohydrate hemiacetals. His research in the field of carbohydrates and vitamin C led to a Nobel Prize in 1937.

3/19 **1943** Birth of Mario Molina. In 1995 he received the Nobel Prize for his work on the chemistry of the atmosphere, especially the formation and decay of ozone.

3/20 **1947** Victor Moritz Goldschmidt dies (*1888). This mineralogist determined atomic and ionic radii in crystals.

3/21 **1932** Birth of Walter Gilbert, who received the Nobel Prize in 1980 for his determination of base sequences in nucleic acids.

3/22 **1924** Death of Siegmund Gabriel (*1851). The general procedure for synthesizing primary amines and amino acids he discovered carries his name.

3/23 **1881** Hermann Staudinger is born (†1965). He created the prerequisites for our understanding of polymeric substances and coined the term "macromolecule." For this work he was awarded the Nobel Prize in 1953. During the First World War he developed artificial pepper and a coffee substitute.

3/24 **1903** Birth of Adolf Butenandt (†1995). Independently of Edward Doisy in 1929 he isolated and crystallized the hormone estrone, and subsequently the additional hormones androsterone, progesterone, and testosterone. This was followed by structure elucidations and syntheses. For this work he was awarded the Nobel Prize in 1939.

3/24 **1917** Birth of John Cowdery Kendrew. He was awarded the Nobel Prize in 1962 for his work on the X-ray structural characterization of myoglobin and hemoglobin.

3/25 **1867** Death of Friedlieb Ferdinand Runge (*1794). In the course of investigating coal tar he isolated aniline, quinoline, and phenol, and he is regarded as the originator of paper chromatography.

3/26 **1847** Birth of Heinrich von Brunck (†1911). As technical director of the Badische Anilin- & Soda-Fabrik he played a crucial role in realization of the synthesis of indigo as well as industrial application of the synthesis of ammonia.

3/26 **1911** Birth of Bernard Katz. He was awarded the Nobel Prize for medicine or physiology in 1970 for his fundamental investigations of chemical processes in the transfer of nerve impulses.

3/26 **1916** Christian Boehmer Anfinsen is born (†1995). For his discoveries related to the structure and mode of action of ribonucleases he was awarded the Nobel Prize in 1972.

3/27 **1845** Birth of Wilhelm Conrad Röntgen (†1923), the discoverer of X-rays.

3/27 **1847** Birth of Otto Wallach (†1931). He pioneered in the field of ethereal oils and terpenes. His name is associated with the *Wallach reactions*, which he developed (Nobel Prize, 1910).

3/28 **1709** J. F. Böttger presents a written report to the chancellory of Saxony regarding the preparation of a white porcelain. In 1710 he was made administrator of the newly created porcelain works in Meissen.

3/29 **1855** Birth of Julius Bredt (†1937). He dedicated himself to the stereochemistry of camphor compounds and in 1902 proposed the rule carrying his name, which says that a bridgehead carbon atom cannot be involved in a double bond.

March

3/30 **1876** Death of Antoine Jérôme Ballard (*1802). He occupied himself with chemical substances found in the sea, and in 1826 he discovered the element bromine.

3/31 **1811** Birth of Robert Wilhelm Bunsen (†1899). His inventions in the field of apparatus include the *Bunsen burner* and the water aspirator. His analytical activities led to the development of iodometry and, in conjunction with Gustav Robert Kirchhoff, spectral analysis, with the help of which cesium and rubidium were discovered.

3/31 **1890** Birth of William Lawrence Bragg (†1971). Together with his father he used the model of reflection of X-rays at crystal planes to lay the groundwork for X-ray diffraction (Nobel Prize in physics, 1915).

April

4/1 **1850** Hans Pechmann is born (†1902). He devised both the *Pechmann coumarin synthesis* and the *Pechmann pyrazole synthesis.*

4/1 **1865** Birth of Richard Zsigmondy (†1929), who built the first ultramicroscope and verified the heterogeneous nature of colloids (Nobel Prize, 1925).

4/1 **1896** The Institute for Serum Research and Serum Testing is established in Berlin, and Paul Ehrlich is called to be its director.

4/2 **1928** Death of Theodore W. Richards (*1868). He determined the precise atomic weights of a large number of elements, for which he received the Nobel Prize in 1914.

4/2 **1953** Francis H. C. Crick and James D. Watson submit their article about the structure of DNA to *Nature.*

4/4 **1915** Death of James Hargreaves (*1834), who developed the *Hargreaves process* for preparing hydrochloric acid from salt and flue gases.

4/5 **1827** Joseph Lister is born (†1912) He concerned himself with the action of chemicals on microbes and introduced the practice of antisepsis.

4/6 **1865** Friedrich Engelhorn establishes the "Badische Anilin- & Soda-Fabrik" in Mannheim (Germany). The first production facilities were developed nearby in Ludwigshafen.

4/6 **1938** The chemist R. J. Plunkett discovers polytetrafluoroethene ("Teflon") at DuPont.

4/7 **1894** Birth of Louis Plack Hammett (†1987). His name is associated with the *Hammett equation* which he formulated in 1935 to describe the influence of substituents on the reactivity of aromatic compounds.

4/8 **1911** Melvin Calvin is born. His investigation of blood-type factors led to isolation of the Rh antigen. For his pioneering work on photosynthesis in the photochemical assimilation of CO_2 he was awarded the Nobel Prize in 1961.

4/9 **1887** Heinrich Hock is born. In the course of his investigations of hydroperoxides he discovered the procedure bearing his name for the synthesis of phenol and acetone by cleavage of cumene hydroperoxide, a process still in use today.

4/10 **1917** Birth of Robert Burns Woodward (†1979). He succeeded in carrying out a number of important total syntheses, including those of quinone (1944), cholesterol and cortisone (1951), and vitamin B_{12} (1976). In 1965 he was awarded the Nobel Prize. Together with Roald Hoffmann he also proposed the *Woodward–Hoffmann rules.*

4/10 **1927** Birth of Marshall Warren Nirenberg. In the 1960s he deciphered the genetic code, for which he was awarded the Nobel Prize in medicine or physiology in 1968.

4/11 **1935** The Wolfen Film Factory submits a patent application for triple-layer color reversal film based on work carried out by Rudolf Fischer around 1910.

4/12 **1884** Birth of Otto Meyerhoff (†1951), who studied carbohydrate metabolism and the chemical processes associated with working muscle. In 1922 he was awarded the Nobel Prize in medicine.

4/15 **1890** Death of Eugène Melchior Péligot (*1811), the discoverer of uranium.

4/16 **1838** Ernest Solvay is born (†1922). He developed a process for the commercial recovery of soda from salt, ammonia, and carbon dioxide, and in 1863 founded the Solvay Works in Brussels.

4/16 **1943** Albert Hofmann discovers the halucinogenic activity of lysergic acid diethylamide (LSD) in tests conducted on himself.

4/18 **1787** A. L. Lavoisier presents the Academy of Sciences in Paris with a systematic nomenclature of chemistry, thereby establishing the basis for a universal technical language.

4/18 **1910** Paul Ehrlich presents a congress report about the antiparasitic activity of "Salvarsan," a chemotherapeutic agent from Hoechst effective against the source of syphilis.

4/19 **1912** Birth of Glenn T. Seaborg, who was awarded the Nobel Prize in 1951 for his discoveries in the field of transuranium elements.

4/20 **1860** Birth of Ludwig Gattermann (†1920). He developed the synthetic procedure named for him for preparing aromatic aldehydes.

4/20 **1899** Death of Charles Friedel (*1832), who together with James Crafts discovered the *Friedel–Crafts reaction.*

4/21 **1774** Birth of Jean-Baptiste Biot (†1862). He applied circular polarization to the determination of sugars and suggested an asymmetric molecular structure as the origin of the phenomenon.

4/21 **1899** Birth of Paul Karrer (†1971), noted for his research into carotenoids, flavins, and vitamins A and B_2 (Nobel Prize, 1937).

4/22 **1858** Birth of Martin Kiliani (†1895). This electrochemist developed the fused-salt electrolysis of aluminum to the point of commercial feasibility.

4/22 **1919** Birth of Donald James Cram. He investigated the stereochemistry in the addition of organometallic compounds to carbonyl groups adjacent to chiral centers, translating his results into *Cram's rule.* His work on host–guest chemistry led to the award of a Nobel Prize in 1987.

4/23 **1823** Anton Dominik Giulini and his nephew Paul establish the Giulini Chemical Works in Ludwigshafen (Germany).

4/24 **1817** Birth of Jean C. de Marignac (†1894), discoverer of the elements gadolinium and ytterbium. He determined the atomic masses of numerous elements and predicted the existence of isotopes.

4/24 **1960** Death of Max von Laue (*1879). He received the Nobel Prize in physics in 1914 for his discovery of X-ray interference by crystals.

4/25 **1900** Birth of Wolfgang Pauli (†1958). For his introduction of the *Pauli exclusion principle* he was awarded the Nobel Prize in physics in 1945.

4/26 **1838** Birth of Wilhelm Kalle (†1919). His company, the Fabrik Kalle & Co. in Wiesbaden, Germany, began operation by producing fuchsin in 1863.

4/26 **1932** Birth of Michael Smith. This 1993 Nobel Prize winner developed a site-specific mutagenesis based on oligonucleotides and applied it to the investigation of proteins.

4/27 **1844** Death of John Dalton (*1766), natural scientist who published the first table of atomic masses and discovered the law of multiple proportions.

4/28 **1903** Death of Josiah Willard Gibbs (*1839). He concerned himself with systems at equilibrium and formulated the *Gibbs phase rule.*

4/29 **1943** Death of Wilhelm Schlenk, Sen. (*1879), known for his work on free radicals.

April

4/30 **1897** Joseph Thomson reports his discovery of the free electron.

May

5/1 **1824** Birth of Alexander Williams Williamson (†1904), who developed the ether synthesis bearing his name.

5/1 **1825** Birth of Johann Jakob Balmer. In 1884/85 he developed the *Balmer formula* for the spectral lines of hydrogen, important as a basis for the development of the Bohr model of the atom.

5/2 **1979** Death of Giulio Natta (*1903), a recipient of the Nobel Prize in 1963 for his work on the chemistry and technology of high polymers.

5/3 **1941** The first successful treatment with penicillin is commenced in the Radcliffe Hospital in Oxford. The patient, who was originally suffering from abscesses, was released in good health on May 15.

5/4 **1777** Birth of Louis Jacques Thenard (†1857). He was the discoverer of *Thenard's blue*, a spinel composed of the oxides of cobalt and aluminum.

5/5 **1892** Death of August Wilhelm Hofmann (*1818), who carried out research in the field of aniline dyes and in 1867 founded the German Chemical Society. His name is associated with the *Hofmann degradation of quaternary ammonium salts*, the *Hofmann degradation of amides*, and also the *Hofmann decomposition apparatus*.

5/6 **1871** Birth of François Auguste Victor Grignard (†1935), the discoverer of the organomagnesium compounds named for him as well as their reaction with carbonyl derivatives. He was awarded the Nobel Prize for this work in 1912.

5/7 **1925** The Deutsche Museum is first opened in Munich.

5/7 **1939** Birth of Sidney Altman, who in 1989 received the Nobel Prize for his discoveries regarding the catalytic activity of RNA.

5/8 **1794** Death via the guillotine of Antoine Laurent Lavoisier (*1743), who played a major role in the establishment of chemistry as a classical science by providing a scientific explanation for the combustion process, redefining the concept of an element, and introducing a rational system of chemical nomenclature.

5/9 **1927** Birth of Manfred Eigen. He investigated extremely rapid reactions in solution and for this purpose developed a number of measuring techniques. In 1955 together with Ade Maeyer he succeeded for the first time in determining the rate of neutralization (Nobel Prize, 1967).

5/9 **1938** Establishment of the Chemischen Werke Hüls GmbH in Marl (Germany).

5/10 **1910** Death of Stanislao Cannizzaro (*1826). In 1853 he discovered the reaction bearing his name for the disproportionation of aldehydes to alcohols and acids.

5/11 **1981** Death of Odd Hassel (*1897), who created a basis for conformational analysis and in particular verified the existence of the chair forms of cyclohexane and decalin. He was awarded the Nobel Prize in 1969.

5/12 **1670** Birth of Friedrich Wilhelm I, elector of Saxony. The sciences were held in especially high regard under his rule, and in 1875 he ordered the establishment of the first European manufacturing facility for porcelain (in Meissen).

5/12 **1803** Birth of Justus von Liebig (†1873). He was among the first chemists in Germany to introduce laboratory work into the chemistry curriculum. Liebig further developed Karl Wilhelm Kastner's ideas for artificial fertilization and characterized a host of compounds, including lactic and formic acids. The *Liebig condenser* and the *Liebig elemental analysis* are reminders of some of his other areas of interest. He was also involved in the development of meat extracts, baking powder, and infant formulas.

5/12 **1884** Hilaire, Count of Chardonnet de Grange, informs the Académie Française that he has succeeded in preparing artificial silk from cellulose. The factory he built was producing about 50 kg of the material per day in 1891.

5/12 **1895** Birth of William Francis Giauque (†1982). He investigated the behavior of substances at extremely low temperatures and discovered the oxygen isotopes ^{17}O and ^{18}O (Nobel Prize, 1949).

5/13 **1842** Ferdinand Fischer is born (†1916). He worked in the field of gas and firing technology and in 1887 founded the Association of German Chemists.

5/14 **1796** The English physician Edward Jenner, who experimented systematically with methods of vaccination, carries out his first smallpox vaccination in Berkeley (Gloucestershire).

5/15 **1899** Birth of William Hume-Rothery (†1968), who concerned himself with solid-state structures and intermetallic compounds and formulated the *Hume-Rothery rule* for intermetallic phases.

5/16 **1940** Death of Otto Dimroth (*1872), for whom the *Dimroth condenser* is named.

5/17 **1861** James Clerk Maxwell presents the first color photographs to the Royal Institution, although the procedure by which they were prepared was of scientific interest only.

5/18 **1889** Birth of Thomas Midgley, Jun. (†1944). His primary interest was in the "knocking" properties of fuels, and in 1922 he discovered the antiknock agent tetraethyl lead.

5/18 **1901** Vincent du Vigneaud is born (†1978). For his isolation, structure determination, and first synthesis of a polypeptide hormone he was awarded the Nobel Prize in 1955.

5/19 **1914** Birth of Max Ferdinand Perutz. His work in the structure determination of proteins with the aid of X-ray analysis led to establishment of the structure of hemoglobin (Nobel Prize, 1962).

5/20 **1860** Birth of Eduard Buchner (†1917), who was awarded the Nobel Prize in 1907 for his discovery and investigation of cell-free fermentation.

5/20 **1895** Carl von Linde (1842–1934) demonstrates in Munich for the first time the liquification of large quantities of air. The yield per hour was about three liters.

5/21 **1934** Birth of Bengt Ingemar Samuelsson. He determined the structure of the prostaglandins and participated in the discovery of the leucotrienes (Nobel Prize in medicine or physiology, 1982).

5/22 **1912** Birth of Herbert Charles Brown, who discovered the addition reaction of diborane to double bonds (now named in his honor) known as hydroboration. For his basic investigations in the field of organoboron chemistry he was awarded the Nobel Prize in 1979.

5/22 **1927** George Olah is born. For his contributions to carbocation chemistry he was awarded the Nobel Prize in 1994.

5/23 **1902** Birth of Rudolf Criegee (†1975), who discovered the cleavage of diols with lead tetraacetate, one of the important reactions in carbohydrate chemistry, and also made contributions to the chemistry of peroxides and ozonides.

5/24 **1686** Birth of Gabriel Fahrenheit (†1736). He constructed the first mercury thermometer calibrated with the scale that bears his name.

5/25 **1877** The Imperial Patent Law is adopted in Germany.

5/26 **1888** Death of Ascanio Sobrero (*1812), the discoverer of nitroglycerin.

5/27 **1857** Birth of Theodor Curtius (†1928). He prepared the first aliphatic diazo compound by treating ethyl α-aminoacetate with sodium nitrite.

5/28 **1906** Death of Rudolf Josef Knietsch (*1854). This chemist developed (at the Badische Anilin- & Soda-Fabrik) an industrial procedure for liquification of chlorine and also the contact process for manufacturing sulfuric acid.

May

5/29 **1829** Death of Humphry Davy (*1778). He discovered the euphoric effect of laughing gas and dedicated himself to the investigation of chemical effects of electrical current. In the course of this work he discovered that chlorine is an element and not a compound of oxygen. His important inventions include (1815) the safety lamp for use in coal mines.

5/30 **1912** Birth of Julius Axelrod, noted for his work on the biosynthesis and inactivation of noradrenalin, the transport substance for nerve impulses (Nobel Prize in medicine, 1970).

5/31 **1918** Death of Alexander Mitscherlich (*1836), a pioneer in the paper industry who developed the sulfite cellulose process.

June

6/1 **1796** Birth of Nicolas Léonhard Sadi Carnot (†1832). In the course of his thermodynamic studies he developed the cyclic process that bears his name.

6/1 **1978** The European Patent Office opens in Munich.

6/3 **1873** Birth of Otto Loewi (†1961). For his investigations of the chemical transmitter for nerve impulses he was awarded in 1936 the Nobel Prize for medicine or physiology.

6/4 **1877** Birth of Heinrich Wieland (†1957), who clarified the constitution of the bile acids and related compounds (Nobel Prize, 1927).

6/5 **1760** Birth of Johan Gadolin (†1852), the discoverer of yttrium. The mineral gadolinite is named in his honor, as is the element gadolinium, isolated by Jean Marignac.

6/6 **1825** Friedrich Bayer (†1880), cofounder of the Farbenfabriken (dye works) Bayer, is born.

6/7 **1896** Birth of Robert Sanderson Mulliken (†1986). Together with Friedrich Hund he developed the molecular orbital theory, and in 1966 was awarded the Nobel Prize.

6/8 **1863** Friedrich August Raschig (†1928) is born. In 1890 he established a factory in Ludwigshafen (Germany) for the production of phenol and cresol, and in 1921 the Raschig ceramics works for manufacturing the packing material for distillation columns that he developed (*Raschig rings*).

6/8 **1916** Birth of Francis Harry Compton Crick, who was awarded the Nobel Prize in 1962 for discovering the double-helix structure of DNA.

6/9 **1812** Birth of Hermann von Fehling (†1885), who discovered *Fehling's test* for sugars.

6/10 **1848** Birth of Ferdinand Tiemann (†1899). He is best known for the synthesis of fragrances, especially vanillin and ionone, some of which were utilized commercially by the Haarmann Company in Holzminden (Germany). Together with Reimer he discovered the *Reimer–Tiemann reaction* for the synthesis of *o*- and *p*-hydroxyaldehydes from phenols with chloroform and potassium hydroxide.

6/11 **1842** Birth of Carl von Linde (†1934), inventor of the ammonia-based refrigeration device and the procedure bearing his name for the liquification of air with the aid of the Joule–Thomson effect. In 1879 he established the firm Linde AG in Wiesbaden (Germany).

6/12 **1899** Birth of Fritz Albert Lipmann (†1986). He concerned himself with the energetics of metabolic processes and recognized the central role played by ATP. In 1953 he was awarded the Nobel Prize in medicine.

6/12 **1920** An article appears in the *Berichte der Deutschen Chemischen Gesellschaft* by Hermann Staudinger entitled "Regarding Polymerization," in which, contrary to current thinking, he postulates the existence of macromolecules. A few years later he verified their existence experimentally.

6/14 **1897** Cyril Norman Hinshelwood is born (†1967). He was awarded the Nobel Prize in 1956 for kinetic studies of gas-phase reactions and interpretation of the mechanisms of chain reactions.

6/15 **1883** The Reichstag adopts the first piece of Bismarck's social legislation, a statute covering health insurance. As a result, the first corporate health coverage was established at the large German chemical companies.

6/16 **1850** Max Delbrück is born (†1919). His scientific activity was of great importance to the fermentation industry.

6/16 **1897** Birth of Georg Wittig (†1987). He discovered the Wittig reaction, with the help of which double bonds could be introduced at specific locations. His pioneering efforts in the chemistry of organophosphorus compounds led to a Nobel Prize in 1979.

6/17 **1940** Death of Arthur Harden (*1865), who studied the process of fermentation and discovered the phosphorylation reaction, which was soon recognized as a fundamental biochemical step. In 1929 he was awarded the Nobel Prize.

6/18 **1865** Birth of Emil Knoevenagel (†1921), who discovered a condensation reaction related to the aldol condensation that now carries his name.

6/18 **1918** Birth of Jerome Karle, who received the Nobel Prize in 1985 for fundamental work on direct methods of crystal structure determination.

6/18 **1932** Birth of Dudley Robert Herschbach. He developed the method of crossed molecular beams, with the help of which dynamic elementary chemical processes could be followed (Nobel Prize, 1986).

6/19 **1783** Friedrich Wilhelm Sertürner, the discoverer of morphine, is born (†1841).

6/19 **1910** Birth of Paul John Flory (†1985). He was awarded the Nobel Prize in 1974 for his theoretical and experimental studies in the field of macromolecular chemistry.

6/20 **1861** Birth of Frederick Gowland Hopkins (†1947), the discoverer of the necessity for supplementary dietary substances later known as "vitamins." In 1929 he was awarded the Nobel Prize in medicine.

6/21 **1929** A patent is issued for the copolymerization of butadiene and styrol with the aid of sodium (German: "Natrium"). The process leads to a rubber that is at least as effective as natural rubber and which became familiar under the name "buna" (for *bu*tadiene–*Na*trium).

6/22 **1839** Louis Daguerre signs an agreement with Alphonse Giroux regarding the manufacture of photographic apparatus. The Giroux Co. became the first photography company in the world that produced not only cameras but also the accessories needed for preparing daguerreotypes.

6/23 **1733** Johann Rudolf Geigy is born (†1793). He founded the Geigy chemical company in Basel (Switzerland) in 1758, which merged with Ciba in 1970.

6/23 **1854** Birth of Rudolf Leuckart (†1889). His name is associated with a number of reactions, including the xanthogenate cleavage or the reductive alkylamination of carbonyl compounds in the presence of formic acid.

6/24 **1900** The city of Mainz (Germany) celebrates the 500th birthday of Johannes Gutenberg. He created interchangeable metal type from an alloy consisting of tin, antimony, and bismuth, thus giving birth to the book-printing industry.

6/25 **1864** Birth of Walter Hermann Nernst (†1941). His most important accomplishments were the formulation of the *Nernst equation*, the *Nernst distribution law*, and the *third law of thermodynamics* (Nobel Prize, 1920).

6/25 **1911** Birth of William Howard Stein (†1980). Together with Stanford Moore he determined the sequence of ribonuclease, for which both received the Nobel Prize in 1972.

6/25 **1921** Friedrich Bergius (1884–1949) reports at a chemical congress in Stuttgart (Germany) the success of his efforts to liquify coal, which acquired importance especially due to increasing motorization. In 1931 he was awarded a Nobel Prize for developments in high-pressure technology.

June

6/26 **1966** "Experimental Breeder Reactor No. I" in Idaho (USA) is declared a national historic landmark after its decontamination. This device was the scene of important milestones in the development of reactor technology: e.g., first electricity generation from nuclear energy, verification of an atomic breeder process, and successful testing of a Na–K alloy as a coolant.

6/27 **1892** Death of Carl Ludwig Schorlemmer (*1834). In 1874 he assumed the first chair ever designated for organic chemistry (at Owens College, Manchester, England). He was also the first to prepare a number of hydrocarbons in pure form and then determine their thermodynamic properties.

6/28 **1825** Birth of Emil Erlenmeyer (†1909). In 1862 he postulated the presence of a double bond in ethene and a triple bond in ethyne. In 1880 he proposed the rule named for him stating that compounds with two or three –OH groups on the same carbon atom are usually unstable, transforming themselves with loss of water into carbonyl and carboxyl systems, respectively. His name has also acquired universal and permanent recognition through the *Erlenmeyer flask* he invented.

6/28 **1927** Birth of Sherwood Frank Rowland. For his investigations into the effects of fluorochlorohydrocarbons on the ozone layer he was awarded the Nobel Prize in 1995.

6/29 **1813** Alexander Parkes is born (†1890). He prepared the first plastic ("Parkesine") from nitrocellulose, albeit one for which no applications were found.

6/30 **1926** Birth of Paul Berg, who introduced segments of animal DNA into bacterial DNA and thus became one of the pioneers in gene manipulation (genetic engineering). Among other things he laid the groundwork for biogenetic production of insulin and interferon (Nobel Prize, 1980).

July

7/1 **1877** The Imperial Patent Office is opened in Berlin, following adoption on May 25 of an Imperial Patent Law. The first German Imperial patent (D. R. P. 1) was issued to the Nuremberg Ultramarine Factory and covered a "process for the production of a red ultramarine dye."

7/1 **1929** Birth of Gerald Maurice Edelman. He investigated the molecular mechanism of binding between an antibody and an antigen, and received a Nobel Prize in medicine or physiology in 1972

7/2 **1862** Birth of William Henry Bragg (†1942). He and his son initiated crystal structure analysis in 1913 through their derivation of *Bragg's law* for the relationship between the wavelength of impinging X-rays and atomic spacings in a crystal lattice (Nobel Prize in physics, 1915).

7/2 **1875** Birth of Fritz Ullmann (†1939). He was the founder and editor of *Ullmann's Encyclopedia of Industrial Chemistry.*

7/4 **1853** Birth of Ernst Otto Beckmann (†1923), who in 1886 observed the intramolecular rearrangement of ketoximes to substituted amides that bears his name. He also developed the *Beckmann thermometer*, which permits very precise temperature measurements (0.01 K).

7/4 **1913** Fritz Klatte describes in his patent application the polymerization of vinyl chloride and processing of the resulting polymer. These findings were first exploited commercially in the early 1930s.

7/5 **1891** Birth of John Howard Northrop (†1987). He concerned himself with enzymes and their purification. In 1941 he succeeded in preparing the first crystalline antibody, the diphtheria antitoxin (Nobel Prize, 1946).

7/6 **1903** Birth of Axel Hugo Theodor Theorell (†1982), the first person to succeed in separating an enzyme reversibly into a coenzyme and an apoenzyme. He received a Nobel Prize for medicine or physiology in 1955.

7/7 **1960** The American physicist T. H. Maiman demonstrates the first laser.

7/8 **1720** Birth of Heinrich Friedrich von Delius (†1791), who introduced chemistry into the academic curriculum.

7/9 **1856** Death of Amadeo Avogadro (*1776). It was he who established that a given volume of any gas under the same conditions (pressure, temperature) contains the same number of molecules (*Avogadro's number*). This permitted the molecular mass of a gaseous compound to be determined.

7/10 **1897** The Badische Anilin- & Soda-Fabrik introduces the first "pure indigo" into the market after 18 years of research and an investment equivalent to the stock value of the entire company.

7/10 **1902** Birth of Kurt Alder (†1958), who together with Otto Diels discovered the diene addition known as the *Diels–Alder reaction* (Nobel Prize, 1950).

7/11 **1895** Adam Miller acquires a patent in Glasgow for the production of threads from gelatin. This achievement represented a significant step along the way to artificial silk.

7/12 **1857** Birth of Amé Pictet (†1937), one of the discoverers of the *Pictet–Spengler reaction* for the synthesis of isoquinolines.

7/12 **1870** John Wesley Hyatt applies for a US patent in which the production of celluloid is fully described for the first time. He developed it as a substitute for ivory in the manufacture of billiard balls.

7/12 **1928** Birth of Elias James Corey, who developed synthetic pathways to such natural products as the prostaglandins and various terpenes (Nobel Prize, 1990).

7/14 **1921** Birth of Geoffrey Wilkinson, who simultaneously with Ernst Otto Fischer established the structure and behavior of ferrocene and other sandwich compounds (Nobel Prize, 1973). He was also the originator of the *Wilkinson catalyst.*

7/15 **1800** Birth of Jean Baptiste André Dumas (†1884). As a public official he was one of the most influential chemists in France. He was responsible for a method for determining the nitrogen content of organic compounds.

7/15 **1871** Birth of Max Ernst Bodenstein (†1942), whose investigations laid the groundwork for the kinetic theory of gases.

7/15 **1921** Birth of Robert Bruce Merrifield. He developed an automated method for peptide synthesis known as the *Merrifield technique* (Nobel Prize, 1984).

7/16 **1876** Alfred Eduard Stock is born (†1946). His pioneering studies of boranes established a new field of chemical research.

7/17 **1821** Birth of Friedrich Engelhorn, founder of the Badische Anilin- & Soda-Fabrik (BASF).

7/17 **1903** Birth of Richard Müller, who independently of Eugene Rochow discovered the *Müller–Rochow synthesis* for the preparation of chlorosilanes, the precursors to silicones.

7/18 **1937** Birth of Roald Hoffmann. He participated in the development of the theory of electrocyclic reactions and together with Woodward proposed the *Woodward–Hoffmann rules* for the maintenance of orbital symmetry (Nobel Prize, 1981).

7/18 **1948** Hartmut Michel is born. He received a Nobel Prize in 1988 for his work on determining the three-dimensional structure of one of the reaction centers in photosynthesis.

7/19 **1838** Death of Pierre Louis Dulong (*1785). Together with Alexis Petit he discovered the *rule of Dulong and Petit,* according to which the molar heat capacity of numerous solid elements is constant at room temperature with a value of ca. 25 $J K^{-1}$.

7/19 **1921** Rosalyn Yalow is born. She developed the radioimmunoassay method and in 1977 received a Nobel Prize in medicine or physiology.

7/20 **1897** Birth of Tadeus Reichstein (†1996). He concerned himself with the isolation and structural characterization of pituitary hormones, including cortisone. In 1950 he was awarded a Nobel Prize for medicine or physiology.

7/20 **1969** Men land for the first time on the moon. The flight of Apollo 11 provided cosmochemists with the first 22 kg of moon rock.

7/21 **1923** Birth of Rudolph Arthur Marcus, who received a Nobel Prize in 1992 for his work on the theory of intermolecular electron-transfer processes.

7/22 **1788** Birth of Joseph Pelletier (†1842). This Paris apothecary received a prize of 10 000 francs from the Paris Academy in 1827 for discovering the quinine bases.

7/23 **1822** Birth of Henry Deacon (†1876). He discovered a method for isolating chlorine from hydrogen chloride by air oxidation in the presence of a copper catalyst. Until the development of the electrolytic procedure this was the most common method of chlorine production.

7/23 **1906** Birth of Vladimir Prelog. His work on the stereochemistry of medium-sized rings led to his being recognized as one of the originators of conformational analysis (Nobel Prize, 1975).

7/24 **1939** The largest British chemical company, Imperial Chemical Industries (ICI), begins producing polyethylene based on the high-pressure polymerization method discovered in 1933.

7/25 **1616** Death of Andreas Libavius (or Libau; *ca. 1550). This physician and chemist compiled the leading chemistry book of his time, *Alchemia collecta,* one of the first to present chemistry in the absence of mysticism and in a comprehensible way.

7/26 **1863** Birth of Paul von Walden (†1957). In 1896 he observed a reversal in the configuration at an asymmetric carbon atom in the course of a substitution reaction, a phenomenon now known as *Walden inversion.*

7/27 **1881** Birth of Hans Fischer (†1945), who successfully synthesized heme and bilirubin and also established the constitution of chlorophyll (Nobel Prize, 1930).

7/28 **1968** Death of Otto Hahn (*1879), who received a Nobel Prize in 1944 for discovering the nuclear fission of heavy atoms.

7/29 **1892** Birth of Walter Reppe (†1969), a chemist at the Badische Anilin- & Soda-Fabrik who developed the techniques of high-pressure acetylene chemistry.

7/29 **1994** Death of Dorothy Crowfoot Hodgkin (*1910). This winner of a 1964 Nobel Prize employed X-ray crystallographic methods to determine the structures of such important biochemical compounds as penicillin, insulin, and calciferol.

7/30 **1841** Bernhard Tollens is born (†1918). He is noted for his investigations in the field of sugar chemistry and from the *Tollens reagent,* used in a sensitive test for aldehydes.

July

7/31 **1825** Birth of August Beer (†1863), who proposed the law of light absorption that bears his name. This relationship later became important for colorimetric determinations.

7/31 **1943** Death of Max Julius LeBlanc (*1865), a physical chemist who developed a series of electrochemical methods, including in 1893 the hydrogen electrode used in pH determinations.

August

8/1 **1863** Friedrich Bayer, a merchant, and the dyer Friedrich Weskott establish a company in Leverkusen (Germany) for the production of aniline dyes under the name "Friedr. Bayer et comp."

8/1 **1885** Birth of György de Hevesy (†1966), who together with Dirk Coster discovered hafnium. For his studies in the use of isotopes for labeling during the investigation of chemical and biochemical reactions he received a Nobel Prize in 1943.

8/2 **1788** Birth of Leopold Gmelin (†1853). He was the founder of the *Gmelin Handbook of Inorganic Chemistry.*

8/3 **1942** Death of Richard Willstätter (*1872), who was awarded a Nobel Prize in 1915 for his investigations of vegetable dyes, especially chlorophyll.

8/4 **1950** Hoechst begins production at its new German penicillin factory, which in principal could meet the needs of the entire West German market.

8/5 **1866** Carl Dietrich Harries is born (†1923). He discovered the addition of ozone to double bonds, with subsequent cleavage of the resulting ozonides

8/6 **1766** Birth of William Hyde Wollaston, who discovered the elements palladium and rhodium. The mineral *wollastonite,* a calcium silicate, was named in his honor.

8/6 **1881** Birth of Alexander Fleming (†1955), the discover of penicillin and winner of a Nobel Prize in medicine in 1945.

8/7 **1972** A German federal law is enacted forbidding the manufacture, import, or application of DDT (4,4′-dichlorodiphenyltrichloroethane). A Nobel Prize in medicine was awarded in 1948 to Paul Müller of Geigy AG for discovering (in 1939) the insecticidal properties of this compound, which had been known since 1874. For decades DDT was the most important insecticide worldwide.

8/8 **1818** Matthias Eduard Schweizer is born (†1860). He discovered that an aqueous solution of tetraamminecopper(II) hydroxide is a solvent for cellulose, thereby preparing the way for the manufacture of cuprammonium silk.

8/8 **1884** George Eastman, the head of Kodak, applies for the first patent for a photographic film based on celluloid.

8/8 **1901** Birth of Ernest Orlando Lawrence (†1958). For his invention of the cyclotron and the scientific results it facilitated – such as the discovery of radioactive isotopes and elements – he was awarded a Nobel Prize in physics in 1939. In his honor, element 103 (discovered in 1961) was assigned the name *lawrencium.*

8/9 **1896** Birth of Erich Hückel (†1980). This theoretical physicist devised the *Hückel rules* for aromaticity in cyclic conjugated systems, and together with Peter Debye he developed the *Debye–Hückel theory* of strong electrolytes.

8/10 **1902** Birth of Arne Wilhelm Tiselius (†1971), who was awarded a Nobel Prize in 1948 for his work on electrophoresis and adsorption chromatography as methods for separating serum proteins.

8/11 **1926** Birth of Aaron Klug. For the development of crystallographic electron microscopy and determination of the molecular structures of proteins, nucleic acids, and their complexes he was awarded a Nobel Prize in 1982.

8/12 **1887** Erwin Schrödinger is born (†1961). He was the cofounder of wave mechanics. In 1926 he proposed the fundamental equation of quantum mechanics that bears his name (Nobel Prize in physics, 1933).

8/13 **1918** Birth of Frederick Sanger. He received a Nobel Prize in 1958 for the final structure determination of insulin (1955), and in 1977 he developed a method for sequencing DNA that led to a second Nobel Prize in chemistry in 1980.

8/14 **1958** Death of Frédéric Joliot (*1900). Together with his wife Irène he discovered artificial radioactivity and shared with her a Nobel Prize in 1935 for the synthesis of new radioactive elements.

8/15 **1875** Birth of Charles August Kraus (†1967). In the course of his research on liquid ammonia as a solvent he demonstrated that the blue color arising upon dissolution of sodium is a consequence of solvated electrons.

8/16 **1849** Birth of Johan Gustav Kjeldahl, who in 1883 published the method known by his name for determining the nitrogen content of organic compounds.

8/16 **1904** Birth of Wendell Meredith Stanley. He isolated the tobacco mosaic virus and thus viral protein in pure form (Nobel Prize, 1946).

8/17 **1893** Walter Noddack is born (†1960). This physical chemist is noted for his work on rhenium as well as for geochemical studies.

8/18 **1916** Founding of the "German Coal-Tar Dye Manufacturers Interest Group," to which two different "big three" triads (Bayer, BASF, and Agfa on the one hand and Hoechst, Casella, and Kalle on the other) as well as Weiler-ter-Meer and the Chemische Fabrik Griesheim belong.

8/18 **1960** The G. D. Searle Drug Co. begins marketing in the United States the first oral contraceptive ("Enovid") designed for public consumption.

8/19 **1830** Birth of Julius Lothar Meyer (†1895). He developed independently of Dimitri Mendeleev the concept of a periodic system of the elements.

8/19 **1839** L. J. Daguerre uses his pension fund to publish a "photographic" secret: a copper plate coated with silver and treated with iodine vapors (AgI) is exposed in a camera obscura. The latent image is developed with mercury vapor and unexposed silver iodide is removed with salt water (later with thiosulfate solution).

8/20 **1779** Birth of Jöns Jacob Berzelius (†1848). He was a determining force in the development of European chemistry for more than five decades. In 1803 he discovered cerium, and between 1807 and 1812 he produced a remarkably precise table of experimentally determined atomic masses. He improved laboratory techniques dramatically through the introduction of such equipment as test tubes and beakers, platinum apparatus, and wash bottles. In 1813 he developed a new set of elemental symbols based on the first letters of the element names as a substitute for the traditional graphic symbols.

8/21 **1901** Death of Adolf Fick (*1829). He derived the laws bearing his name, which describe the quantitative course of diffusion.

8/24 **1888** Death of Rudolph Julius Emanuel Clausius (*1822), who formulated the second law of thermodynamics and introduced such concepts as "mean free path" and "collision number" into the kinetic theory of gases. He also modified Clapeyron's equation and coined the term "entropy" as a measure of the orderliness of a system.

8/25 **1867** Death of Michael Faraday (*1791). He discovered benzene in 1824, magnetic induction in 1831, and the *Faraday effect* in 1845, and made important contributions to the development of electrochemistry. With *Faraday's laws* he established a relationship between current flow in an electrolysis and the amount of substance deposited, and he discovered the principle of self-induction. His name is associated not only with concepts but also with a physical quantity.

8/26 **1668** Friedrich Jakob Merck acquires the Engel Apothecary Shop in Darmstadt (Germany), which ultimately develops into the corporate entity E. Merck AG.

August

8/26 **1856** British patent number 36 140 is issued for mauvein, the world's first artificial dye, which was synthesized quite by accident by the young English chemist Perkin.

8/27 **1874** Birth of Carl Bosch (†1940), who developed high-pressure technology on an industrial scale and together with Haber perfected the *Haber–Bosch process* for the synthesis of ammonia from its elements (Nobel Prize, 1931).

8/28 **1841** Death of Johann August Arfvedson (*1792), the discoverer of lithium.

8/29 **1868** Death of Christian Friedrich Schönbein (*1799), who discovered ozone and nitrocellulose and is considered the founder of geochemistry.

8/30 **1852** Birth of Jacobus Henricus van't Hoff (†1911). In 1901 he became the first Nobel Prize winner in chemistry, recognized for discovering the laws of chemical dynamics and osmotic pressure in solution. He was also one of the founders of stereochemistry.

8/30 **1884** Birth of Theodor Svedberg (†1971). With his construction of the ultracentrifuge it became possible to separate proteins. His name has been associated with the sedimentation constant, which specifies the rate at which particles migrate in the centrifugal force field of an ultracentrifuge (Nobel Prize, 1926).

8/31 **1842** Birth of Adolf Pinner (†1909), who together with Richard Wolffenstein clarified the structural formula of nicotine. He also developed the *Pinner reaction.*

September

9/1 **1873** Birth of Ragnar Berg (†1956), one of the fathers of modern nutrition. He recognized the importance of trace elements in living systems and showed that many metabolic diseases are a consequence of nutritional shortcomings.

9/2 **1836** Death of William Henry (*1774), who discovered the proportionality between the amount of gas dissolved in a liquid and the pressure exerted by that gas above the liquid, now know as *Henry's law.*

9/2 **1853** Birth of Wilhelm Ostwald (†1932). He discovered among other things the *Ostwald rule, Ostwald's law of dilution,* and the *Ostwald process* for preparing nitric acid. He was awarded a Nobel Prize in 1909 for his studies of catalysis as well as chemical equilibrium and reaction rates.

9/3 **1869** Birth of Fritz Pregl (†1930), who developed both qualitative and quantitative microanalysis as well as the microelemental analysis of organic compounds, which produces reliable results from milligram quantities of material.

9/4 **1913** Birth of Stanford Moore (†1982). Working together with William Stein he established the constitution of ribonucleases and investigated their biochemical mode of action, for which he was awarded a Nobel Prize in 1972.

9/4 **1967** The revised German patent law takes effect, according to the terms of which not only processes but also chemical substances themselves can be protected.

9/5 **1961** A German detergent law specifies that after October 1, 1964, detergents and cleansing agents must contain only ingredients that are at least 80% biodegradable.

9/6 **1906** Birth of Luis Frederico Leloir (†1987). He was awarded a Nobel Prize in 1970 for his discovery of sugar nucleotides and their role in carbohydrate synthesis.

9/7 **1829** Birth of August Kekulé von Stradonitz (†1896). He developed during the 1860s his valence and structural theories (including a structural formula for benzene) in conjunction with recognizing the tetravalency of carbon.

9/7 **1917** Birth of John Warcup Cornforth. He was awarded a Nobel Prize in 1975 for his studies of the stereochemistry of enzyme-catalyzed reactions.

9/8 **1894** Death of Hermann von Helmholtz (*1821). His recognition of electromotive force as a form of work permitted the establishment of a relationship between it and thermodynamics, which in turn furnished a powerful impetus for the further development of physical chemistry.

9/8 **1918** Birth of Derek Harold Richard Barton, noted for his studies of the conformation of organic compounds and their successful application to such natural products as steroids and alkaloids (Nobel Prize, 1969).

9/9 **1913** The first reactor based on the Haber–Bosch process is activated by the Badische Anilin- & Soda-Fabrik in Oppau (Germany).

9/10 **1797** Birth of Carl Gustav Mosander (†1858), who engaged in studying the rare earths and discovered the elements lanthanum, erbium, and terbium.

9/10 **1850** Birth of Karl Heumann (†1894). He developed two syntheses of indigo, which were later implemented on an industrial scale by the Badische Anilin- & Soda-Fabrik and Hoechst.

9/11 **1967** The lunar lander Surveyor 5 transmits back to earth a soil analysis from the moon: 58% oxygen, 18.5% silicon, 6.5% aluminum, and a total of 13% sulfur, iron, cobalt, and nickel.

9/12 **1897** Birth of Irène Joliot-Curie (†1956). Together with her husband she discovered artificial radioactivity, and she was awarded a Nobel Prize in 1935 for the synthesis of new radioactive elements.

9/12 **1909** Bayer AG becomes the first company worldwide to receive a patent for the production of synthetic rubber.

September

9/13 **1886** Birth of Robert Robinson (†1975). This 1947 Nobel Prize winner investigated natural products from plants and made important contributions to the structural elucidation of numerous alkaloids, including morphine, narcotine, and strychnine. His name is associated with a number of reactions, including the *Robinson anellation*, a combination of the Michael and aldol reactions which can be used for the construction of anellated rings.

9/13 **1887** Birth of Leopold Ruzicka (†1976). It was he who formulated the isoprene rule, which has simplified the assignment of structures to natural products. He also succeeded in preparing androsterone in the first synthesis of a sex hormone (Nobel Prize, 1939).

9/14 **1909** A patent application is submitted in Germany in Fritz Haber's name for the high-pressure synthesis of ammonia.

9/15 **1794** Birth of Heinrich Emanuel Merck (†1855). This pharmacist and chemist accomplished the commercial preparation of numerous medicinal agents, including morphine and codeine, and therefore founded in 1927 in Darmstadt (Germany) the pharmaceutical company that bears his name.

9/15 **1854** Birth of Traugott Sandmeyer (†1922), who discovered the *Sandmeyer reaction* for the synthesis of aryl halides and pseudohalides from the corresponding amines.

9/16 **1893** Birth of Albert Szent-Györgyi (†1986). In 1928 he succeeded for the first time in isolating vitamin C (ascorbic acid) in pure, crystalline form in the course of his investigations of the biological combustion process (Nobel Prize in medicine or physiology, 1937).

9/17 **1936** Death of Henry Le Châtelier (*1850), who formulated the principle of least constraint with respect to the influence of external parameters on chemical equilibrium, now known as *Le Châtelier's principle.*

9/18 **1907** Birth of Edwin Mattison McMillan (†1991), co-discoverer of neptunium and plutonium (Nobel Prize, 1951).

9/20 **1842** Birth of James Dewar (†1923). In the course of his work on gas liquification he developed the *Dewar flask*, which facilitated the storage and investigation of substances at low temperature. The now familiar thermos bottle is based on the same principle.

9/20 **1946** The "Society of German Chemists" is founded in Göttingen to continue the traditions of its predecessor organizations: the "German Chemical Society" and the "German Chemists Club."

September

9/22 **1956** Death of Frederick Soddy (*1877). Together with Ernest Rutherford he discovered radon and recognized it to be one of the noble gases. With Alexander Russel and Kasimir Fajans he formulated the law of displacement for elemental transformations in radioactive processes (Nobel Prize, 1921).

9/23 **1882** Death of Friedrich Wöhler (*1800), whose synthesis of urea constituted the birth of synthetic organic chemistry.

9/24 **1905** Birth of Severo Ochoa (†1993). He successfully isolated the polynucleotide phosphorylases, permitting extracellular biosynthesis of RNA strands (Nobel Prize in medicine, 1959).

9/25 **1986** Death of Nikolai Semenov (*1896). He was awarded a Nobel Prize in 1956 for elucidating the mechanisms of chain reactions.

9/26 **1876** Fritz Henkel establishes the Henkel Corporation in Düsseldorf (Germany), which in 1907 introduced "Persil®," the world's first self-acting cleansing agent.

9/27 **1818** Birth of Adolf Wilhelm Hermann Kolbe (†1884), noted for his industrially applicable synthesis of salicylic acid from metal phenolates and carbon dioxide.

9/27 **1857** Birth of Ludwig Wolff (†1919), whose name is associated with the *Wolff rearrangement* and the *Wolff–Kishner reduction.*

9/28 **1852** Birth of Henri Moissan (†1907). He is known for his investigations of fluorine as well as development of the electric furnace, with which he was able to synthesize a host of carbides, silicides, hydrides, and borides (Nobel Prize, 1906).

9/29 **1861** Birth of Carl Duisberg. He joined the board of directors of Bayer in 1912 and promoted the development of the I. G. Farben conglomerate. He is thus regarded as one of the founders of the large-scale German chemical industry.

9/29 **1920** Birth of Peter Dennis Mitchell (†1992). He was awarded a Nobel Prize in 1978 for development of the chemiosmotic theory as a way of explaining biological energy-transfer processes.

9/30 **1939** Birth of Jean-Marie Lehn, who received a Nobel Prize in 1987 for developing macropolycyclic compounds with highly selective structure-specific interactions.

9/30 **1943** Birth of Johann Deisenhofer, who received a Nobel Prize in 1988 for determining the three-dimensional structure of a photosynthetic reaction center.

October

10/2 **1852** Birth of William Ramsay (†1916), who was awarded a Nobel Prize in 1904 for discovering the noble gases helium, neon, argon, krypton, and xenon, and for situating them within the periodic table.

October

10/2 **1907** Birth of Alexander Robertus Todd (†1997), whose life work was the structure elucidation and synthesis of biologically important phosphorus compounds. Among other things he carried out total syntheses of ADP and ATP (Nobel Prize, 1957).

10/3 **1904** Birth of Charles Pedersen (†1989), who synthesized macrocylic crown ethers and investigated their complexes with ions (Nobel Prize, 1987).

10/4 **1918** Birth of Kenichi Fukui. In 1952 he developed the theory of frontier orbitals. For his quantum-mechanical investigations of chemical reactivity he was awarded a Nobel Prize in 1981.

10/5 **1889** Birth of Dirk Coster (†1950), who in 1922 together with György de Hevesy discovered hafnium.

10/6 **1807** Humphry Davy for the first time prepares potassium metal by the electrolysis of molten potassium hydroxide. He also succeeded in preparing sodium by the same method.

10/7 **1885** Birth of Niels Bohr (†1962). Starting with Rutherford's notions and through the application of quantum conditions he developed the atomic model that bears his name (Nobel Prize in physics, 1922).

10/8 **1883** Birth of Otto Heinrich Warburg (†1970). He studied among other things the metabolic process in tumors. For his work on cell respiration he was awarded in 1931 a Nobel Prize in medicine or physiology.

10/8 **1906** The hairdresser Karl Ludwig Nessler introduces his new process for giving "permanent waves."

10/8 **1917** Birth of Rodney Robert Porter (†1985). He made important contributions to the structure determination of antibodies and in 1970 established the complete structure of immunoglobulin G (Nobel Prize, 1972).

10/9 **1852** Birth of Emil Hermann Fischer (†1919), who was awarded a Nobel Prize in 1902 for his groundbreaking synthetic studies in the field of carbohydrate and purine chemistry. His name is associated with a number of name reactions, including the *Fischer phenylhydrazine synthesis* and the *Fischer indole synthesis*, as well as with the *Fischer projection formulas* for carbohydrates.

10/10 **1731** Birth of Henry Cavendish (†1810), who discovered hydrogen. He also demonstrated that water is not an element, but a compound of hydrogen and oxygen.

10/10 **1897** Felix Hoffmann, working at Bayer, describes in his laboratory notebook the compound acetylsalicylic acid (ASS), which he prepared for the first time in chemically pure form. In 1899 ASS was introduced into the market as a pain reliever with the trade name "Aspirin®." This was the first product ever to be marketed in tablet form.

10/11 **1884** Birth of Friedrich Bergius, who initiated experiments in the conversion of coal, tar, and oil into gasoline and also demonstrated the commercial feasibility of coal hydrogenation (Nobel Prize, 1931).

10/12 **1792** Birth of Christian Gottlob Gmelin (†1860). He contributed to the development of analytical chemistry in Germany and in 1828 succeeded in synthesizing ultramarine.

10/13 **1965** Death of Paul Hermann Müller (*1899). He first recognized the insecticidal activity of DDT, and for this was awarded a Nobel Prize in medicine in 1948.

10/14 **1885** Birth of Murray Raney (†1966), who prepared the first *Raney catalysts* for hydrogenation and dehydrogenation. The most readily accessible of these is *Raney nickel.*

10/15 **1845** Prof. Justus von Liebig (1803–1873) acquires a British patent for his method of preparing artificial fertilizer from Chile saltpeter.

10/16 **1846** The physician John Collins Warren in Boston attempts the first major surgery under anesthesia by administering ether prior to operating on a patient with a throat tumor.

10/17 **1936** The age of color photography begins with a public presentation of "Agfacolor® new color-reversal film" for slides and prints. Kodak introduced its "Kodachrome®" film at almost the same time.

10/18 **1906** Death of Friedrich Konrad Beilstein (*1828), who in 1880–1882 introduced his comprehensive *Handbook of Organic Chemistry.*

10/19 **1937** Death of Ernest Rutherford (*1871). He developed the model of the atom that bears his name, formulated the law of decay for radioactive substances, and in 1919 accomplished the first artificial nuclear reaction, whereby nitrogen was transformed into oxygen by bombardment with α-particles (Nobel Prize, 1908).

10/20 **1891** Birth of James Chadwick. In 1932 he discovered the neutron by irradiating beryllium with α-particles. In 1935 he was awarded a Nobel Prize in physics.

10/21 **1803** John Dalton's atomic theory is described in an article bearing this date, a publication in which for the first time quantitative data are presented regarding atoms (relative atomic weights). This paper contains the basic premises underlying modern atomic theory.

10/21 **1822** Adolph Strecker is born (†1871). He was the originator of both the *Strecker synthesis* and the *Strecker degradation* of amino acids.

10/22 **1986** Death of Albert Szent-Györgyi von Nagyrapolt (*1893). He concerned himself with the constituents of the pepper and isolated for the first time vitamin C. His investigations into the biological mechanism of oxidation led to an understanding of the processes accompanying muscle contraction (Nobel Prize in medicine or physiology, 1937).

10/23 **1871** The pharmacist Ernst Schering founds Schering AG in Berlin.

10/23 **1875** Birth of Gilbert Newton Lewis (†1946), who developed a theory of chemical bonding that distinguished between polar and nonpolar bonds. In 1916 he introduced *Lewis structures* for representing schematically the bonding relationships in molecules, and in 1923 he expanded the concept of acids and bases.

10/24 **1965** Death of Hans Meerwein (*1879). In order to explain certain rearrangements of camphor derivatives he postulated the formation of carbocations and then demonstrated their presence experimentally. His name is associated with the *Wagner–Meerwein rearrangement*, the *Meerwein reaction*, the *Meerwein arylation*, and the *Meerwein–Ponndorf–Verley reduction*.

10/25 **1929** This day has gone down in history as "Black Friday," when prices on the New York Stock Exchange fell by as much as 90%. The resulting economic crisis saw the number of employees of I.G.-Farbenindustrie reduced by 45%.

10/26 **1939** The Nobel Prize Committee decides to recognize Gerhard Domagk for his discovery of the antibacterial activity of "Prontosil®" – a sulfonamide – although Domagk was unable to accept the award until 1947.

10/27 **1894** Birth of Sir John Edward Lennard-Jones (†1954). He was concerned with problems in theoretical chemistry and became known for his work on interatomic forces.

10/28 **1914** Richard Laurence M. Synge is born (†1994). In 1952 he was awarded a Nobel Prize for his discovery of partition chromatography.

10/30 **1975** Death of the physicist Gustav Hertz (*1887), who received a Nobel Prize in physics in 1925 for discovering the excitation of atoms by electron impact.

10/31 **1832** Birth of Walter Weldon (†1885), who developed the no longer important *Weldon process* for the preparation of chlorine from hydrochloric acid.

10/31 **1835** Birth of Adolf von Baeyer (†1917), who studied organic dyes and hydroaromatic compounds. He also elucidated the structure of indigo and succeeded in synthesizing the substance (Nobel Prize, 1905).

11/1 **1869** Birth of Raphael Eduard Liesegang (†1947), discoverer of the *Liesegang rings.* These periodic precipitation events are analogous to the periodic structures that arise in oscillating reactions.

11/2 **1966** Death of Peter Josephus Wilhelm Debye (*1884). He received a Nobel Prize in 1936 for his contributions with respect to molecular structure and introduction of the dipole theory. Together with Paul Scherrer he discovered a process permitting crystal structures to be determined not only from single crystals but also crystalline powders: the *Debye–Scherrer process.*

11/4 **1806** Birth of Karl Friedrich Mohr (†1879). Among other things he introduced the *Mohr balance* for determining the density of liquids as well as the *Liebig condenser* for carrying out laboratory experiments under reflux. He also initiated the use of *Mohr's salt,* iron(II) sulfate, in quantitative analysis as a standard for potassium permanganate solutions.

11/4 **1854** Birth of Paul Sabatier (†1941), who discovered the catalytic activity of finely divided metals, especially nickel, in the hydrogenation of unsaturated hydrocarbons (Nobel Prize, 1912).

11/4 **1902** Birth of Otto Bayer (†1982), a chemist at the Bayer Company and the discoverer of a polyaddition process patented in 1937 between diisocyanates and diols leading to polyurethanes.

11/5 **1879** Death of James Clerk Maxwell (*1831), who formulated the basic equations of electrodynamics that bear his name and postulated the existence of electromagnetic waves. His investigations into the kinetic theory of gases led to the *Maxwell–Boltzmann distribution of velocities.*

11/6 **1822** Death of Claude Louis Berthollet (*1748). He introduced chlorine as a bleaching agent and was the first to employ potassium chlorate in explosives. His name is associated with the *berthollides* – a collective term for intercalated mixed crystals, intermetallic compounds, and other nonstoichiometric systems.

11/7 **1867** Birth of Marie Curie-Sklodowska (†1934), the founder of radiochemistry. She coined the term "radioactivity" and in 1897 discovered the natural radioactivity of thorium as well as the elements polonium (together with her husband Pierre Curie) and radium (with P. Curie and G. Bémont). She was awarded a Nobel Prize in physics in 1903 and a Nobel Prize in chemistry in 1911.

11/7 **1888** Birth of Chandrasekhara Raman (†1970), the discoverer of the *Raman effect.* With the aid of the spectroscopic technique based on this effect it became possible to characterize certain aspects of the bonding relationships in molecules (Nobel Prize in physics, 1930).

November

11/8 **1854** Birth of Johannes Robert Rydberg (†1919). His name is associated with the *Rydberg constant* that appears in the formula for series equations in atomic spectroscopy and in Moseley's law.

11/8 **1895** Wilhelm Conrad Röntgen discovers a penetrating type of radiation (now known as X-rays) in the course of investigating a cathode ray tube, the applicability of which to medicine he quickly recognizes. On November 22 he prepared the first X-ray image of his wife's hand (Nobel Prize in physics, 1901).

11/9 **1897** Birth of Ronald George Norrish (†1978), the developer of the *Norrish reactions*, photochemical transformations of carbonyl compounds. He also developed flash photolysis and in 1967 received a Nobel Prize for his investigations of the kinetics of very fast reactions.

11/10 **1918** Birth of Ernst Otto Fischer, who received a Nobel Prize in 1973 for his fundamental studies of sandwich compounds. He later turned his attention to carbene and carbyne complexes.

11/11 **1986** In reaction to an industrial accident at the Sandoz facilities in Basel (Switzerland) the German Association of Chemical Industry and the Federal Environmental Ministry reach agreement on a catalog of measures to be taken in the interest of improving safety conditions in chemical storehouses and increased water-quality protection.

11/12 **1842** Birth of John William Rayleigh (†1919), who together with William Ramsay discovered argon. He also explained the blue color of the sky as a consequence of light scattering by air molecules. Rayleigh's name is familiar from *Rayleigh radiation* and *Rayleigh's law* (Nobel Prize in physics, 1904).

11/13 **1937** The basic patent is issued for production of the plastic polyurethane (invented by Otto Bayer).

11/14 **1891** Birth of Frederick Grant Banting (†1941), the discoverer of insulin. He was awarded a Nobel Prize in medicine or physiology in 1923.

11/15 **1919** Death of Alfred Werner (*1866), the father of coordination chemistry and winner of a Nobel Prize in 1913.

11/16 **1945** The United Nations Educational, Scientific, and Cultural Organization (UNESCO) begins its work. This was the first worldwide institution designed to promote international cooperation in education, science, and culture.

11/17 **1953** Karl Ziegler applies for a patent of a procedure developed under his direction for atmospheric-pressure polymerization of ethene in the presence of organoaluminum and transition metal compounds.

11/18 **1789** Birth of Louis-Jacques Daguerre (†1851), the inventor of photography.

11/19 **1887** Birth of James Batcheller Sumner (†1955), who succeeded for the first time in crystallizing an enzyme (Nobel Prize, 1946).

11/20 **1945** Death of Francis William Aston (*1877). He used the mass spectrograph he developed to discover many isotopes of nonradioactive elements, and derived from this work the law of unit atomic masses (Nobel Prize, 1922).

11/21 **1895** Birth of Josef Mattauch (†1976). In 1934 and 1941 he proposed the isobar rules that specify there is no pair of stable isotopes with nuclear charges that differ by a single unit, and that for every odd atomic weight there is only a single stable nucleus.

11/22 **1981** Death of Hans Adolf Krebs (*1900). His research on metabolism led to discovery of the ornithine and citric acid cycles (Nobel Prize in medicine, 1953).

11/23 **1837** Johannes Diderik van der Waals is born (†1923). His name has been given to intermolecular forces that are responsible for the deviation of real gases from ideal behavior.

11/23 **1904** The Badische Anilin- & Soda-Fabrik and Bayer form a cooperative alliance, to which Agfa is added on December 10. This "triumvirate" is confronted by a duo consisting of Hoechst and Casella, joined in 1907 by Kalle & Co.

11/24 **1859** Charles R. Darwin's famous work *On the Origin of Species by Means of Natural Selection, or the Preservation of Favoured Races in the Struggle for Life* appears in London. The theory presented here of change, adaptation, competition, and survival in the course of the evolution of animal species contradicted the theological teachings of creation by God.

11/25 **1877** The "Organization for Protection of the Interests of the German Chemical Industry" is founded in Frankfurt am Main. This was the precursor to the Union of Chemical Industries (Verband der Chemischen Industrie, VCI).

11/26 **1898** Birth of Karl Ziegler (†1973). His name is associated with a number of discoveries. Among the best known are the *Ziegler–Natta catalysts,* organometallic compounds that catalyze the stereospecific polymerization of olefins at atmospheric pressure (Nobel Prize, 1963).

11/27 **1903** Birth of Lars Onsager (†1976), who proposed the laws of reciprocity in the thermodynamics of irreversible processes, for which he was awarded a Nobel Prize in 1968.

11/27 **1968** Death of Lise Meitner (*1878) a physicist who, together with Otto Hahn, in 1918 discovered the element protactinium, and in 1938 along with O. R. Frisch described as "nuclear fission" a process observed by Hahn and Strassmann in the course of irradiating uranium with neutrons that resulted in the formation of alkaline earth elements.

November **11/29** **1915** Birth of Earl Wilburne Sutherland (†1974), who in 1971 received a Nobel Prize in medicine or physiology for his research on metabolism.

11/29 **1936** Birth of Yuan Tseh Lee. He was awarded a Nobel Prize in 1986 for his research on the dynamics of elementary chemical reactions.

11/30 **1915** Birth of Henry Taube, who clarified the mechanisms of electron transfer, especially with respect to metal complexes (Nobel Prize, 1983).

December **12/1** **1947** Death of Franz Fischer (*1877), who in 1925 together with Tropsch developed the *Fischer–Tropsch process* for the synthesis of hydrocarbon products from coal.

12/1 **1949** The chemist Fritz Stastny records an entry in his research notebook at the Badische Anilin- & Soda-Fabrik that will lead to the discovery of "Styropor®": a reaction mixture inadvertently left in a drying oven had foamed up on the following day.

12/2 **1859** Birth of Ludwig Knorr (†1921). His work was dedicated to the synthesis of heterocycles, and his name is still associated with such reactions as the *Knorr pyrazole synthesis* and the *Knorr pyrrole synthesis.*

12/2 **1942** With the first self-sustained controlled nuclear reaction, Enrico Fermi proves at Columbia University in New York that it is possible to build and operate a nuclear reactor. Uranium served as the fuel.

12/3 **1900** Birth of Richard Kuhn (†1967), who was awarded a Nobel Prize in 1938 for his work on carotenoids and vitamins.

12/3 **1910** Georges Claude demonstrates for the first time – at the Paris Auto Show – the neon lighting he developed.

12/3 **1933** Paul Crutzen is born. He was awarded a Nobel Prize in 1995 for his work on atmospheric chemistry.

12/4 **1893** Death of John Tyndall (*1820), noted for investigating the chemical consequences of light and the *Tyndall effect*, which bears his name.

12/5 **1831** Birth of Hans Heinrich Landolt (†1910). He and Richard Börnstein were responsible for the *Physical-Chemical Tables* that remain an important reference work today.

12/6 **1778** Birth of Joseph-Louis Gay-Lussac (†1850). Together with Alexander von Humboldt he determined the correct ratio of hydrogen to oxygen in water. He also made an important industrial contribution through introduction of the *Gay-Lussac tower* for the absorption of nitrous gases in the lead-chamber process for making sulfuric acid.

12/6 **1920** Birth of George Porter, who together with Ronald Norrish developed spectroscopic methods for the investigation of rapid reactions (Nobel Prize, 1967).

12/7 **1951** The Farbwerke Hoechst is the first of the former I.G.-Farben companies to be reestablished.

12/8 **1947** Birth of Thomas Robert Cech, who was awarded a Nobel Prize in 1989 for discovering the catalytic activity of ribonucleic acids.

12/8 **1970** Death of Christopher Kelk Ingold (*1893). He introduced such concepts as "electrophile," "nucleophile," and "mesomerism." His name is also associated with the *Cahn-Ingold-Prelog rules,* which permit the specification of absolute configuration at chiral centers.

12/9 **1742** Birth of Carl Wilhelm Scheele (†1786), discoverer of the elements oxygen and chlorine.

12/9 **1868** Birth of Fritz Haber (†1934). He was awarded a Nobel Prize in 1918 for the laboratory-scale method he developed for the synthesis of ammonia from its elements. Bosch at the Badische Anilin- & Soda-Fabrik succeeded in transforming this into an industrial-scale process. Haber also directed the first use of chlorine as a chemical weapon in 1915 at Ypres.

12/9 **1919** Birth of William Nunn Lipscomb. He studied the structure of boranes and was able to resolve questions regarding their chemical bonding, for which he was awarded a Nobel Prize in 1976.

12/10 **1896** Death of Alfred Nobel (*1833). By adding kieselgur to the explosive oil nitroglycerin he managed to prepare a safe detonating material: dynamite. In his will he made provisions for the establishment of the Nobel Foundation and the annual award of prizes. The first Nobel Prizes were awarded on the fifth anniversary of his death in 1901.

12/11 **1909** Death of Ludwig Mond (*1839). He developed several techniques that bear his name, including the *Mond process* for preparing pure nickel via the intermediate nickel tetracarbonyl. Mond was cofounder of the corporation Brunner Mond, Ltd., which later became ICI.

12/12 **1850** Death of Germain Henri Hess (*1802). He was the formulator of *Hess's law,* a special case of the law of conservation of energy (which was at that time unknown), permitting heats of reaction to be determined for processes that are not directly measurable.

12/13 **1780** Birth of Johann Wolfgang Döbereiner, who concerned himself with questions related to catalysis. In the process he discovered that spongy platinum will ignite a mixture of hydrogen and oxygen at room temperature, whereupon he constructed the *Döbereiner lighter*. In 1829 he grouped a number of elements into triads "on the basis of their analogy" (Li–Na–K, Ca–Sr–Ba, S–Se–Te, Cl–Br–I), an important step in the direction of the periodic table.

12/14 **1900** Max Planck publishes in Berlin his essay "Radiation from Black Bodies," thereby laying the cornerstone for the development of quantum theory.

12/15 **1852** Birth of Henri Becquerel (†1908), who in 1896 discovered natural radioactivity (Nobel Prize in physics, 1903).

12/16 **1921** In the course of experimenting with catalysts, Fritz Winkler discovers the principle of the fluidized-bed process, which is then adapted for the industrial-scale preparation of synthesis gas. Other applications were also developed, however, including the roasting of sulfide-containing ores.

12/17 **1493** Birth of Paracelsus (Philippus Aureolus Theophrastus Bombastus von Hohenheim, †1541), a physician and natural philosopher credited with founding pharmaceutical chemistry.

12/17 **1908** Willard Frank Libby is born (†1980). He developed the method for determining the age of an object based on the carbon isotope ^{14}C, which has been frequently utilized in geology and archeology (Nobel Prize, 1960).

12/18 **1856** Birth of Joseph John Thomson (†1940). In the course of his investigations into the transport of electricity through gases he discovered the free electron (Nobel Prize in physics, 1906).

12/19 **1930** Death of Conrad Willgerodt (*1841). It was he who developed the *Willgerodt reaction* for transforming ketones into amides and salts of carboxylic acids.

12/19 **1951** The Farbenfabriken Bayer AG (Leverkusen, Germany) becomes the second of the I.G. Farben successors to be reestablished.

12/20 **1890** Birth of Jaroslav Heyrovsky (†1967). He was awarded a Nobel Prize in 1950 for inventing and developing polarography as an analytical method.

12/20 **1971** Death of Stefan Goldschmidt (*1889), who in 1920 succeeded in preparing the first organic nitrogen radical, diphenylpicrylhydrazyl.

12/21 **1805** Birth of Thomas Graham (†1869). He was the cofounder of colloid chemistry and his name is associated with *Graham's salt*, a polymeric sodium metaphosphate.

12/22 **1838** Birth of Vladimir Markovnikov (†1904). In his dissertation, which was published in 1870 in *Liebig's Annalen*, he proposed the rule for *Markovnikov addition* of hydrogen halides to unsymmetrically substituted alkenes.

12/23 **1722** Birth of Axel Fredrik Cronstedt, the discoverer of nickel (†1765). He also succeeded in distinguishing between galenite (PbS) and graphite, which until then had been assumed to be chemically identical.

12/24 **1818** Birth of James Prescott Joule (†1889). His name is used to designate a unit of energy and is also associated with the *Joule–Thomson effect*, the cooling that real gases undergo upon expansion, a phenomenon he discovered along with William Thomson.

12/25 **1876** Adolf Windaus is born (†1959). He demonstrated the structural relationship between cholesterol and the bile acids. He also recognized ergosterol as provitamin D and clarified the mechanism for the photochemical rearrangement to vitamin D_2 (Nobel Prize, 1928).

12/25 **1904** Birth of Gerhard Herzberg. He concerned himself with the electronic structure and geometry of molecules and advanced astrochemistry by developing spectroscopic methods for detecting atoms and molecules in space (Nobel Prize, 1971).

12/26 **1825** Birth of Felix Hoppe-Seyler (†1895), the founder of modern physiological chemistry and originator of the term "biochemistry." His research laid the groundwork for the physiology of respiration as well as the chemistry of the pigment in blood.

12/27 **1822** Louis Pasteur is born (†1895). He accomplished through fractional crystallization the first separation of a racemate into its enantiomers. His investigations into the phenomenon of fermentation led to the development of microbiology and also an important method for preserving foods. Indeed, *pasteurization* has been a familiar term for nearly 120 years. His research on immunity led to preventive vaccination using weakened strains of bacteria.

12/27 **1846** Birth of Wilhelm Michler (†1889), whose name is still associated with the ketone tetramethyldiaminobenzophenone, an important intermediate in the synthesis of textile dyes.

12/28 **1818** Carl Remigius Fresenius is born (†1897). This analytical chemist developed methods for determining the composition of mineral waters and metal alloys. In 1861 he established the journal *Zeitschrift für analytische Chemie.*

12/28 **1895** The brothers Lumière opened the first movie theater in Paris using films prepared on the basis of celluloid.

12/28 **1944** Birth of Kary Banks Mullis. This 1993 winner of a Nobel Prize developed the polymerase chain reaction (PCR).

December

12/29 **1800** Birth of Charles Goodyear (†1860), an American who discovered the principle of vulcanization.

12/30 **1691** Death of Robert Boyle (*1627), who discovered that the pressure and volume of a gas are inversely proportional (*Boyle–Mariotte law*) and is considered to be one of the originators of both the atomic theory and the concept of an element.

12/31 **1877** Louis Cailletet reports his successful liquification of nitrogen and air to the Paris Academy of Science.

12/31 **1924** Effective with the end of the current business year, the following companies declare themselves united as the I.G. Farbenindustrie Aktiengesellschaft: Agfa, Badische Anilin- & Soda-Fabrik, Bayer, Casella, Griesheim Elektron, Hoechst, Kalle, and Weiler ter Meer. The individual companies thus give up their independence to become simply branches of I.G. Farben.

Name
Index

Subject
Index

Subject
Index

Subject Index